INTEGRATED CIRCUIT MANUFACTURABILITY

IEEE Press
445 Hoes Lane, P.O. Box 1331
Piscataway, NJ 08855-1331

IEEE Press Editorial Board
Roger F. Hoyt, *Editor-in-Chief*

J. B. Anderson	S. Furui	P. Laplante
P. M. Anderson	A. H. Haddad	M. Padgett
M. Eden	R. Herrick	W. D. Reeve
M. E. El-Hawary	S. Kartalopoulos	G. Zobrist
	D. Kirk	

Kenneth Moore, *Director of IEEE Press*
John Griffin, *Acquisition Editor*
Marilyn G. Catis, *Assistant Editor*
Denise Phillip, *Production Editor*

Cover design: Caryl Silvers, *Silvers Design*

Technical Reviewers

Phil Allen, *Georgia Institute of Technology*
Randall Collica, *Digital Equipment Corporation*
Christian Landrault, *LIRMM*
Gary S. May, *Georgia Institute of Technology*

Books of Related Interest from IEEE Press

INTEGRATED CIRCUITS FOR WIRELESS COMMUNICATIONS
Edited by Asad Abidi, Paul Gray, and Robert Meyer
1998   Hardcover   544 pp   IEEE Order No. PC5716   ISBN 0-7803-3459-0

HIGH-TEMPERATURE ELECTRONICS
Edited by Randall Kirschman
1998   Hardcover   1392 pp   IEEE Order No. PC5735   ISBN 0-7803-3477-9

DELTA-SIGMA DATA CONVERTERS: Theory, Design, and Simulation
Edited by Steven R. Norsworthy, Richard Schreier, and Gabor C. Temes
1997   Hardcover   512 pp   IEEE Order No. PC3954   ISBN 0-7803-1045-4

**MONOLITHIC PHASE-LOCKED LOOPS AND CLOCK RECOVERY CIRCUITS:
Theory and Design**
Edited by Behzad Razavi
1996   Hardcover   512 pp   IEEE Order No. PC5620   ISBN 0-7803-1149-3

SEMICONDUCTOR MEMORIES: Technology, Testing, and Reliability
Ashok K. Sharma
1997   Hardcover   480 pp   IEEE Order No. PC3491   ISBN 0-7803-1000-4

MULTIMEDIA TECHNOLOGY FOR APPLICATIONS
Edited by Bing Sheu and Mohammed Ismail
1998   Hardcover   552 pp   IEEE Order No. PC5645   ISBN 0-7803-1174-4

Books are to be returned on or before the last date below.

INTEGRATED CIRCUIT MANUFACTURABILITY

The Art of Process and Design Integration

Edited by

José Pineda de Gyvez
Department of Electrical Engineering
Texas A&M University

Dhiraj Pradhan
Department of Electrical Engineering
Stanford University
(on leave from Texas A&M University)

IEEE Circuits and Systems Society, *Sponsor*

The Institute of Electrical and Electronics Engineers, Inc., New York

This book and other books may be purchased at a discount from the publisher when ordered in bulk quantities. Contact:

IEEE Press Marketing
Attn: Special Sales
445 Hoes Lane, P.O. Box 1331
Piscataway, NJ 08855-1331
Fax: 1-732-981-9334

For more information on the IEEE Press,
visit the IEEE home page: http://www.ieee.org/

© 1999 by the Institute of Electrical and Electronics Engineers, Inc.,
3 Park Avenue, 17th Floor, New York, NY 10016-5997

*All rights reserved. No part of this book may be reproduced in any form,
nor may it be stored in a retrieval system or transmitted in any form,
without written permission from the publisher.*

Printed in the United States of America

10   9   8   7   6   5   4   3   2   1

ISBN 0-7803-3447-7
IEEE Order Number: PC4481

Library of Congress Cataloging-in-Publication Data

Integrated circuit manufacturability : the art of
   process and design integration / edited by José Pineda
   de Gyvez, Dhiraj Pradhan.
         p.   cm.
   "IEEE Circuits and Systems Society, sponsor."
   Includes bibliographical references and index.
   ISBN 0-7803-3447-7 (alk. paper)
   1. Integrated circuits—Design and construction—
Data processing. 2. Metal oxide semiconductors,
Complimentary—Design and construction—Data
processing. 3. Computer-aided design. 4. Integrated
circuits—Testing. I. Pineda de Gyvez, José.
II. Pradhan, Dhiraj K. III. IEEE Circuits and Systems
Society.
TK7874.I4713   1998
621.3815—dc21   98-6335
                                              CIP

Para Ricardo y Samantha por la alegria y amor que me dan

—José Pineda de Gyvez

Contents

Preface xiii

Chapter 1 Introduction 1
José Pineda de Gyvez

    References 7

Chapter 2 Defect Monitoring and Characterization 9
Eric Bruls

- **2.1 Market Developments 9**
- **2.2 Price 9**
- **2.3 Quality and Reliability 10**
- **2.4 IC Manufacturing Defects 11**
  - 2.4.1 IC Development Flow 12
  - 2.4.2 The Manufacturing Process 15
  - 2.4.3 Defect Mechanisms 19
- **2.5 Defect Monitoring 21**
  - 2.5.1 Global Defects 22
  - 2.5.2 Local Defects 24
- **2.6 Defect Modeling 29**
  - 2.6.1 Global Defects 29
  - 2.6.2 Local Defects 31
- **2.7 Summary 39**
- **2.8 Exercises 40**
  - **References 41**

vii

Chapter 3 Digital CMOS Fault Modeling and Inductive Fault Analysis 43
Manoj Sachdev

3.1 Introduction 43
    3.1.1 Quality and Reliability Awareness 44
    3.1.2 Role of Testing in Quality Improvement 44
3.2 **Objectives of Fault Modeling** 45
3.3 **Levels of Fault Modeling** 46
    3.3.1 Logic-Level Fault Modeling 47
    3.3.2 Transistor-Level Fault Modeling 53
    3.3.3 Layout-Level Fault Modeling 60
    3.3.4 Functional-Level Fault Modeling 60
    3.3.5 Delay Fault Models 61
    3.3.6 Leakage Fault Models 63
    3.3.7 Temporary Faults 64
3.4 **Inductive Fault Analysis** 65
    3.4.1 The Defect–Fault Relationship 66
    3.4.2 IC Design and Layout-Related Defect Sensitivity 68
    3.4.3 Basic Concepts of IFA 69
    3.4.4 Practical Experiences with IFA 71
    3.4.5 The IFA: Strengths and Weaknesses 77
3.5 **Summary** 77
    References 78

Chapter 4 Functional Yield Modeling 85
Gary C. Cheek and Geoff O'Donoghue

4.1 Introduction 85
4.2 **Basic Yield Statistics: Random Defects** 86
    4.2.1 Yield Model Derivations 89
4.3 **Classes of Yield Models** 94
    4.3.1 Class I Yield Models 94
    4.3.2 Class II Yield Models 95
    4.3.3 Class III Yield Models 96
    4.3.4 Class IV Yield Models 99
4.4 **Yield Model Components** 99
    4.4.1 Defect Density Term in Yield Models 100
    4.4.2 Area Term in Yield Models 103
    4.4.3 Probability of Fail 104
    4.4.4 Computation of Critical Area 106
    4.4.5 The Y_0 Term in Yield Models 107
4.5 **Applications of Functional Yield Models** 108
    4.5.1 Low-Yield Cutoff and Chip Costing 109
    4.5.2 Spatial Yield Distributions: Y_0 110
    4.5.3 Yield Distributions 110
    4.5.4 Critical Area in Product Design: SRAM Example 111

        4.5.5 Critical Area in Yield Calculation 113
        4.5.6 Use of Yield Models for Scaling Applications 113
4.6 **Summary 115**
4.7 **Exercises and Solutions 115**
    References 118

Chapter 5   Critical Area and Fault Probability Prediction 121
D. M. H. Walker

5.1 **Introduction 121**
5.2 **Theoretical Background 124**
5.3 **Contamination to Defect Mapping 125**
5.4 **Defect to Fault Mapping 126**
    5.4.1 Geometrical Methods 126
    5.4.2 Monte Carlo Methods 132
    5.4.3 Combined Methods 135
    5.4.4 Three-Dimensional Defects 136
    5.4.5 Spatial Clustering Within Chips 137
    5.4.6 Circuit Model Issues 137
5.5 **Hierarchical Defect to Fault Mapping 138**
    5.5.1 Identification of Nonoverlapping Layout Areas 139
    5.5.2 Hierarchical Circuit Extraction of Nonoverlapping Layout 140
    5.5.3 Hierarchical Defect to Fault Mapping of Nonoverlapping Layout 141
    5.5.4 Global Fault Reporting 141
5.6 **Fault to Failure Mapping 142**
5.7 **Applications 142**
    5.7.1 Yield Prediction 142
    5.7.2 Redundancy Analysis 143
    5.7.3 Test Generation 144
    5.7.4 Process Diagnosis and Monitoring 144
    5.7.5 Design for Manufacturability 145
5.8 **Summary and Research Directions 146**
5.9 **Exercises and Solutions 147**
    References 148

Chapter 6   Statistical Methods of Parametric Yield and Quality Enhancement 157
Maciej Styblinski

6.1 **Problems and Methodologies of Statistical Circuit Design 158**
6.2 **Circuit Variables, Parameters, and Performances 158**
    6.2.1 Designable Parameters 159
    6.2.2 Random Variables 159

6.2.3 Circuit (Simulator) Variables 159
  6.2.4 Circuit Performance 160
6.3 **Statistical Modeling of Circuit (Simulator) Variables 161**
  6.3.1 Passive Discrete RLC Elements 161
  6.3.2 Passive Integrated RLC Elements 162
  6.3.3 Single Active Device Modeling for Discrete and Integrated Circuits 162
  6.3.4 Global and Local (Mismatch) Models for Integrated Circuits 164
6.4 **Acceptability Regions 166**
  6.4.1 Methods of Acceptability Region Approximation 168
6.5 **Parametric Yield 171**
6.6 **Indirect Methods of Yield Enhancement 175**
  6.6.1 Simplicial Approximation-Based Design Centering 176
  6.6.2 Worst-Case Distance-Driven Design Centering 177
  6.6.3 Performance Space-Oriented Design Centering 178
6.7 **Statistical Methods of Yield Optimization 180**
  6.7.1 Problem Classification 180
  6.7.2 Large-Sample versus Small-Sample Methods 181
  6.7.3 Using Standard Deterministic Optimization Algorithms 182
  6.7.4 Large-Sample Heuristic Methods for Discrete Circuits 183
  6.7.5 Large-Sample, Derivative-Based Methods for Discrete Circuits 184
  6.7.6 Large-Sample, Derivative-Based Method for Integrated Circuits 188
  6.7.7 Small-Sample, Stochastic Approximation-Based Methods for Discrete Circuits 191
  6.7.8 Small-Sample, Stochastic Approximation Methods for Integrated Circuits 196
  6.7.9 Case Study: Process Optimization for Manufacturing Yield Enhancement 199
6.8 **Design for Quality 201**
  6.8.1 Generalized Formulation of Yield, Variability, and Taguchi Circuit Optimization Problems 202
  6.8.2 Propagation of Variance Method 205
6.9 **Conclusion 210**
  References 210

Chapter 7 Architectural Fault Tolerance 217
S. K. Tewksbury

7.1 **Introduction 217**
  7.1.1 Use of Known-Faulty Components When Manufacturing an Electronic System 218
  7.1.2 Large-Area Integrated Circuits 219
  7.1.3 In-Service Failures 220
  7.1.4 Difficulty of Repair 221
  7.1.5 Selective Fault Tolerance 221

7.2 **Local Fault Tolerance** 223
    7.2.1 Modular Redundancy 223
    7.2.2 Error Correcting Codes 227
    7.2.3 Algorithm-Based Fault Tolerance 232
7.3 **Global Reconfiguration** 234
    7.3.1 Reconfigurable Arrays of Computation Cells 235
    7.3.2 Two-Dimensional Arrays 241
    7.3.3 Self-Reconfiguration Algorithms for Two-Dimensional Arrays 243
    7.3.4 Replacement of Full Rows (Columns) by Spare Rows (Columns) 246
7.4 **Physical Switch Technologies for Reconfiguration** 248
    7.4.1 Electronically Programmable Reconfiguration 249
    7.4.2 Physical Restructuring of Interconnections 255
7.5 **Summary** 260
    References 260

Chapter 8   Design for Test and Manufacturability 269
Dhiraj Pradhan and Adit Singh

8.1 **Introduction** 269
    8.1.1 The Basic Problems of Testing 269
8.2 **Testing for Stuck-at Faults** 270
    8.2.1 Limitations of the Stuck-at Fault Model 272
8.3 **Test Coverage and Defect Levels** 274
8.4 **Quality Screening Based on Defect Clustering** 276
    8.4.1 Binning for Low Defect Levels 277
    8.4.2 Analysis 278
    8.4.3 Test Transparency Function 281
    8.4.4 Numerical Results 283
8.5 **Applications** 284
    References 285

Chapter 9   Testing Solutions for MCM Manufacturing 287
Yervant Zorian

9.1 **Introduction** 287
9.2 **MCM Testing Problem** 289
    9.2.1 Wafer and Bare Die Test 290
    9.2.2 Substrate Testing 292
    9.2.3 MCM Assembly Testing and Repair 292
9.3 **A Structured Testability Approach** 293
    9.3.1 The Bare Dies Test Procedure 294
    9.3.2 The Assembled MCM Test and Diagnosis Procedure 294
9.4 **Chip-Level Structured Testability Inclusion** 298

9.5 Module-Level Testability Needs 300
9.6 Conclusion 301
    References 301

Index 305

About the Editors 315

Preface

Integrated circuit (IC) manufacturability has received a great deal of attention in the last few years. As technological processes are advancing toward submicron resolution features—with higher transistor integration on silicon taking place—the need to foresee the ease and feasibility of fabrication of IC designs is becoming a must.

IC manufacturability is no longer a practice that belongs exclusively to industry. Today it is a flourishing area of academic research in which systematic solutions are sought for yield-related problems. This presents a dramatic departure from the previous practice of ad hoc research to address such problems. The result of this interaction is the emergence of methodologies for yield and fault prediction, taking into account existing manufacturing conditions. It is important to emphasize "existing manufacturing conditions" because IC design can no longer be seen as a task using ideal nominal values, but instead as a task where process variations and manufacturing disturbances must be considered. Despite the emphasis on manufacturability, traditional academic curricula do not include special-purpose courses in this field. In all likelihood, this stems from the fact that in common engineering practices, IC design and semiconductor process technology are two distinct and isolated domains. Typically, the design engineer is more accustomed to, say, behavioral and electrical simulations, while the process engineer delves into the physical and chemical components of the technological process. As a result, electrical engineering-related courses are also separate domains. In one domain, for instance, all the courses pertain to IC design (e.g., Digital and Analog IC Design, Very Large-Scale Integration Systems, Logic Synthesis), and in process domain there are courses such as Microelectronic Circuit Fabrication and Microelectronic Device Design.

The traditionally separated areas of IC design and semiconductor manufacturability have been integrated in this new text. This work addresses the study of process and design variables in order to determine the ease and feasibility of fabrication—or manufacturability—of integrated circuits. The book consists of four main sections:

(1) basic processing technology and related disturbances, (2) functional yield prediction, (3) layout defect-sensitivity analysis, and (4) manufacturing fault analysis/debugging. Obviously, given the nature of the topics, some of them could be a book by themselves. Actually, this would correspond to the previously mentioned separation of fields by domains.

The purpose of *Integrated Circuit Manufacturability: The Art of Process and Design Integration* is to link the four sections—for example, to present the impact of process disturbances on the IC performance; to study the feasibility of fabrication through yield prediction and estimation; and to present designs for manufacturability techniques. As process technology and circuit design are still separate domains, a middle strategy is pursued; that is, we seek a compromise between theory and practice, as well as extent of theoretical coverage. The use of computer-aided design tools throughout an academic course is strongly recommended. This is an important aspect of the learning process, for it provides an almost "turnkey" solution to specific application domains. As progress is made during the course, time and resources can be dedicated to understanding and managing more complex and practical problems, rather than putting all efforts into small "classic" examples.

The four core sections of this textbook present the student with practical issues that are normally applied in industry and are usually required by quality, product, and design engineering departments. In fact, the book is a response to the industry's continuous need for qualified engineers in manufacturing positions. Ideally, these engineers have knowledge in defect engineering, circuit design, testing, and failure analysis, and are capable of coordinating and monitoring production activities to ensure that the product meets functional and performance requirements. However, only rarely does the recently graduated engineer possess all this knowledge. Quite often, industry must also incur expensive training programs.

The book consists of nine chapters carefully written by leading authorities in the field. The order of the book is meant to advance progressively from semiconductor processing to electrical design and from electrical design to system architecture. Chapter 1 provides an overview of the book. Chapter 2 deals with the environmental conditions prevailing in the manufacturing line. In particular, it examines methods that characterize defects, and it takes this defect information and translates it into electrical fault models using the inductive fault analysis technique, which is discussed in Chapter 3. Chapter 4 addresses the feasibility of fabrication of the design, taking into account the environmental conditions of the manufacturing line. Chapter 5 discusses the way ICs can be designed to tolerate defects using methods such as critical area analysis, design rule optimization, and process-oriented monitoring. Chapter 6 is aimed at quality control problems of circuit designs; it includes topics such as performance variability minimization and sensitivity reduction w.r.t. process variations. Chapter 7 focuses on tolerance, redundancy, and testability issues at the system level, while Chapter 8 concentrates on design for testability and testing practices at the circuit level. Finally, Chapter 9 presents testing methodologies for multichip modules (MCMs), including topics such as the burn-in of bare dies.

Each chapter is general and is not a review or survey of research topics only. Included are tutorial-like exercises, for the book is intended to serve as a textbook.

Most of the material addressed here is covered in a course on IC Manufacturability taught at Texas A&M. This course is certainly not ultimate; rather, it is meant to serve as an introduction to manufacturing issues from a circuit and production point of view. While the course is functional, our experience has shown that more work is needed to develop a complete curriculum in manufacturing covering formal aspects of yield and testability (e.g., the need of an adequate textbook). Usually, manufacturing courses are related to issues in quality control and originate mostly from an industrial engineering perspective. This course departs from this tradition by giving attention to electrical engineering aspects and by involving students in the use of CAD tools. In fact, use of CAD tools was an important component of this course. These tools give students the ability to work in different areas without having to be experts in any of them. The techniques enable novice students to grasp complex issues such as extraction of manufacturing yield and wafer yield management, or extraction of critical areas and realistic sets of faults. It is hoped that this textbook will complement the needs of a more thorough manufacturability curriculum.

José Pineda de Gyvez
Department of Electrical Engineering
Texas A&M University

1

Introduction

José Pineda de Gyvez

As the complexity of integrated circuits advances toward submicron technologies with high transistor densities in the order of 10^6 elements, the integration between process technology and system design becomes a relevant issue. This book addresses the study of process and design variables in order to determine the ease and feasibility of fabrication—or manufacturability—of contemporary Very Large-Scale Integration (VLSI) systems and circuits. The book will introduce the reader to key aspects in today's design practices. The four topics shown in Figure 1.1—yield modeling, defect and fault modeling, testing, and fault tolerant architectures—are intended to link common interests to both design and process engineers. The whole manufacturing process from circuit to silicon will be examined, with each subject supporting each other. For instance, the use of yield modeling will be examined to foresee the IC's feasibility of fabrication. This will be followed by the study of manufacturing defects and their impact on circuit design, and will be complemented with the implementation of strategies for fault tolerant architectures.

Integrated circuit (IC) manufacturability has received a great deal of attention in recent years. As technological processes are advancing toward submicron resolution features—with higher transistor integration on silicon taking place—the need to foresee the ease and feasibility of fabrication of IC designs is becoming essential. IC manufacturability is no longer a practice that belongs exclusively to industry [1]. Today it is a flourishing area of academic research in which systematic solutions are sought for yield-related problems. This presents a dramatic departure from the previous practice of ad hoc research to address such problems. The result of this interaction is the emergence of methodologies for yield and fault prediction, taking into account existing manufacturing conditions.

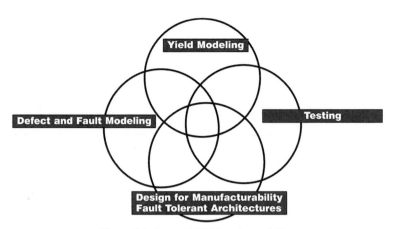

Figure 1.1 Areas of IC Manufacturability

Defects can be classified as local and global. The global class concerns the disturbances that affect complete regions of a wafer, whereas the local class concerns disturbances that are peculiar to only one IC. Spot defects belong to the class of local disturbances. As the IC pattern resolutions tend to shrink more and more, the impact of spot defects on the layout geometry plays a more important role in yield losses. Traditional approaches for layout verification concentrate on validating design rules imposed by the technological process. However, they do not verify the robustness of the design when it is exposed to defects in a real manufacturing environment. In order to perform this verification task, it is necessary to capture the design's *critical areas* [2]. The so-called critical areas are the places in the layout where spot defects can induce an incorrect behavior of the IC. For instance, a spot defect that creates a *bridge* between two patterns can induce a *short circuit* in the design. A figure of merit that measures the design's "vulnerability" is obtained as the ratio of the critical area for a given defect size to the total layout area. This figure of merit is known as *defect sensitivity*. Recall that semiconductor yield is the probability of manufacturing ICs without faults. Therefore, yield can be predicted by determining the defect sensitivity of a given layout design and by studying the stochastic behavior of defects in a given manufacturing environment.

We cannot talk about defects and put testing aside. Testing is a field that should profit from knowledge of defect behavior. Traditionally, testing is practiced at a convenient level of abstraction without considering the real causes of the fault. It was only a couple of years ago that this way of thinking changed. By considering a *realistic list of faults*, better quality test vectors can be obtained [3]. A simple way to handle fault-modeling complexity is to support several levels of abstraction in the description of a fault. For example, a system designer is interested in fault models that describe the faults in the architectural modules of the design, perhaps Multichip Modules (MCMs) rather than the faults in the IC layers. For each level of abstraction the fault models can be described with certain primitives appropriate to that level.

One such possible fault hierarchy is the following. At the highest level of abstraction, the *engineering* faults describe the functional faults of module units such as MCM's dice and provide an architectural description of the fault. *System*-level faults are concerned with module units such as Programmable Logic Devices (PLDs) and Arithmetic Logic Units (ALUs), providing a behavioral description of the unit. The *logic* level of abstraction describes internal faults in terms of logical expressions. At the next lower level, the *circuit* abstraction describes the electrical faults of the design. This description provides lists of faulty nodes and elements such as transistors. Finally, the *physical* level describes the process-related faults containing information such as defective patterns and process incongruities.

The four topics of this volume present the reader with practical issues that are normally applied in industry and are usually required by quality, product, and design engineering departments. In fact, the material hereby presented is a response to industry's continuous need for qualified engineers in manufacturing positions. Ideally, these engineers have knowledge of defect engineering, circuit design, testing, and failure analysis, and are capable of coordinating and monitoring production activities to ensure that the product meets functional and performance requirements. However, recently graduated engineers rarely possess all this knowledge. Quite often industry must also incur expensive training programs.

In Chapter 2, the complete development flow of an IC is divided into three parts. During the design phase, requirements and specifications have to be generated and translated into a design. In the second phase, this design has to be implemented on actual silicon. And finally, since the manufacturing processes are rather complex and susceptible to the occurrence of defects, the products have to be tested as to their correctness. The intrinsic quality of the actual manufacturing process should be as high as possible. Obviously, this benefits the economical viability of the products because the percentage of good devices will increase, thereby reducing the cost per device. However, the impact of the processing quality on the ultimate quality and reliability of products is also very high. Although a test procedure after the production phase is used to determine whether a device satisfies the requirements, this can only be of limited scope. Hence, it is important to gather information on the status of a production process with respect to its specifications.

On the basis of the Metal Oxide Semiconductor (MOS) processing steps, a short introduction will be given on the physical phenomena that result in defects in a wide meaning of the word. Global defects often cause some of the electrical parameters to lie outside the specified process window. Since these defects influence a large area of a wafer, these kinds of defects can be easily detected by means of Process Control Monitors (PCMs). These monitors are located either on a few positions of a wafer or they are placed in the scribe-lines between the products. These PCMs are used as a first filter in the test procedures. When the measured parameters lie outside the specified process window, the whole wafer is rejected without measuring the separate products. In a stable production environment and for a mature process, the occurrence of global defects is very rare, but if they do occur they can be detected rather easily.

The occurrence of local defects poses a more serious problem. These defects can occur anywhere on a wafer and can be related to local process disturbances or to environmental influences (like dust). The occurrence of these defects has to be minimized for creating a viable manufacturing line. For the purpose of optimizing the process conditions, specific defect monitors are used. The application of defect monitors is most valuable when the information about the status of the production line is fed back to the process engineers as soon as possible. This has resulted in the development of various short-loop defect monitors that focus on the detection of local defects in specific production steps. Currently, the most critical production steps are related to the gate-oxide and the interconnect between the various active devices. For these production steps, we will give an overview of various defect monitors used in the semiconductor world.

The information collected through defect monitors can be used not only to improve the process quality, but also to give a prognosis for the expected production yield. For that purpose, layout information of the product has to be combined with a modeled representation of the defects. In principle, three aspects of these defects have to be modeled. First, the number of defects for each layer of the design has to be known. Second, a design's susceptibility to specific defects depends on the size of those defects. Third, a model has to be given for the fact that defects tend to occur in clusters on the wafers, instead of randomly distributed. Finally, Chapter 2 concludes with an introduction of the translation of defects into their electrical influence on the circuit, varying from resistive contacts and leaky capacitors to parasitic diodes. A correct translation from a geometrical level to an electrical level is important for generating effective test procedures. These imperfections of the manufacturing process necessitate testing the manufactured integrated circuits. The abstraction of defect to fault semantics for IC testing is covered in Chapter 3.

Functional yield models have been developed in an effort to predict the number of good dice per wafer that will result from a manufacturing process. The ability to accurately predict the yield of a manufacturing process is crucial to the management of business operations, cost containment, and profit maximization. Functional yield models have traditionally been used to evaluate scaled circuitry, to maintain a competitive position through yield improvement activities, and to provide a market analysis tool. The purpose of Chapter 4 is to provide an overview of functional yield models and their application to mixed-signal VLSI circuitry. The first section of this chapter reviews the definition of functional yield and its application in the industrial environment. The basic mathematics of several key models are presented, and some of the limitations of these models are outlined. The basic components of a functional yield model include the critical area of the circuit layout and the defect density. These key components are defined, and various methods to measure the basic physical model parameters (such as defect sensitive area, defect distribution, and defect density) are described using vehicles such as defect monitors, particle counting systems, and analysis of circuit layouts. Application of yield model and analysis techniques are described with industrial examples. The special conditions that apply to analog circuitry that could have significant parametric yield loss content are also described. The separation of yield loss components, such as defect versus parametric, is demonstrated through the use of critical area, a defect-based testing methodology, and a deterministic yield model. The methods that have been developed to decrease product reliability hazards owing to functional yield degradation utilizing critical area analysis are addressed in detail. A section describes the use of functional yield models to contain and manage business costs and details the key components of a yield management environment. These components include the ability to optimize chip area to maximize the total number of dice per wafer during circuit design, an accurate metric for manufacturing efficiency, and a testing strategy. The development of deterministic yield models is particularly important when working with wafer fab foundries and test houses. This last section of the chapter presents examples of new products that are being developed within the above yield management environment.

Chapter 5 focuses on techniques for computation of catastrophic fault probabilities and functional yield based on knowledge of the product mask artwork, manufacturing process flow, and statistics of the process disturbances. The fault and yield results are then used in design, manufacturing, and test applications. The software that performs the fault and yield computation acts as a virtual fabrication line in that it stimulates the production of a population of chips with distribution of functional yield. The term *yield stimulation* is often applied to this process. For many applications, the probability of functional circuit faults is also important; thus, tools targeted as functional fault probability, as well as functional yield computation, are described.

Chapter 6 presents various methodologies developed in the integrated circuit Design for Quality (DFQ) area, involving performance variability minimization, manufacturing yield optimization, and sensitivity reduction with respect to manufacturing process variations and environmental conditions such as temperature. Recently, practical application of DFQ methodologies has been recognized as critical for the competitiveness of the U.S. electronics industry. The following main classes of DFQ methodologies are addressed in the chapter: Traditional, including statistical design centering, parametric yield optimization, worst-case design, sensitivity minimization (leading to performance variability minimization), element tolerancing and tuning, and so on, all based on the use of various statistical or deterministic optimization methods [4].

The Taguchi approach is based on the design of experiments and on Taguchi's Off-Line Quality Control philosophy. It has recently been popularized in the United States, owing to the well-known success of the Japanese in its practical applications. A new approach generalizing the notion of manufacturing yield and combining yield and variability minimization into one coherent methodology is presented. Methodologies for IC long-term reliability improvement, related to IC element "aging" and drift of element values in time (e.g., due to such phenomena as hot-electron effects), are addressed. A new formulation of drift reliability optimization problem is presented. Practical optimization examples are discussed, utilizing a combination of traditional, Taguchi, and novel-generalized DFQ techniques.

Chapter 7 focuses on the importance of VLSI tolerant architectures. Fault tolerance is important to the system designer (as opposed to the IC designer) since the overall system may need to tolerate failures that arise during the system operation. In some cases, the failure may be due to a manufacturing defect (e.g., early failure when notches into interconnections cause locally excessive current densities.) However, many "environmental" failure mechanisms also confront the correct operation of an entire system. Transients on power lines may cause transient faults in the operation of the circuitry. Data errors may arise owing to metastability effects during asynchronous transfers between two parts of the system. Connectors may cause "weak" signals to be transferred between ICs. Transient errors may arise due to alpha particles. When developing a fault tolerant architecture at the system-design level, the specific cause of the failure may be less important than the functional behavior induced by the failure. For example, faults may be persistent, transient, or intermittent. Architectural fault tolerance generally emphasizes such function behavior rather than the specific mechanism causing the fault. There does remain a dependence on the number of simultaneous faults that can be corrected. Some approaches can only correct a single fault, whereas others can correct multiple faults. The fabrication-induced faults have a particular importance because they are intrinsic mechanisms resident within the individual ICs and are not addressable by system designers, whereas the "environmental" faults can generally be addressed at the system-design level.

Addition of fault tolerance leads to additional circuitry and possibly lower speeds. In the case of earlier generations of IC technologies, the amount of circuitry that could be placed on a single IC was so limited that the overhead of fault tolerance generally could not be tolerated. However, a surprisingly extensive amount of research and exploratory development has for several years been directed at developing circuits with many times the functional complexity of an IC by using the entire area (or a large portion of the area) of the monolithic circuits. The probability of a manufacturing fault is virtually 100%, requiring such Wafer Scale Integration (WSI) circuits to incorporate fault tolerance to obtain a functional circuit. Several of these earlier WSI functions have presaged the volition of ICs to today's VLSI (and soon Ultra Large Scale Integration [ULSI]) circuits. Since the total amount of circuitry in an IC has increased, the fraction of circuitry needed for fault tolerance has decreased and fault tolerance has become a more realistic manufacturing approach. A similar change is seen in the area of testing, in which the circuitry for built-in testing used to be excessive for earlier generation ICs but it is now a decreasing fraction of the total area with each successive generation of IC technology. Just as design for testing (using the Joint Test Association Group [JTAG] standard, for example) has become an increasingly necessary functional feature of VLSI ICs, so, too, may fault tolerance. The connection is particularly useful since the built-in testing provided by standards such as JTAG target general fault mechanisms during the operation of the system, with manufacturing defects merely a subset of the total fault mechanisms that are covered.

Chapter 8 reviews approaches that generate correct results for faulty circuits. Some, though not all, mechanisms causing circuits to fail have their origins in defects that are introduced during fabrication of the circuit. In such cases, fault tolerance may help extend manufacturing limits (e.g., the area of a monolithic circuit) of ICs. However, fault tolerance is a capability of particular architectures realized by an IC fabrication process rather than a general approach to enhanced manufacturability of general ICs in a given fabrication process.

Fault tolerance is not an alternative to high yield as IC technologies evolve. Substantial progress along the "learning curve" to a well-controlled, high-yield fabrication process is generally necessary before we can proceed to the next-generation, more technologically challenging, fabrication process. Chapter 8 begins with an overview of the VLSI test problem and test generation algorithms. In particular, a review of the classical D, PODEM, and FAN algorithms, along with more recent heuristics-based approaches such as learning and transitive closure, have been employed to speed up test pattern generation. Also covered are test generation for sequential circuits with partial or no scan access to flip-flop states, testing for bridges, opens and delay faults, and Built-in-Self-Test (BIST) strategies, the IEEE 1149 test standard. Finally, the effectiveness of testing is studied in detail—that is, the relationship between yield, test coverage, and product quality (defect levels or field reject ratios).

The efficiency (minimum additional IC area and speed degradation) with which fault tolerance can be added is a strong function of the specific function architecture, rendering fault tolerance a viable direction for only a subset of IC functions produced on a given fabrication line. For example, the regular organization of memory arrays is well matched to highly effective reconfiguration based on spare rows and columns to replace a faulty element with a functional element. Microprocessor architectures, on the other hand, are more irregular and less favorable for efficient addition of reconfiguration to bypass defective circuitry. For such reasons, fault tolerance is a specialized approach that can extend the ability of an IC technology to implement particular functions, given the underlying IC technology.

Today Multichip Modules consist of complex and dense VLSI devices mounted into packages that allow almost no physical access to internal nodes [5, 6]. The complexity and cost associated with testing and diagnosing MCMs is one of the major obstacles to their current usage. Chapter 9 discusses a set of test solutions for MCM manufacturing. It is meant to cover state-of-the-art test solutions for MCMs and to provide comprehensive knowledge of high-quality test techniques for MCM manufacturing. In particular, emphasis is placed on approaches to testing and burn-in of bare dies; schemes to test unpopulated MCM substrates in manufacturing; advanced techniques to test, diagnose, and repair assembled modules in order to improve MCM assembly yields; and finally, solutions for reusing MCM testability techniques beyond assembly test during subsequent manufacturing levels, that is, board, system, and field-level tests.

REFERENCES

[1] J. Pineda de Gyvez and G. Cheek, "Special Issue on Advanced Yield Modeling," *IEEE Transactions on Semiconductor Manufacturing,* vol. 8, no. 2, May 1995.

[2] C. H. Stapper, "Modeling of Integrated Circuit Defect Sensitivities," *IBM Journal of Research & Development,* vol. 27, no. 6, pp. 549–557, Nov. 1983.

[3] F. J. Ferguson and J. P. Shen, "A CMOS Fault Extractor for Inductive Fault Analysis," *IEEE Transactions on Computer-Aided Design,* vol. 7, no. 11, pp. 1181–1194, Nov. 1988.

[4] S. W. Director, "Optimization of Parametric Yield," *IEEE International Workshop on Defect and Fault Tolerance on VLSI Systems,* pp. 1–19, Nov. 1991.

[5] S. K. Tewksbury, "Wafer-Level System Integration," New York: Kluwer Academic Publishers, 1989.

[6] A. Flint, "Testing Multichip Modules," *IEEE Spectrum,* pp. 59–62, Mar. 1994.

2
Defect Monitoring and Characterization

Eric Bruls

The rapidly increasing level of integration and the decreasing life cycle of electronic systems and integrated circuits demand great flexibility on the part of IC suppliers. They have to be able to supply the market at the right time with the right devices in order to achieve a return on the capital invested in development and production. This requires close examination of the various factors that will determine whether the required profit margin will be achieved.

2.1 MARKET DEVELOPMENTS

The timing of the introduction of ICs on the market with respect to the life cycle of that particular type has a great influence on its potential success. Depending on the type of IC involved, insight into developments in the field of the systems can also be essential. A delay in the assumed acceptance rate of a new system can cause considerable problems because reaching the breakeven point for an investment is dependent on the time between investment and return on the invested capital.

Typically, however, the time to market for an IC tends to be too long. Launching an IC onto the market too late will mean serious loss of market share for the competitors. In such a situation, something innovative will have to be offered in order to regain the lost ground.

2.2 PRICE

Depending on the kind of product involved and the stage of the life cycle for that particular product, the price can be more or less fixed or can be left almost entirely up to the manufacturer. The price of ICs for industrial applications, for example, in workstations or factory control can be relatively high because of the high system costs. ICs for application in consumer goods, for example, in televisions and kitchen appliances, are continuously under great price pressure because of the small profit margins and price sensitivity of this market.

If a company is the first on the market at the beginning of an IC's life cycle, the price can be determined relatively freely. Later, after the competitors introduce similar products, the price will decrease to a certain saturation level.

Important aspects that determine the cost price of an IC are as follows:

- **Design Effort** These costs are very significant in low-volume products. However, for mass-produced products, these costs are less important.
- **Silicon Area** Based on the technology and production line used for manufacturing, a price per square centimeter of processed silicon can be computed. The larger the area of a design, the higher the price of the final product.
- **Production Yield** Because the production of defective devices reflects on the price of correct ICs, the yield of the manufacturing line is an important factor in determining the price of mass-produced ICs.
- **Package Costs** The costs of the package itself and of packaging a device are relatively high. Minimizing the number of defective ICs being packaged and optimizing the packaging yield will reduce the total costs.
- **Test Costs** In view of the increasing complexity of ICs, it is becoming more difficult to test whether such a circuit is functionally correct, and an increasing amount of time has to be spent on expensive general-purpose Automatic Test Equipment (ATE). Presently, test-related costs may constitute a significant part of the total costs for an IC (even up to 50%, especially for complex mixed-signal ICs).

2.3 QUALITY AND RELIABILITY

The customer requirements are not limited to functionality and price of the ICs. In view of the complexity of the manufacturing process and the resulting circuits, inevitably a small fraction of the devices that reach the customer will fail to meet the specifications. The term *quality* is used as a measure for the number of defective products shipped to the customer. The quality of a product is expressed in parts per million (PPM), that is, the average number of failing devices per million shipped to the customer. In the mid-1990s, a quality level of 100 PPM or better was required for complex VLSI chips, and the quality requirements will continue to rise.

The reliability of the product is also important. A well-known phenomenon is the "bathtub"-like failure rate with respect to time as shown in Figure 2-1 [8]. During normal operation of the device, the geometries are placed under electrical stress. Devices without any production-based deformation can withstand this stress, but devices with marginal deformations that do not affect the functionality may cause a relatively high level of early-life failures or infant mortality. After this initial

2.4 ■ IC Manufacturing Defects

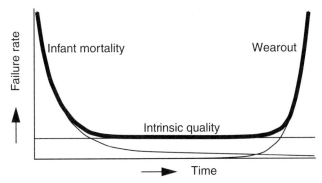

Figure 2-1 A traditional bathtub reliability curve.

period, the failure rate falls to a constant level throughout a large part of the IC's lifetime. When nearing the end of this period, the failure rate again increases due to wearout mechanisms. For specification reasons, the failure rate is recomputed for a standard amount of time (10^9 seconds) and expressed as Failures in Time Standardized (FITS). In addition, with the increasing level of integration and cost constraints, product reliability more and more implies requirements for Wafer Level Reliability. This implies very stringent requirements for the integrity of the production process.

This chapter discusses the occurrence of defects in the IC manufacturing process. First, the IC manufacturing process is introduced in order to give the reader a better understanding of the various defect mechanisms. Next, the chapter discusses how information about the occurrence of defects can be gathered in a production environment in order to optimize the entire process. Finally, the information gathered is used to model the defects by means of a number of characteristics.

2.4 IC MANUFACTURING DEFECTS

The whole development and manufacturing flow has an impact on the ultimate quality and reliability of an integrated circuit. A device's failure to operate properly can have two different causes: either design or manufacturing-related. Making a high-quality product requires close interaction between three areas of interest: design, manufacturing process, and testing. As indicated in Figure 2-2, a poor link in this chain may limit the achievable overall result. Some of these interactions are well understood and widely applied, whereas others are still under development. For example, the design rules originating from the process limitations clearly form a limiting condition for the design. It is probably just as clear, though not as widely applied, that information about the processing and statistics relating to defects can be used to direct the test efforts.

Design Failures. The increase in the complexity of ICs has made it a very tedious task to ensure that the formally specified functionality indeed implements the requirements at system level. Problems with this translation can cause system

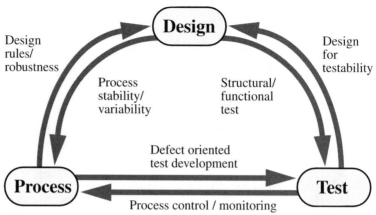

Figure 2-2 Interaction between the various areas of interest in producing high-quality ICs.

failures that are difficult to pinpoint and might be found only after introduction on the market. Only extensive design validation at all levels of development can help to detect such problems before the market introduction.

Another frequently occurring design-related failure involves timing problems. The processing is specified to achieve parameters within specific margins. Because of the complexity of modern ICs, critical paths will exist in the logic circuitry, which only just satisfy the required performance. However, the accuracy of the simulations is limited due to technical as well as economical restrictions. Furthermore, it is not possible to simulate all possible environmental combinations (e.g., temperature) and processing conditions. Such design failures are normally found during the characterization phase and can be solved by a redesign.

Manufacturing Defects. By means of the validation and characterization step in the IC development flow, a level of confidence can be achieved concerning the correctness of the design. However, manufacturing defects are a source of continuous concern. A manufacturing defect can occur at any stage of the manufacturing process, causing the device to fail the specifications. On the one hand, it is important to reduce the number of manufacturing defects (increase production yield) by strict process control strategies, but, on the other hand, effective test techniques remain necessary in order to achieve the required quality level.

The following section presents an introduction to the complete IC development and manufacturing flow and pinpoints the various mechanisms that contribute to the occurrence of a defect.

2.4.1 IC Development Flow

Implementing an idea on system level in one or more ICs involves many small steps. One way of representing this development is illustrated in Figure 2-3, which shows the flow from high-level to low-level specifications, the manufacturing, and the various levels of testing.

2.4 ■ IC Manufacturing Defects

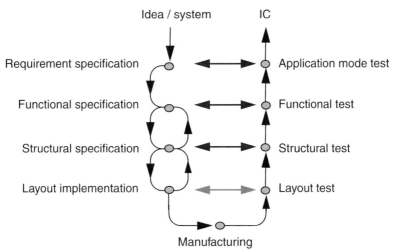

Figure 2-3 IC implementation flow.

Design. The development starts with an idea for a system. Once the various ICs are partitioned off, a requirement specification can be written for each IC. This is often not a formal description using a specific hardware description or programming language, but rather is a plain textual description of the high-level behavior of the complete IC.

The second step is to translate this requirement specification into a formal functional description. This level describes the functionality of the complete IC on a level of, for example, adders, multipliers, and registers. For this purpose, formal hardware description languages, such as VLSI Hardware Description Language (VHDL), or normal programming languages (such as Pascal, C) are used. The functional specification can be simulated to verify the correctness. However, this part of the development flow is a weak point and remains a source of concern because formal verification is impossible owing to the lack of a formal requirement specification.

The formal description at the functional level can be used to generate the structural specification (gate-level description) and layout implementation fairly quickly. It is common practice to apply synthesis tools for these translation steps, whereas verification of the correctness is possible as a result of the formal approach.

Manufacturing. The resulting layout specification has to be implemented on silicon. For this purpose each layer of the layout is considered separately, and several processing steps are required for each of them. A brief description of the manufacturing process is given in the following section. It is important to notice here that, owing to the complexity and sensitivity of various processing steps to disturbances, an IC will likely contain a defect causing it to fail one or more specifications. Depending on the size of the chip and the stage of development of the IC as well as the process, the percentage of fault-free processed VLSI circuits (i.e., the production yield) can have virtually any value. For complex innovative ICs and immature processes it may be as low as 10% or even less, while for mass-produced ICs in a mature process the production yield may be as high as 95% or even more.

Testing. To ensure that the manufactured ICs perform the specified tasks, various tests have to be applied. Depending on the product development stage (e.g., prototype or mass production), this test can be very extensive or just minimal to check the correctness of the production process. The three types of tests that can be distinguished are:

1. **Validation Test** To check whether the device actually implements the specified functionality and can be applied in the complete system. Once the design has been shown to be correct, it does not need to be tested again.
2. **Characterization Test** To check under which conditions the IC is able to perform according to the requirements. This includes checks on process variability, temperature variations, and power supply changes. This is done for a small sample size and only needs to be carried out once for a design and a given process.
3. **Production Test** To check that no abnormal processing conditions or local disturbances occurred during the manufacturing process, causing a particular IC to malfunction. Because such a defect can occur on any device, each IC has to be tested for its correctness with respect to the processing.

Each of these three tests can be performed at various levels of abstraction, which are indicated in Figure 2-3. At the lowest level of abstraction, we can test whether the layout as specified by the designer is actually realized on silicon, bearing in mind certain process margins. Despite the direct relationship between a test and the detection of defects, this level of testing is only starting to be applied because of the relatively computation-intensive technique of fault list generation and test vector generation.

For production testing, the structural level is applied in most situations. Because of the available fault models, which are an abstraction of the electrical impact of the defects on the corresponding level of design description, and corresponding efficient algorithms for test vector generation and evaluation, structural tests offer an excellent opportunity to derive a metric for the effectiveness of a test set. For the application of these structural tests, the scan technique often is used as simple controllability and observability improvement for the sequential circuit elements. During the 1980s and early 1990s, the stuck-at fault model was the de facto standard for a large part of the industry in classifying test effectiveness. However, the limited accuracy of the available fault models, especially the stuck-at fault model, in representing the behavior of actual defects and the limitations of the static behavior of a scan-based test is becoming more evident, and it is becoming more and more difficult to meet the ever-increasing quality requirements. Additional and alternative test methods are considered to overcome these problems. These vary from voltage stress measurements to supply current measurements in the quiescent mode of a circuit (I_{DDQ} or I_{SSQ}).

Some defects cannot be detected by structural tests. As a result of a local processing disturbance, the signal propagation delay between two points on the IC might be increased, causing the device to fail the functional specification, although the structure of the connections is not corrupted. Therefore, some tests also need to be carried out at the functional level.

On top of this level, it might be necessary to test the operation of the device in the way it is supposed to function in the system for which it is designed. This is known as an *application mode test*. Here the issue to be checked is not whether the IC implements a certain algorithm, but whether the implementation of that algorithm in combination with the other part of the system performs the expected tasks. In order to obtain enough controllability over the various conditions that might need to be varied, an application mode test is preferably applied on an ATE instead of a bench implementing the application. However, this implies some requirements with respect to the availability of the right input signals and processing of the output signals. Application mode testing is generally applied only for validation and characterization purposes. For production testing, this approach is too costly and will be used only if no other solution exists.

2.4.2 The Manufacturing Process

Given the design of a circuit, the structure of the various layers has to be copied onto the IC. The basic technique used to transfer the designed structure onto the surface of the silicon wafer is photolithography, which is discussed later in this section. For this purpose, the structure of each layer is first copied on a glass plate by selectively deleting an opaque layer (e.g., chromium, by means of an e-beam). This glass plate containing the pattern of one layer of the layout is called a mask.

The number and types of layers that have to be copied onto the surface of the semiconducting material are dependent on the technology used. The most well-known technologies are:

- Bipolar
- NMOS
- CMOS
- BiCMOS (combined bipolar and CMOS)

An in-depth discussion of these technologies with respect to their application areas and limitations is outside the scope of this book. However, the techniques applied for implementing an IC are similar for all of them, and a short overview of some basic techniques is given here as background knowledge to enable the reader to understand the various defect mechanisms.

The main types of layers and geometries used for implementing integrated circuits are [11]:

- **Doped Areas** The operation of ICs is based on the various combinations of differently doped semiconducting material. To achieve a specific doping profile, an N- or P-type dopant material is introduced into the semiconducting bulk, which typically is silicon. For this purpose, two different techniques can be applied: implantation and diffusion.

Implantation

In this technique, an electric field is used to accelerate specific ions in the direction of the silicon wafer. As a result of the high kinetic energy, the ions enter the silicon where they gradually lose their energy due to collisions with atoms. A disadvantage of this technique is the damage caused in the crystal lattice of the silicon due to the collisions; this damage needs to be repaired by means of an annealing step. Nevertheless, this technique is preferred because of the good control it offers over the dopant profile, which is determined by the amount of ions per square centimeter and the kinetic energy of the ions. A masking material like silicon dioxide or resist is required to achieve implantation in specific places only.

Diffusion

To diffuse dopant material into the silicon, a layer of dopant material is deposited on top of the wafer. Where the dopant layer is in contact with the silicon, the dopant will diffuse into the silicon wafer when the temperature is high enough. To obtain a deeper junction or to reduce the surface concentration of dopants, it is possible to apply a second high-temperature step, known as a "drive-in," after the dopant layer has been separated from the silicon. A localized barrier like silicon dioxide can be used to achieve local diffusion. For manufacturing VLSI circuits in modern IC processes, this technique is no longer applied.

- **Oxide Layers** Throughout the whole IC processing, oxidation is an important and frequently encountered technique, either to create an insulating layer or to provide a masking layer for other processing steps. Again two techniques can be distinguished.

Thermal Oxidation

When the bare silicon wafer is exposed to oxygen or water vapors at an elevated temperature, a silicon dioxide film is formed. The rate of oxidation is controlled by the chemical reaction between the silicon and the oxidant as well as by the diffusion of the oxidant through the silicon dioxide film. This technique gives a high-quality insulation layer. To apply localized oxidation, a masking material is applied. This masking material is resistant to the influence of oxygen and water vapors and can withstand high temperatures. Application of thermal oxidation requires a temperature of around 1000°C, which makes it inappropriate when metal (aluminum) has already been deposited, which has a melting point of approximately 550°C.

Deposition

For thermal oxidation, bulk material is used as one of the reactants. To create an insulation layer after deposition of a conducting layer, the deposition technique has to be used. Generally, the Chemical Vapor Deposition (CVD) technique is used; in this approach the deposited material is created by a chemical reaction in the vapor phase on the surface of the wafer or in

its vicinity. These processes take place under reduced pressure or in a vacuum. Unwanted oxide again has to be etched away.
- **Polysilicon and Metal Patterns** The gates of the transistors and the connections between the various electrical elements are formed in these layers. Conducting layers are created mostly by depositing a thin layer of the required material. The CVD technique tends to be used for polysilicon layers, while for metal layers the Physical Vapor Deposition (PVD) technique is applied, where the deposited material does not react chemically during the deposition. The patterns are created by etching away the redundant material after a photolithographic step to duplicate the patterns.

Throughout the implementation of each of these layers, various separate processing steps are required in order to realize the designed patterns. Because a large part of the randomly occurring defects is related to the photolithographic step used for transferring the patterns from the masks to the chip, some aspects will be discussed here on the basis of the processing for the metal layer, as shown also in Figure 2-4.

(a) The base material, a layer of silicon dioxide, is shown. For simplicity, this layer is depicted as one smooth plane. However, in reality, significant height variations can exist as a result of previous processing steps.

(b) On top of the oxide, a layer of metal is deposited by means of a PVD process. For metal layers, this is generally done by sputtering.

(c) Deposition of a photosensitive film, the photoresist, is done using a spin-on technique in which the photoresist is deposited in liquid form and uniformly distributed over the wafer by spinning it around. After a heating step, the photoresist becomes solid.

(d) The mask derived from the designed layout is used to copy the correct structure onto the photoresist. In the example, the photoresist is assumed to be positive, which means that exposed parts will become soluble in the developing solution. Consequently, on those parts of the wafer that are exposed to the light, the metal can ultimately be removed. Modern IC technologies typically apply positive photoresists because of the higher resolution compared with existing negative photoresists.

(e) Now the photoresist can be developed, and exposed parts are removed by the developing solution.

(f) The remaining photoresist structure protects part of the metal layer during the etching step. The unprotected metal parts are removed.

(g) The remaining photoresist no longer serves any purpose and is removed, leaving the original silicon dioxide layer and on top of it a metal layer patterned according to the layout.

At this stage, a significant statement can be made about the defects. It can be assumed that the masks that are used for the photolithographic steps are correct. These masks undergo an automatic check by means of a die-to-die or a die-to-database comparison, which means that any defect will very likely be detected. Consequently, it is unlikely that any part of the photoresist which is not intended

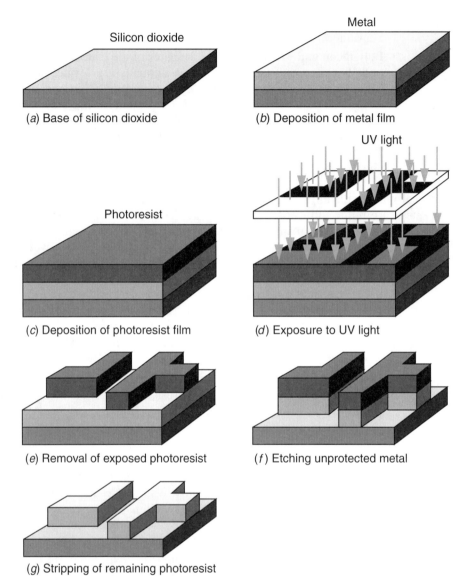

Figure 2-4 Various processing steps for defining the metal structures.

to be exposed to the light will actually be exposed. Translating this to the wafer the structure is supposed to be copied to, we find that it is unlikely that too much material will be etched away. The exception to this phenomenon might be some pollution in the photoresist itself, which is removed by the developing solution. In such a situation, the underlying material might be etched away without requiring the photoresist to be exposed.

However, any obstacle for the light, for example, a dust particle on top of the photoresist, will prevent part of the photoresist from being exposed and will thus cause some additional material of the deposited layer to remain on the final IC. Therefore, the likelihood that an extra material defect will occur is much higher than for a missing material defect.

2.4.3 Defect Mechanisms

In view of the decreasing minimum feature sizes and the complexity of the total manufacturing process, many things can cause the produced ICs to fail the specifications. The most important mechanisms that lead to the occurrence of defects are discussed briefly below [1, 9, 10].

1. **Wafer Defects** The bulk material that forms the basis for all devices is the bare silicon wafer. Although techniques for producing these wafers have improved considerably, problems such as contamination and micro-cracks cannot be eliminated completely. By causing a shift in basic parameter values, contamination as well as micro-cracks, can result in problems concerning the performance of the IC or elements that lie in the affected area.

2. **Human Errors** The impact of human interaction with the IC manufacturing process was identified long ago as a significant and uncontrolled source of defects. In addition to pollution in the air caused by humans, which may result in defects on the wafers, more major problems can occur—for example, scratches over wafers as a result of careless handling, or in some cases complete process steps forgotten or even steps done twice as a result of poor administrative discipline or badly organized logistics. In mass-production sites in particular, the human factor is being reduced as much as possible by ever-increasing automation.

3. **Equipment Failure** There is also the risk of equipment failure, whether or not the equipment is automated. Two problems here are similar to those caused by human error. First, equipment also can contaminate the air, be it a rubber belt for transport of the wafers, or a vacuum clock for evaporation of metal, which also adheres to the side-walls and can fall off at a later point in time. These contaminations in the air can fall on a wafer and cause defects on any position. Second, if the equipment is not tuned correctly, handling of the wafers by machines can also cause problems. These problems, if present, often have a systematic behavior, which makes it important to identify and solve them as soon as possible.

   Equipment failures are the main defect mechanism in modern manufacturing lines. They can be minimized by strict maintenance planning. The intervals between maintenance are decided by weighing up the costs arising from the maintenance and the possible reduction in production yield if it is postponed. The maintenance costs consist of the time in manpower and nonoperational hours of the equipment, but also of a short, maintenance-related reduction in yield as a result of the disturbance to the operational conditions.

4. **Environmental Impact** For a large part of production time, the wafers are in contact with the air inside the production building. Any contamination in this air which is the same size as or bigger than the minimum feature size of the process can cause functional problems if it falls on top of a wafer. Huge investments are necessary to reduce the number of particles in the air to an acceptable level, but the presence of particle sources inside the manufacturing building implies that it is impossible to achieve an environment without particles. Therefore, it will remain necessary to test each IC for the presence of defects and the resulting faulty behavior.
5. **Process Instabilities** Some of the processing steps applied are critical and susceptible to variations in process conditions. For example, in CVD techniques, the locally deposited amount of material and the variations in layer thickness are largely dependent on the regularity of the gas flow. Turbulent gas flows, which may be caused by obstacles in the flow or around the injection point, will result in strong thickness variations and thus parametric differences over the wafer. Other process instabilities, like the furnace temperature during an oxidation step, cause some parameters to change more globally over a wafer.

With respect to the area affected by a defect, two classes of defect mechanisms can be distinguished: global defects and local defects [10].

Global Defects. As indicated earlier, some of the defect mechanisms affect large areas or even a complete wafer. All circuitry in the affected area is influenced by the global defect in a similar way, taking into account the variability of the effect. Examples of global defects are mask misalignment causing a different parametric behavior, line registration errors (lines too wide or narrow as a result of variations in the etching time), and different implantation levels causing a shift of transistor parameters. Detection of such defects might be done by monitoring certain parameters at a few locations of a wafer.

Local Defects. Contrary to the global defects, local defects have an impact on only a small area. Typical examples are dust particles or pollution by chemicals affecting the photolithographic process, scratches due to handling, and cracks. These defects may cause more ICs to be defective on one wafer, but this can only be confirmed by measuring the ICs themselves.

Another aspect that can be used to categorize the defects concerns the electrical impact, the "fault." As for the area-related division, it is not always clear in which category a specific fault falls, and evaluating a defect on its electrical impact is even more difficult because of the circuit dependency. Nevertheless, it is possible to identify two different classes of faults, which are the parametric faults and the functional faults.

Parametric Faults. Some defects tend to change only the parametric behavior of the resulting circuit, such as a reduced threshold voltage or increased resistances of connections. Although these effects can be caused by either global or local defects, the nature of these defects in an actual processing line and the susceptibilities of the various processing steps imply that most global defects result in parametric faults. These kinds of faults can be difficult to detect in an IC because they may only cause the device to fail some performance-related specifications. A very specific test is therefore required.

Functional Faults. The other type of defects tends to cause a catastrophic failure in the behavior of the device, ranging from a logic failure of an output under certain limited input conditions to completely incorrect operation of the device, regardless of the input signals. A simple example of such a functional fault is a circuit of which the output is connected through a short with one of the supply lines. The output level will virtually remain constant, and thus the IC cannot perform the specified functionality.

This classification of the defects and their electrical impact forms the basis of any defect monitoring and characterization work and the test strategy applied for their detection. There is one additional aspect concerning the interaction between defects and the impact on the IC. A local as well as a global defect is defined as an out-of-spec difference between the designed and the actually implemented geometry. Not all of these defects will cause a significant change in the geometry (be it a dopant profile or some metal patterns) of the IC with respect to the parameters or the connectivity. In particular, the defects that cause a deviation that is only a little larger than the allowed process variations may not affect the electrical behavior of the IC in any stage of its lifetime and thus are less important considerations. When a defect clearly changes the geometry of the IC with respect to process parameters or connectivity, it will be called a hard defect. A defect that does not cause a clear geometric change is called a soft defect. The defect modeling section of this chapter will present a more formal definition of hard and soft local defects.

2.5 DEFECT MONITORING

Given that the occurrence of defects during the IC manufacturing process cannot be prevented completely, it remains necessary to minimize the impact of major defect mechanisms and thus the number of defects. Because the costs of the defect-free devices is almost inversely proportional to the yield of the production process, this defect reduction is an important aspect of making the IC manufacturing business viable. However, a situation in which no defects are caused throughout the manufacturing process will not be reached as a result of economical considerations. In such a situation, it will be beneficial to increase the level of integration to implement more and faster circuitry on a single IC, using the available manufacturing processes again to the limits. The sooner the process and defect control programs are effective and result in a mature production process, the sooner new technologies can be developed and applied.

A reduction in the number of defects is possible only if enough information is available about the up-to-date condition of the process and the processing steps that cause most of the defects. A straightforward method to gather this information seems to be evaluation of the defective products [19]. However, this approach has some serious disadvantages.

- **Fault Diagnosis** In a production environment, the aim is to detect defective devices as soon as possible, while incurring minimum costs. Consequently, after a device fails a test, it is not tested any further. However, even if all results of the complete production test are available, it is virtually impossible to identify the cause of the defect. On the one hand, not every defect will cause a uniquely identifiable electrical behavior (e.g., two nets lying next to each other in the polysilicon as well as the metal layer can be shorted together in both layers, resulting in the same fault behavior). On the other hand, the computation of the fault signature for each defect will require unrealistically long simulation times for even a relatively small IC.
- **Long Feedback Loop** To make any decisions for process control purposes, it is important to have up-to-date information on the state of the process. However, the complete processing required for integrated circuits generally takes about four to six weeks. This means that problems that occurred in the beginning of the processing may only be detected after a month has passed. By that time, all products may contain the same defect and a whole month's production of ICs could be lost. Obviously, this situation is intolerable, and alternatives must be found.

Considering these issues, defect monitoring techniques will have to be able to identify the defect location and cause in an easy way. Furthermore, the whole time span between the moment of data acquisition through the processing and the moment the interpreted data on defects becomes available should be as short as possible. A variety of such defect monitoring techniques for global as well as local defects are discussed in the remaining part of this section.

2.5.1 Global Defects

The occurrence of a global defect in a large-volume production line can be very costly, because within a relatively short time period a whole batch or even several batches can be affected. To detect and solve the presence of such global defects effectively, the quality of the processing is monitored regularly throughout and after the manufacturing phase.

Process Control Monitors. A well-known method to gather this kind of information is based on specially designed test modules, also called Process Control Monitors (PCMs) [23]. These PCMs consist of basic structures, such as single transistors, single lines of conducting material, and chains of via contacts. In most cases, each wafer with products also contains about five PCMs distributed over the wafer to monitor the correctness of the processing. Figure 2-5 shows such a configuration of products and PCMs on a wafer. An increasingly common alternative being used nowadays for VLSI circuits is to place the PCMs in the scribe-lines

2.5 ■ Defect Monitoring

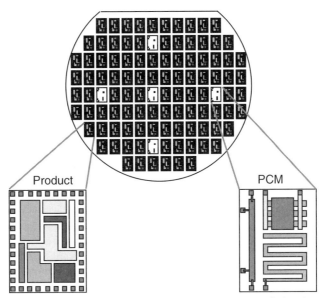

Figure 2-5 Configuration of products and PCMs on a typical wafer.

between the products, which are necessary to be able to cut the wafer into separate ICs.

At some stages throughout the manufacturing process, the process quality can be checked by carrying out some in-line measurements on the PCM structures. This can, for example, be an automated inspection of the realized linewidth by means of a light or Scanning Electron Microscope (SEM)-based measurement. Each PCM can contain for each layer a set of patterns that are representative of the structures in the product and can thus be used for this visual inspection.

After the wafer has been processed completely, some simple electrical measurements can be used to determine whether all electrical parameters, such as sheet resistance of the conducting layers and threshold voltage of the transistors, are within the specified range. Only when all parameters satisfy the requirements can the wafer be said to have been processed correctly, and dies subsequently will be tested on their correctness.

Parameter Monitoring. Another technique complementary to the PCMs is the parameter monitoring approach [16]. Whereas the PCMs are used to measure rather low-level parameters, it is also possible to translate these parameters to a higher-level behavior that can be monitored. Because most global defects cause a parametric failure, a performance-related test of a well-chosen monitor design can show the presence of certain parametric variations.

A circuit that is quite popular with some manufacturers for these applications is the ring-oscillator. The resulting oscillation frequency is dependent on the parametric characteristics of all the building blocks. However, the results of this technique tend to be susceptible to misinterpretation. Several questions need to be answered before the performance of such a parameter monitor can be used safely to check the parameters and thus the performance of a product.

- Is the performance of the monitor correlated well enough to the performance of the product? A whole set of conditions may influence the performance of the monitor, while the performance of the product might be influenced by other parameters or at least in a different manner.
- How greatly do the parameters vary over the wafer? Can we be sure that the parameter values do not vary significantly for the product and the monitor?
- On top of the global defects, which cause global parameter variations, the parameters vary locally in a random way as a result of nondeterministic process circumstances. Is the monitor as sensitive to these random variations as the product?

Unless these questions have been considered before the application of a parameter monitoring approach at a high level, the possible impact is uncertain. Misinterpretation of the measurement results can result either in defective products passing the tests and thus reducing the quality level or in good products failing the tests and thus increasing the manufacturing costs unnecessarily. Without prior investigation of these items, this technique is not recommended.

2.5.2 Local Defects

A prerequisite to produce ICs with high quality and yield is the cleanliness of the base material, that is, the silicon wafers. Particles on top of the wafers, when not removed, might impact the processing and electrical behavior of the product. To ensure that the number of wafer pollution-induced local defects is minimized, the wafer vendor specifies a maximum number of particles on the wafer that are larger than a specific size threshold. These particles can be counted by means of a (laser) light inspection technique [14]. If this number is too high, either the quality of the incoming wafers will have to be improved or the cleaning steps will have to be made more effective.

Experience shows that most processing-related local defects tend to occur in a few layers of the complete process. For modern CMOS IC processes, defects in the gate-oxide and interconnect layers form the vast majority of all defects. This can also be explained from a processing point of view, as will be discussed further in this section. For monitoring the occurrence of these defects, two alternative approaches can be applied. The *in-line monitoring technique* makes use of different inspection techniques for test wafers and products at various stages in the manufacturing process. The *defect monitoring technique* applies specially designed structures that are optimized to simplify the detection of specific frequently occurring defects.

In-line Monitoring. One of the most important techniques for achieving a high production yield and good product quality is the in-line monitoring approach. At various stages in the process, after the implementation of known critical pro-

2.5 ■ Defect Monitoring

cessing steps, the result is inspected. Two different inspection techniques are applied for this purpose.

1. **Surfscan** This technique applies a bundle of light and evaluates the reflections in order to count the number of particles on a surface. Mainly for inspection of layer deposition equipment, this technique is very efficient and provides fast feedback. The technique is mainly applied for (though not limited to) inspection of unpatterned wafers. In Figure 2-6, an example of such a surfscan is shown for a wafer with many particles.

2. **Image Evaluation** By means of a manual or automated (e.g., a KLA™) image inspection system, it is possible to check the occurrence of local defects on patterned wafers. By applying this inspection on a sample basis at a few critical points in the production process—for example, after processing of the poly layer and each of the metal layers—the current status of part of the processing line can be monitored effectively.

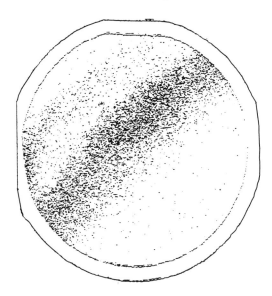

Figure 2-6 Surfscan of a wafer with a haze of contaminants; the majority is much smaller than 1 μm. The regular pattern implies that a systematic physical phenomenon causes the particles.

Both techniques focus on the essential part of the total quality improvement chain, that is, the quality of each processing step separately. Only when the quality of these basic steps is controlled well enough will the result of the total processing stand a chance of satisfying the overall requirements.

Gate-oxide Monitors. The formation of the gate-oxide layer is a very critical process step, which is susceptible to many contaminations and process disturbances. Because for any CMOS process the thickness of the gate-oxide is the smallest dimension in the complete process and lies in the order of only 10 nm for modern processes, nonideal processing conditions are very likely to cause problems in this layer. For correct operation of transistors, a good insulating gate-oxide layer is required. Short-circuits between the channel and the gate of the transistor through the gate-oxide may result in parametric as well as functional faults. The fact that gate-oxide shorts have been found to be unstable and to degrade with time [18] means it is important to minimize the number of these defects.

Only simple test structures are required to detect information about the contamination problem causing gate-oxide shorts. A typical test structure is a combination of gate-oxide capacitors of various sizes. The gate-oxide serves as the insulator, while the active doped area and the polysilicon serve as the two plates. Defects causing a poor quality gate-oxide layer can be detected easily by a simple current leakage measurement. Furthermore, the size of the capacitance is a good measure for the thickness of the gate-oxide layer, which is also a very important parameter (and is also monitored in the PCMs).

In order to obtain some background information on the cause of the defects, various sizes and shapes are used for the plates of the capacitors. Whereas some defect mechanisms are related to the area of the gate-oxide (contamination in the air), others are strongly correlated to the length of the edge of the gate-oxide (impact of polysilicon structure development). By applying capacitors with different shapes and extreme ratios on these variables, differences in the number of defects can indicate the main defect mechanism.

The production of these test structures requires only two masks (polysilicon and active) and the corresponding processing steps. Contrary to the PCM structures, which are placed on each product wafer, one or several monitor wafers contain only a repetition of this gate-oxide monitor in order to satisfy the requirements for a defect monitor. Fault diagnosis has become feasible through some electrical measurements (catastrophic versus parametric defects, area versus edge-related), while the time taken to gather the information has been reduced drastically by applying only two masks for these special monitor wafers.

Interconnect Monitors. The other important source of defects is the interconnect layers, such as polysilicon and metal. To form the conducting layer, a large amount of material is deposited on top of the wafer and afterward removed again partially. Consequently, during this process step a large number of particles are generated, and these form a possible threat as contamination particles. These particles are removed as much as possible in various cleaning steps, but with increasing contours throughout the manufacturing process, the cleaning becomes less effective. Therefore, the combined effect of an increased number of particles generated by the process and the reduced effectiveness of cleaning steps make the interconnect layers susceptible to the occurrence of extra material defects that might cause short circuits.

Detection of defects in the interconnect layers can be translated to the detection of short and open circuits. These faults can be detected by means of simple resistance measurements when the right test structure is chosen. Based on this condition, various test structures have been developed, of which Figure 2-7 shows three basic types [2, 12, 17, 24]. In all three types, all shaded patterns are implemented in one layer of polysilicon or metal, which requires only one mask for such a monitor. Therefore, the requirement of a short feedback loop of the information gathered is satisfied. Defects are detected through the analysis of resistance measurements.

2.5 ■ Defect Monitoring

These measurements can be done easily by probing the wafers. For placing the probe-needles, small pads have to be connected to the ends of a meander structure and to the base of each comb structure. These pads do not, like bond-pads for real devices, contain protection and driver elements, but only need to contain a plane of conducting material in the same layer as the core of the test structure.

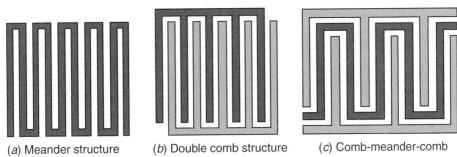

(a) Meander structure   (b) Double comb structure   (c) Comb-meander-comb

Figure 2-7 Basic types of test structures for interconnect layers.

For each of these structures a number of generic aspects can be considered during the design phase.

1. The area of such a defect monitor should be dependent on the expected number of defects per unit area. In order to minimize the number of measurements, the area should be as large as possible. However, to reduce the probability of more than one defect falling in a monitor, the permitted area for a monitor is limited. A first-order analysis satisfies for determining the optimum monitor die size, because exact defect data, of course, are not available when the monitor is being designed.

   Assume, for example, a process line that should be able to produce products with an area of half a square cm and a yield of 75%. For that particular process line, approximately half of the faults are known to occur in the metallization process, equally distributed over the two metal layers. To obtain accurate information about the character and number of defects, the yield of the monitor should not be lower than about 50%. However, we want to obtain accurate data not only when the process is at target, but also when the defect density is increased. A factor 2 more defects than targeted should be accounted for at least. For a single-layer metal monitor that should have a yield of approximately 50% for a defect density twice as high as targeted, the maximum area of one test structure is thus approximately 2 square cm. Whether the resulting resistances to be measured lie within the specifications of the equipment has to be checked.

2. The feature sizes in these structures determine the sensitivity to the occurrence of electrical faults as a result of the defects. Application of more structures in one monitor with various feature sizes gives additional information in this respect about the behavior of the defects and the defect mechanisms.

Depending on the planned technology strategy, various combinations of feature sizes can be applied for a monitor. Assume, for example, a process with design rules of 1.0 μm for the spacing and width of metal lines. When trying to gather some information concerning the capabilities of the present process line, it would be interesting to include some modules containing feature sizes of 1.2 μm and 0.8 μm. When large differences in defect density exist for the 1.2 μm and the 1.0 μm modules, the specified design rules use the capabilities of the process line to the limits and a process shrink still is out of the question. When only a large difference in defect density is obtained between the 1.0 μm and 0.8 μm modules, the currently applied design rules lie well within the capabilities of the process and attention could be spent on developing a shrunk version of the process. When no large differences in defect density are obtained at all, either the quality of the processing is very poor (high defect density) or the presently applied design rules are too wide and don't make full use of the processing capabilities (low defect density).

3. The meander structure (a) provides the opportunity to detect open circuits through an increased resistance of the meander from one end to the other. The double comb structure (b) can be used to detect short circuits, also by means of resistance measurements between the two combs. Any extra conducting material connecting both combs will reduce the resistance significantly. Finally, the comb-meander-comb structure (c) combines both previous types, enabling the detection of both open and short circuits with one monitor structure.

The meander structure can be used to detect the presence of defects easily, but it does not provide any diagnostic information about size or location. The double comb and the comb-meander-comb structure in addition can be used to provide some information concerning the size of short circuits. A small defect only connects two wires together and causes a single conducting path between both combs. A larger defect can connect more fingers of the combs together, creating two or more parallel connections and a corresponding reduction in the resulting

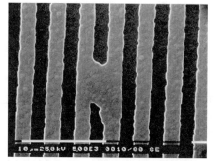 

(*a*) A small short between two lines.   (*b*) A large short between multiple lines.

Figure 2-8 Two SEM photographs of extra material defects detected by such interconnect monitors. The light structures are conducting material.

resistance between them. For the comb-meander-comb structure, it is even possible to distinguish between situations with one, two, or three defects. Figure 2-8 shows the effect of spot defects on an array of lines.

2.6 DEFECT MODELING

Defect monitoring activities are a valuable technique in improving the quality of the manufacturing process. Process engineers may use such monitors to determine defect densities and to identify the physical phenomena causing the defects. They don't require a more abstract defect model. For yield estimation purposes and for defect-oriented test strategies, however, such an abstract defect model is necessary. Therefore, the subject of modeling defects has received much attention throughout the years. In view of these applications, this section discusses some modeling aspects for global and local defects, respectively.

2.6.1 Global Defects

Modeling global defects is not an easy task. The parameter values implemented over a wafer are the result of the permitted random process variations combined with the impact of any defects [16]. An example of such a combined impact is shown in Figure 2-9. The difference between the impact of global defects and allowed process variations is less clear than for local defects.

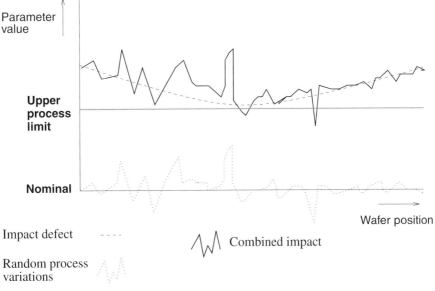

Figure 2-9 Parameter value affected by random process variations and a global defect.

The relative relationship between the amplitude of the random process variations, the impact of the global defect, and the distance between the nominal and extreme processing values determines whether the impact of a global defect will be identified without a doubt or whether the local variations will leave some room for uncertainty.

As can be seen in Figure 2-9, the effective parameter value still lies within the specified process window at two places. On the one hand, this may mean that a product on that location can still be functionally correct, although on the basis of the PCM results the whole wafer will be rejected. This phenomenon causes correct devices to be discarded as being defective, which is also referred to as a Type I test error [26]. On the other hand, even when all PCM results are within specification, the parameter values of all products do not necessarily have to be correct. If the parametric behavior of the ICs is not tested in another way, some defective devices might pass the tests, which is called a Type II test error.

Both types of test errors have a significant impact on the economic viability of the production of ICs by increasing the production costs (Type I) and reducing the quality (Type II). They should therefore be avoided as much as possible. The present approach to solving this problem lies in a very rigorous control of the process variations with respect to the design robustness. The Statistic Process Control (SPC) programs aim at reducing and optimizing these process variations through statistical evaluation techniques.

This approach is reflected in the requirements for the CP and CPK factors [13]. The CP factor is a measure of the process's capability to meet the specified requirements. The CPK factor is a measure of the process's capability to achieve parameter values that lie in the center of the specification range. They are defined as:

$$CP = \frac{USL - LSL}{6\sigma} \tag{2-1}$$

$$CPK = \text{minimum}\left\{\left(\frac{USL - \mu}{3\sigma}\right), \left(\frac{\mu - LSL}{3\sigma}\right)\right\} \tag{2-2}$$

where USL = upper specification limit
LSL = lower specification limit
σ = standard deviation of process parameters
μ = mean value of process parameters

Improving both factors will reduce the probability of Type I or II test errors. One approach to achieve this reduction is to improve the robustness of designs and thus increase the specification window. An alternative approach is to obtain better control over the process parameters in order to reduce the process spread and improve the centering capability. Some examples of parameter distributions, specification limits, and the corresponding CP and CPK factors are indicated in Figure 2-10. Whereas typically values for CP and CPK are required to be larger than 1.3, discussions are in progress as to whether a value of 2 (corresponding to the 6-sigma design approach) would be required to ensure high-level quality and reliability.

2.6 ■ Defect Modeling

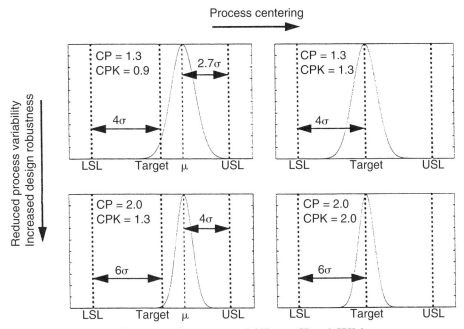

Figure 2-10 Impact of parameter variability on CP and CPK factors.

2.6.2 Local Defects

In general, a local defect can be described as a significant difference between the designed patterns and the structures actually implemented on silicon. The term *significant* indicates that the difference between both structures is larger than is allowed on the basis of the process window. To generalize this description on a higher level of abstraction, the following two assumptions are made.

1. Each defect is assumed to be a certain amount of extra or missing conducting material. Independent of the defect mechanism and the real type of material, such a characterization can be done without loss of general validity. It is only when one comes to the electrical modeling of the defect that the exact type of material may be important.
2. Each defect is assumed to have a circular shape. In reality, no defect is exactly circular, but most defects on an IC tend to have a circular shape. The first reason for this lies in the fact that the distribution of shapes for the dust particles affecting the processing has a natural peak for circular forms. In addition, the surface tension of materials used in the IC processing tends to make all forms circular.

The three main characteristics of a defect described in the remaining part of this section are the defect density, the defect size distribution, and the cluster behavior. In addition to the characterization of the defects themselves, the impact of the defects is determined by the combined defect and layout geometry. A discussion of hard and soft defects will therefore conclude this section.

Defect Density. The defect density specifies for each layer and defect type (extra and missing material) the number of defects occurring per unit area. In computing this information from data gathered with a defect monitor, it has to be considered that not all defects will be detected. Depending on the defect size distribution and the sensitivity of the design (see the discussion on hard versus soft defects in this section), the number of detected defects has to be corrected by a certain factor to obtain the real defect density number.

Semiconductor companies guard their process defect density data, and hence it is difficult to present such data here. However, Table 2-1 gives an overview of some typical relative defect densities for various defect types, where the defect density of extra material defects in the metal layers has been set arbitrarily to 100. The extra conducting material defects may result in short circuits, either inside one layer (metal, polysilicon, or doped silicon) or between two layers (pinhole through an oxide layer). Missing conducting material defects may cause open circuits. For the thick-oxide layer, this corresponds to a via-contact being broken, whereas for the gate-oxide no equivalent defect exists because no contact-holes are designed in those areas.

TABLE 2-1 Some Typical Relative Defect Densities for a CMOS Process.

	Relative Defect Density	
Layer	Extra Conducting Material	Missing Conducting Material
Metal	100	1
Polysilicon	50	1
Thick-oxide	2	5
Gate-oxide	20	—
Doped area	1	1

The aim of Table 2-1 is not to provide accurate numbers for a specific manufacturing line, because this would not be useful. Each manufacturing line has its own characteristics and defect density numbers. Rather, the table seeks to provide an impression of the overall ratio in the number of defects for each type.

For extraction of the defect density numbers from measured data, two approaches can be used which give different results.

1. When using the defect density data only for process control purposes, the variability and relative numbers form a solid basis for corrective actions. In this case, a kind of engineering defect density can simply be derived by dividing the total number of defects detected by the total silicon area monitored.
2. When extracting defect density data for application in yield prediction or other, equation-based analyses, it is necessary to derive the value according to the meaning it has been given through these equations. Although the

defect density number is a characteristic of the defect model representing all defects, the defect monitor only detects those defects that cause a fault (i.e., an abnormal electrical behavior). Therefore, the probability that a defect will cause a fault has to be taken into account when computing the official defect density. The amount of correction is dependent on the sensitivity of the layout and the defect size distribution function. This will be explained in the next section, which discusses the extraction of the defect size distribution parameters from the data measured. Consequently, the official defect density value can be obtained only after the defect size distribution function has been extracted.

Defect Size Distribution. The size of defects is a very important consideration. It has already been assumed that defects are circular deformations of the designed geometry. The size of these defects will, of course, have a significant impact on their possible electrical effect. A set of equations typically used for describing the probability density for the various defect sizes is given by [7, 21]:

$$S(x) = c \frac{x^q}{x_m^{q+1}} \quad \text{for} \quad 0 \le x \le x_m \tag{2-3}$$

$$S(x) = c \frac{x_m^{p-1}}{x^p} \quad \text{for} \quad x_m \le x \le x_M \tag{2-4}$$

$$S(x) = 0 \quad \text{for} \quad x_M \le x \tag{2-5}$$

$$c = \frac{(q+1)(p-1)}{(q+p) - (q+1)(x_m/x_M)^{p-1}} \quad \text{for} \quad p \ne 1 \tag{2-6}$$

$$c = \frac{(q+1)}{1 + (q+1)\ln(x_M/x_m)} \quad \text{for} \quad p = 1 \tag{2-7}$$

where $S(x)$ = probability density for the size of the defects
x = size of the defect
x_m = location of peak in probability density function
x_M = size of largest defect
p, q = shape parameters of the model to be fitted

Equations 2-3 to 2-7 give a general representation of the defect size distribution function, which is a probability density function. Figure 2-11 shows the defect size distribution function for typical values of parameter p. For most situations, the following remarks are applicable with respect to these equations.

- The parameter x_M represents the largest possible defect that may occur on the device. In practice, this parameter is large enough to be replaced by infinity in the equations, thus simplifying the defect size distribution function.
- The presence of the peak in the defect size distribution function is not based on a measured peak. Because the defect size distribution function is a probability density function, integration of this function over all defect sizes should equal 1. Considering the observed x^{-p}-like behavior of the defect

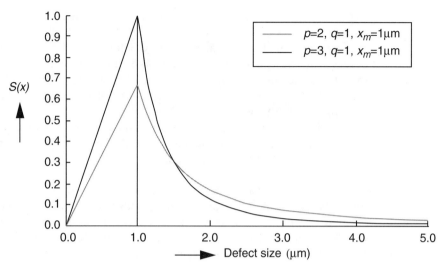

Figure 2-11 Typical defect size distribution functions.

size distribution above a given threshold, this relationship conflicts with that requirement when applied to all small defects. The values of parameters q and x_m are chosen rather arbitrarily and are not related to observed physical behavior. To avoid these arbitrary choices affecting any results derived from this model, x_m is chosen to be smaller than the minimum feature size in the design. For reasons of simplicity, q often equals 1.

The strong increase of defects for smaller sizes can be explained to be natural for two reasons. First of all, when considering the pollution in the air and the particles generated by the processing, it is more likely that a small particle will be encountered than a really large particle. The mechanism that creates a large particle is very likely to create a large number of small particles at the same time. Second, the air in the manufacturing building is filtered and cleaned continuously. However, this cleaning system is designed to remove particles of sizes larger than a specific threshold. The smaller the particle, the less effective the cleaning system will become.

What remains to be determined from the defect data gathered is the actual value of p, the only parameter in the defect size distribution function which is directly coupled to the physical characteristics of the defects. Whereas for short circuits (extra material defects) it is relatively easy to gather information on their size through automated electrical measurements and data processing, for open circuits (missing material defects) this information is much more difficult to collect. A possible solution would be automated visual inspection of the defects, but this is relatively time-consuming work. Therefore, considering the relatively small defect density for missing material defects, the same size distribution function is often assumed for missing material defects as is derived for extra material defects in the same layer.

The exact shape of the size distribution function is very dependent on the processing conditions. However, based on published data and experience, most manufacturing lines can be said to achieve values for parameter p between 2 and 3 (of which the corresponding curves are indicated in Figure 2-11) [3, 21]. The value of p is one aspect that indicates the maturity of the process line. The smaller the value of p, the slower the increase in number of defects for smaller sizes. Such a small increase indicates that the process can be shrunk without encountering a large increase in the number of faults affecting the yield of the products dramatically. In case the value of p is relatively high, strict defect and process control should be applied first to improve the processing yield and quality significantly.

Extraction of parameter p can be done only by a recursive computer-based fit algorithm. Fitting the parameter directly to the measured distribution of the defect size will neglect the influence of the varying probability for a defect to cause a fault and thus be detected. Assume, for example, a double comb test structure for detection of extra material defects causing a fault. The size of the faults is determined by an additional automatic inspection system. The feature sizes of the test structure are s and w for the spacing and the width, respectively. Extra material defects with sizes smaller than s cannot cause a short circuit and thus will not be detected. Defects with sizes larger than $2s + w$ will always cause a short circuit, independent of the exact location on the circuit, and thus are guaranteed to be detected. For defects with sizes between these two limits, the probability of causing a short circuit and thus being detected increases linearly from 0 to 1. Therefore, for fitting parameter p of the defect model, the detected size distribution for the faults first has to be corrected for this detection probability.

Assuming that the inspection system provides a distribution for the fault sizes with only a limited resolution, a correction factor needs to be computed for each size interval. The correction factor equals the ratio between the integration of the defect size distribution function for that size interval and the integration of the defect size distribution function multiplied by the fault probability function. A simple least mean square algorithm can be applied to fit the value of p to the obtained size distribution data. Because the correction factor for the measured size distribution data is dependent on the value of parameter p, an iteration process is needed to fit the value of the defect size distribution parameter to the observed fault size distribution.

Cluster Behavior. An aspect that deals with the overall impact of the occurrence of defects concerns the tendency of defects to occur in clusters instead of being randomly distributed over the wafer [22, 25]. Although this information is not important for determining the electrical impact of a defect on the circuit, it can be very significant when computing the yield on the basis of the defect data.

This cluster behavior causes the Poisson yield model, assuming independent occurrence of defects, to provide yield numbers that are too pessimistic when compared with the actual yield numbers. More than is the case for randomly occurring defects, a device containing at least one defect is likely to contain more defects. Although a quantitative explanation for this behavior is unknown, the relationship between defects through the defect mechanisms (like edge of the wafer sensitivity to handling, equipment disturbance, and local gas turbulence) at least gives an explanation for the existence of cluster behavior.

Two different approaches can be used to extract a parameter for the cluster behavior.

1. The simplest approach is to apply a defect monitoring technique that provides information concerning the number of defects in one monitor. As mentioned in the previous section, based on the comb-meander-comb structure, this is quite possible.
2. If this information is not available, it has to be created artifically. For that purpose, we consider a wafer containing products or monitors that have all been tested. Next, these devices can be grouped together, for example, as indicated in Figure 2-12. This fact again provides us with information concerning the yield of the groups and the number of defective devices per group. Evaluation of this information will give the required parameter for modeling the cluster behavior.

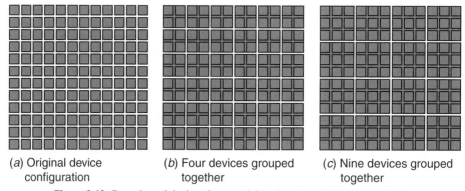

(a) Original device configuration (b) Four devices grouped together (c) Nine devices grouped together

Figure 2-12 Grouping of devices for examining the cluster behavior of defects.

The value and meaning of the cluster parameter depend on the way this parameter is applied. A primary field of application of this cluster parameter is yield prediction. Value and meaning of the cluster parameter in this case are dependent on the yield model. A rather popular yield model is the Negative Binomial Model given in Equation 2-8 [20].

$$Y = \left(1 + \frac{\lambda}{\alpha}\right)^{-\alpha} \quad (2\text{-}8)$$

where Y = yield
λ = average number of faults per device (not defects!)
α = cluster parameter

Note that the meaning of λ in this equation is the average number of faults per device, not the number of defects. Only those defects that have an electrical impact on the device, causing it to fail the specifications (functionality, performance, or reliability), are called a fault. The parameter α in this Negative Binomial Model is a measure for the clustering of defects. The smaller α is within the restriction that it has to be positive; the stronger defects tend to occur in clusters. Typically, α is found to have values between 0.3 and 0.8 [3].

Extraction of the α parameter from measured data can be based on the number of faults in each circuit or monitor. The Negative Binomial Model for the yield is derived from the gamma distribution resulting in the following equation, which describes the distribution of the number of defects per circuit:

$$p(x, \lambda) = \frac{\Gamma(\alpha + x)(\lambda/\alpha)^x}{x!\Gamma(\alpha)(1 + \lambda/\alpha)^{x+\alpha}} \tag{2-9}$$

$$\Gamma(\alpha) = \int_0^\infty e^{-t}t^{\alpha-1}\,dt = (\alpha - 1)\Gamma(\alpha - 1) \tag{2-10}$$

where $p(x, \lambda)$ = probability for x faults occurring in one circuit
λ = average number of faults per device (not defects!)
$\Gamma(\alpha)$ = gamma function

Deriving explicit expressions for x between 0 and 3 gives the following equations:

$$p(0, \lambda) = \frac{1}{(1 + \lambda/\alpha)^\alpha} \tag{2-11}$$

$$p(1, \lambda) = \frac{\lambda}{(1 + \lambda/\alpha)^{1+\alpha}} \tag{2-12}$$

$$p(2, \lambda) = \frac{\lambda^2(\alpha + 1)}{2\alpha(1 + \lambda/\alpha)^{2+\alpha}} \tag{2-13}$$

$$p(3, \lambda) = \frac{\lambda^3(\alpha + 2)(\alpha + 1)}{6\alpha^2(1 + \lambda/\alpha)^{3+\alpha}} \tag{2-14}$$

For fitting the α parameter to the measured data, a simple least mean square algorithm can be applied for comparing the measured data and the theoretical distribution and for selecting the best value. In this search procedure, only parameter α is unknown, because the average value of faults per device can be computed by dividing the total number of detected faults by the total number of devices. A maximum value of 3 for the number of faults per device to be included in the search algorithm is high enough for realistic situations. In practice, the vast majority of the monitors will contain 0, 1, or 2 faults. If the average number of faults per monitor becomes much higher, the evaluation technique loses accuracy, and thus a smaller area should have been chosen.

Hard Versus Soft Defects. When discussing the defects related to the IC manufacturing process, each difference between the designed structure and the implemented structure, larger than is permitted by the process variations, is considered. However, not all of these defects will affect the electrical behavior of the device, neither at zero-hour testing or throughout the lifetime. Considering the defects that may have an impact on the electrical behavior, we can distinguish two groups, based on the defect and layout geometry [4, 5, 8].

- **Hard Defect** A defect that physically shorts two or more lines together or disconnects one signal line into two disconnected parts. Most hard defects cause a functional fault that is detectable by simple DC tests.
- **Soft Defect** A defect that reduces the distance between two lines but does not yet connect them together (a near short) or a defect that locally reduces the width of a line (a near open). Soft defects may affect the performance of the device only under specific conditions or may have a significant impact only after degradation under the influence of operational stresses, causing reliability problems. Although the relationship between the reduced space between two lines as a result of an extra material defect and the degradation probability is not quantitatively known, a defect is defined to be a soft defect when the space is larger than 0 but smaller than d_{max}. The meaning of d_{max} is also explained in Figure 2-13. Similarly, a missing material defect is defined as a soft defect when it does not break a line completely but reduces the width to less than w_{max}.

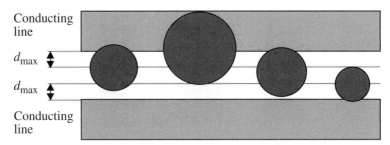

Figure 2-13 Examples of soft defects: the space between lines is reduced to d_{max} or smaller as a result of extra material defects, but the lines are not shorted together.

Given the process quality and the defect model, the percentage of defects having a significant impact on the operation of the device, either at zero-hour or during the lifetime, is dependent on the layout of the circuit. The density of the layout determines how sensitive the circuit will be to the effect of the defects. Based on the layout of the product, it is possible to compute, for each defect size, what the probability will be that such a defect will cause a hard defect. This information for each defect size forms the sensitivity function [6, 15].

Combining the sensitivity function and the defect size distribution function, we can evaluate the relationship between the number of hard and soft defects. This may be a critical factor because it reveals the relationship between defects that cause "easy" to detect faults and defects that may cause "hard" to detect faults affecting the reliability. Figure 2-14 compares for two typical values of parameter p from the defect size distribution function the number of hard and soft defects. For the layout, a regular double comb monitor structure has been assumed with

2.7 ■ Summary

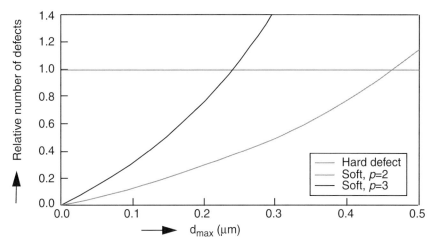

Figure 2-14 Relative number of hard and soft extra material defects.

minimum feature sizes of 1 μm. In both cases, the number of hard defects has been normalized to 1.

From these curves it can be concluded that even for relatively small values of d_{max}, the number of soft defects is significant. In addition, for increasing values of p the number of soft defects increases rapidly. The high number of soft defects does not imply that all devices containing such a defect will ultimately fail functionality prematurely. Only a small part of these defects will probably degrade throughout the lifetime. Nevertheless, more soft defects can be related directly to a reduced level of reliability. The significance of this impact depends on the other aspects that determine the reliability of an IC, such as electromigration and hot-electron degradation.

Although the research on reliability issues of integrated circuits has not yet provided a fixed relationship between soft defects and reliability, this topic has been addressed in certain MIL standards. The requirements with respect to reliability in these standards are very high as a result of the critical application areas and thus cannot be transferred directly to all ICs in general. According to those standards, the spacing between two lines should be at least 50% of the minimum design rule. Similarly, the width of a line should be at least 50% of the designed dimension. In view of such requirements, further research into reliability issues will cover a very interesting and important subject.

2.7 SUMMARY

This chapter presents an overview of the various types of defects causing an IC to fail its specifications. Based on economic consequences of aspects such as price, quality, and reliability, it is shown that minimization of the number of defects is essential in achieving a competitive position on the market. Whereas an IC's failure may be caused by either the design or the manufacturing process, this chapter focuses on manufacturing.

A variety of defect mechanisms can be identified as contributing to the total number of defects in the manufacturing process. These defect mechanisms are identified as wafer defects, human errors, equipment failures, environmental impact, and process instabilities. The resulting defects can be classified according to various metrics. With respect to the geometrical impact, we can distinguish between global and local defects as well as between hard and soft defects. In view of the electrical impact of defects, parametric and functional faults can be identified.

The number of defects can be reduced only when enough information is gathered showing the relationship with the defect mechanism. Defect monitoring strategies for global as well as local defects are presented. For monitoring global defects, PCMs and a parameter monitoring strategy can be applied. For local defects, special monitoring structures for the detection of gate-oxide and interconnect defects are discussed.

For purposes of yield prediction, these defects have to be described by a defect model. For global defects, the model should be related to the difference between the expected and the actual process performance and spread, although a true model is not given. For local defects, specific modeling aspects such as defect density, defect size distribution function, and cluster behavior are covered. Finally, the model for local defects is applied to derive an expression for the relationship between hard and soft defects. From this relationship we can conclude that the number of soft defects forming a reliability hazard is relatively high and thus deserves to be studied in more detail.

2.8 EXERCISES

2.1. Indicate the difference and relationship between the quality and reliability of an IC.

2.2. What is the difference in impact of design failures and manufacturing defects with respect to the test strategy and the various stages of testing throughout the IC development flow?

2.3. Explain the difference in likelihood between missing and extra material defects which are lithography-related in modern IC technologies.

2.4. PCMs as well as the so-called parameter monitoring aim at detecting global defects. Describe the basic difference between both approaches.

2.5. Enumerate the three characteristics of local defects which need to be considered for modeling purposes.

2.6. Consider only the regular core of a double comb test structure with spacing s and width w. (a) Derive the expression for the probability that an extra material defect causes a short circuit by integrating the layout sensitivity function with the defect size distribution function. Make use of the relationships ($x_M = \infty$). (b) Compute the fault probability for $s = w = 1 \ \mu m$, ($x_m = s$), $q = 1$, and $p = 3$. (c) Determine for this fault probability the maximum area of a double comb test structure when the average number of faults per monitor is maximal 1 and the defect density $D_o \leq 12 \ cm^{-2}$. (d) Determine with single-digit accuracy the value of cluster parameter α of the Negative Binomial Yield model when the yield of such a double comb monitor is 58%. (e) Answer questions b and c again when $w = 0.5$ μm instead.

2.7. A relationship exists between the value of parameter p of the defect size distribution function and the quality of the manufacturing line. Describe what this relationship will look like.

2.8. What would be the impact of stronger clustering of defects on the manufacturing yield? And how does defect clustering impact manufacturing yield when increasing the area of an IC?

REFERENCES

[1] P. Borden, "The Nature of Particle Generation in Vacuum Process Tools," *IEEE Transactions on Semiconductor Manufacturing,* vol. 3, no. 4, pp. 189–194, Nov. 1990.

[2] E. M. J. G. Bruls, F. Camerik, H. Kretschmann, and J. A. G. Jess, "A Generic Method to Develop a Defect Monitoring System for IC Processes," *Proceedings IEEE International Test Conference,* pp. 218–227, 1991.

[3] E. M. J. G. Bruls, "Characterization of Defects in Integrated Circuits: Resources, Models and Applications," Report of Eindhoven University of Technology, Department of Electrical Engineering, ISBN 90-5282-190-9, May 1992.

[4] E. M. J. G. Bruls, "Reliability Aspects of Defect Analysis," *Proceedings of the IEEE European Test Conference*, pp. 17–26, 1993.

[5] H. G. Claudius, "Practical Defect Reduction in an MOS IC Line," *Microcontamination,* vol. 5, no. 4, pp. 47–52, Apr. 1987.

[6] A. V. Ferris-Prabhu, "Modeling the Critical Area in Yield Forecasts," *IEEE Journal of Solid-State Circuits*, vol. SC-20, no. 4, pp. 874–878, Aug. 1985.

[7] A. V. Ferris-Prabhu, "Defect Size Variations and Their Effect on the Critical Area of VLSI Devices," *IEEE Journal of Solid-State Circuits*, vol. SC-20, no. 4, pp. 878–880, Aug. 1985.

[8] H. H. Huston and C. P. Clarke, "Reliability Defect Detection and Screening During Processing—Theory and Implementation," *Proceedings of the IEEE International Reliability Physics Symposium,* pp. 268–275, 1992.

[9] J. A. Lange, "Sources of Semiconductor Wafer Contamination," *Semiconductor International*, pp. 124–128, Apr. 1983.

[10] W. Maly, A. Strojwas, and S. Director, "VLSI Yield Prediction and Estimation: A Unified Framework," *IEEE Transactions on CAD,* vol. CAD-5, no. 1, pp. 114–130, Jan. 1986.

[11] W. Maly, *Atlas of IC Technologies; An Introduction to VLSI Processes,* Menlo Park, CA; Benjamin/Cummings Publishing Co., 1987.

[12] W. Maly, M. Thomas, J. Chinn, and D. Campbell, "Characterization of Type, Size and Density of Spot Defects in the Metallization Layer," pp. 71–90 in *Yield Modelling and Defect Tolerance in VLSI*, eds. W. Moore, W. Maly, and A. Strojwas, Philadelphia: Adam Hilger, 1988.

[13] P. Mullenix, "The Capability of Capability Indices with an Application to Guard-banding in a Test Environment," *Proceedings of the IEEE International Test Conference,* pp. 907–915, 1990.

[14] K. Okamoto and S. Yoshitome, "Wafer Inspection Technology for Submicron Devices," *Integrated Circuit Metrology, Inspection, and Process Control,* vol. 1087, pp. 524–531, 1989.

[15] J. Pineda de Gyvez and J. A. G. Jess, "On the Definition of Critical Areas for IC Photolithographic Spot Defects," *Proceedings of the IEEE European Test Conference,* pp. 152–158, 1989.

[16] M. M. A. van Rosmalen, E. M. J. G. Bruls, K. Baker, and J. A. G. Jess, "Parameter Monitoring: Advantages and Pitfalls," *Proceedings of the IEEE International Test Conference,* pp. 115–124, 1993.

[17] H. R. Sayah and M. G. Buehler, "Comb/Serpentine/Cross-bridge Test Structure for Fabrication Process Evaluation," *IEEE Proceedings on Microelectronic Test Structures,* vol. 1, no. 1, pp. 23–28, Feb. 1988.

[18] J. M. Soden and C. F. Hawkins, "Test Considerations for Gate Oxide Shorts in CMOS ICs," *IEEE Design and Test of Computers,* vol. 3, no. 4, pp. 56–64, Aug. 1986.

[19] J. M. Soden and R. E. Anderson, "IC Failure Analysis: Techniques and Tools for Quality and Reliability Improvement," *Proceedings of the IEEE,* vol. 81, no. 5, pp. 703–715, May 1993.

[20] C. H. Stapper, F. M. Armstrong, and K. Saji, "Integrated Circuit Yield Statistics," *Proceedings of the IEEE,* vol. 71, no. 4, pp. 453–470, Apr. 1983.

[21] C. H. Stapper, "Modeling of Defects in Integrated Circuit Photolithographic Patterns," *IBM Journal of Research Developments,* vol. 28, no. 4, pp. 461–475, July 1984.

[22] C. H. Stapper, "Correlation Analysis of Particle Clusters on Integrated Circuit Wafers," *IBM Journal of Research Developments,* vol. 31, no. 6, pp. 641–650, Nov. 1987.

[23] S. Swaving, A. Ketting, and A. Trip, "MOS-IC Process and Device Characterization within Philips," *IEEE Proceedings on Microelectronic Test Structures,* vol. 1, no. 1, pp. 180–184, Feb. 1988.

[24] C. W. Teutsch and D. C. Drain, "Combining Electrical Defect Monitors with Automatic Visual Inspection Systems," *Integrated Circuit Metrology, Inspection, and Process Control,* vol. 1087, pp. 189–199, 1989.

[25] R. M. Warner, Jr., "Applying a composite model to the IC yield problem," *IEEE Journal of Solid-State Circuits,* vol. SC-9, no. 3, pp. 86–95, June 1974.

[26] R. H. Williams, R. G. Wagner, and C. F. Hawkins, "Testing Errors: Data and Calculations in an IC Manufacturing Process," *Proceedings of the IEEE International Test Conference,* pp. 352–361, 1992.

3

Digital CMOS Fault Modeling and Inductive Fault Analysis

Manoj Sachdev

3.1 INTRODUCTION

Imperfections in the manufacturing process necessitate testing the manufactured Integrated Circuits (ICs). The fundamental objective of testing is to distinguish between good and faulty ICs. This objective can be achieved in several ways. Earlier, when the ICs were relatively less complex, this objective was achieved through functional testing. Functional tests are closely associated with IC functionality. Therefore, these tests are comparatively simple and straightforward. A 4-bit binary counter can be tested exhaustively only in $2^4 = 16$ test vectors. However, later as the complexity of the fabricated ICs increased, it was discovered that the application of the functional test was rather expensive as regards test resources and inefficient in detecting the manufacturing process imperfections (or defects as they are popularly known). For example, a digital IC with 32 inputs and having only combinational logic, a modest complexity by today's Very Large-Scale Integration (VLSI) standard, will require $2^{32} = 4,294,967,296$ test vectors for exhaustive functional testing. If these vectors are to be applied at the rate of 10^6/sec, it will take 71.58 min to test a single IC for exhaustive functional testing. The test becomes even longer if the IC contains sequential logic. Obviously, this test solution is too expensive to be practical.

The test problem is further compounded by the rapid developments of CAD tools in the areas of IC design and manufacturing which have helped engineers design and fabricate complex ICs. Such tools for test and testability analysis became visible only when testing was recognized as a bottleneck in achieving increasingly important quality and reliability goals. The packaging limitation puts severe additional constraints on the testing of complex ICs. For example, over the last decade packaging capabilities have increased only by an order of magnitude. The state of the art today is a 500- to 600-pin package compared to a 50- to 60-pin package nearly a decade ago. On the other hand, in the same period, the device integration on an IC has increased by an order of 3 or more. Effectively, the depth of logic that is to be accessed from the primary pins increased for each successive generation of chips. In other words, controllability and observability objectives became much more difficult to achieve for modern ICs. As a result, test vectors became longer and added to the test cost. During the same period, the cost of general-purpose Automatic Test Equipment (ATE) increased significantly. State-of-the-art ATE costs as much as a few million dollars. An expensive ATE and the longer test vectors pushed the test costs to an unacceptable level.

3.1.1 Quality and Reliability Awareness

Following the invention of the transistor in the 1950s, the semiconductor industry grew into diverse applications areas, ranging from entertainment electronics to space applications. Computers and telecommunications are other notable applications. Regardless of the application area, the quality and reliability demands for semiconductor devices increased significantly [1, 2]. This requirement is not difficult to understand. Today, our whole lifestyle depends heavily on reliable, continuous operation of electronics to which ICs belong. Many conveniences have been taken for granted. However, it has taken a painstakingly long effort to ensure quality and reliability in ICs. It is a well-known rule of thumb that if it costs x dollars to test a defective component at chip level, it will cost x^2 dollars at board level and x^3 dollars at system level to test, diagnose, and replace the same defective component. Therefore, economically it makes a lot of sense to build a system with high-quality components. Pulat and Streb [3] added numbers to this hypothesis to make it clear: *Imagine that a process step that provides 99% product yield generates 10,000 defective parts per million items produced. If 30 process steps are required to make a product, each with 99% yield, assuming independence, the overall product yield will be only 74%.* Hence, modest failure rates at the component level may result in significant likelihood of failure at the board or system level. The increasing system complexities require still better quality from IC suppliers so as to make economic products.

As systems became complex, their maintenance and repair became costlier. Often, specialists are required for such functions. Therefore, reliable system operation over its lifetime became an absolute requirement. These developments led to slogans such as Design for Quality, Design for Reliability, and Six-Sigma Design.

3.1.2 Role of Testing in Quality Improvement

Design, fabrication process, and test are three major activities in the development of an IC. It is futile to believe that the overall quality of any IC can be built considering only the design or the process or the test alone. In other words, a robust design, a well-controlled process, and an effective test together result in a quality product. The role of design and process in building IC quality and reliability has been investigated in depth and is the focus of continuous investigations [4].

In this chapter, we focus on testing which has an equally important role to play in IC quality improvement. Testing is the last check-post before the product is shipped to its destination: in other words, it is the last opportunity to prevent a faulty product from being shipped. Pulat and Streb [3] stressed the need of component (IC) testing in Total Quality Management (TQM). In a large study conducted over three years and encompassing 71 million commercial grade ICs, Hnatek [2] reported the differences in quality seen by IC suppliers and users. One of the foremost conclusions of the study was that IC suppliers often do not carry out enough testing. How thorough must functional testing of digital ICs be to guarantee adequate quality? Is fault grading necessary? If yes, what must be the single-stuck at fault coverage? These were the objectives of a study conducted by McCluskey and Buelow [5]. The result of their theoretical analysis as well as experimental evidence indicated that Boolean production test fault coverages of greater than 99% are necessary for manufacturing high-quality ICs. However, 1% escape of faults will cause 10,000 PPM (parts per million), which is a long way from the industry's aim of sub-100 PPM.

This chapter is divided into three sections: the objectives of fault modeling; digital CMOS fault modeling; and Inductive Fault Analysis (IFA). In Section 3.21, various levels (abstractions) of modeling are demonstrated with examples. The advantages and disadvantages of different fault modeling levels are also illustrated with examples. Only Complementary Metal Oxide Semiconductor (CMOS) technology is considered because of its overwhelming popularity in implementing digital logic functionality. Its popularity derives from its integration capability, extremely low power consumption, high performance, and simple logic implementation. Section 3.4 describes the basic assumptions and steps of IFA. Some industrial examples are shown which demonstrate the potential of IFA in test quality and test economics improvement. IFA's strengths and weaknesses are also described.

3.2 OBJECTIVES OF FAULT MODELING

The exponential increase in the cost of functional testing has led to tests that are not functional in nature, but rather aim at detecting possible faulty conditions in ICs. The Circuit Under Test (CUT) is analyzed for faulty conditions, and tests are generated to test for such conditions. Like any other analysis, this fault analysis also requires a model (or abstraction) that can represent likely faults in ICs with an acceptable level of accuracy. This type of model is called the fault model, and this type of testing is known as structural testing. The name "structural test" has two origins. First, the testing is carried out to validate the design's structural composition rather than its functionality; second, the test methodology has a structured basis, that is, the fault model for test generation. In fact, the concept of structural testing dates back to the 1950s. In one of the first papers on the subject, Eldred proposed a methodology that would test whether or not all tubes and diodes within a gating structure were operating correctly [6]. However, structural testing gained popularity in the 1970s and 1980s when structural Design for Test (DFT) methodologies like Scan Path and Level Sensitive Scan Design (LSSD) [7, 8] emerged for digital circuits. These DFT methods became popular because their application could change distributed sequential logic into a big unified shift-register for testing purposes [7, 8]. As a result, the overall test complexity is reduced [9]. Owing to these techniques, the test generation and the fault grading for complex digital circuits became a possibility.

Breuer and Friedman [10] described fault modeling as an activity concerned with the systematic and precise representation of physical faults in a form suitable for simulation and test generation. Such a representation usually involves the definition of abstract or logical faults that produce approximately the same erroneous behavior as the actual physical defects. Here, it is important to distinguish between a defect and a fault. A defect is physical in nature, and a fault is its representation. Therefore, a fault can also be defined as follows: *A fault is the electrical impact of a physical defect at an appropriate level of abstraction.* Fault models played a pivotal role in the success of structural testing, which must test for all modeled faults. Structural testing has some notable advantages over functional tests, especially the following.

- The effectiveness of the structural test is quantifiable. It is possible to ascertain what percentage of the modeled faults have been tested by a given test suite. This percentage coverage is popularly known as test coverage. Thus, it allows the user to establish a relationship between test coverage and quality of tested ICs.
- The test generation for structural tests is considerably simpler than functional test generation for a complex CUT. Computer-Aided Design (CAD) tools (e.g., Automatic Test Pattern Generation [ATPG] and Fault Simulators) ensure faster and effective test generation.

The underlying assumptions of the structural test are that the design is essentially correct and that its functionality on silicon has already been verified and characterized. It is the nonideal manufacturing process that introduces defects in ICs. These defects cause faults that result in erroneous IC behaviors. Moreover, faults in structural testing are defined only for fault simulation purposes and do not represent how likely a fault is compared to another one. Structural testing further assumes the time invariance of a fault. Time variant or temporary faults are not considered for structured test generation. Lastly, it is assumed that a fault has a local impact. For example, a fault causes the output of a NAND gate to be always logic high. This assumption derives from the fact that a fabrication process line is regularly monitored; hence, the global defects are controlled early in the production environment. As a result, the vast majority of defects to be tested for are local in nature.

3.3 LEVELS OF FAULT MODELING

Numerous possibilities exist to represent faults for the purpose of fault simulation. However, most of them are categorized according to the level of abstraction for fault modeling. The level of abstraction is essentially a tradeoff between the fault model's ability to accurately represent an actual physical defect and the speed of processing the fault in a fault simulation environment. For example, behavior (functional) level fault modeling is the fastest but the least accurate. On the other hand, layout level (IFA) fault models are most accurate but require enormous computational resources. Most of the fault models can be classified according to the following level of abstraction.

3.3.1 Logic-Level Fault Modeling

Initial work on fault modeling concentrated on logic-level fault modeling. It is assumed that the faulty behavior due to defects can be mapped onto the incorrect Boolean functionality of basic gates in the CUT. Circuits were simple, and feature sizes were relatively large. Therefore, fewer defects could cause fatal faults. In the early days of the semiconductor industry, the whole industry was battling to solve complex design and process-related problems, and paid little attention to IC testing. The yield of the early ICs was poor, primarily because of equipment and technological hurdles rather than manufacturing process defects. Furthermore, limited knowledge concerning the origins of different defects, their impact on circuit behavior, and testing forced researchers to adopt many simplifying assumptions.

The implementation details of the logic gates are not considered while modeling faults at the logic level. The logic-level fault modeling has some notable advantages. The Boolean nature of the fault model allows the usage of powerful Boolean algebra for deriving tests for complex digital circuits [11]. The gate-level representation of the fault behavior resulted in technology-independent fault model implementations and test generation. Technology-independent tests increased the portability of designs to different technologies.

3.3.1.1 Stuck-At Fault Model. The Stuck-At Fault (SAF) model is the most commonly used logic-level fault model. Poage [12] was one of the first to propose the SAF model. It became a popular fault model in the 1960s owing to its simplicity, and it is still widely used in academic research as well as in the industry. Its simplicity derived from its logical behavior. These faults are mapped onto the interconnect (or nets) between the logic gates. Thus, often they are also referred to as the Stuck-Line (SL) fault model [13, 14]. Under the faulty condition, the affected line is assumed to have a permanent (stuck-at) logic 0 or 1 value that cannot be altered by input stimuli. Figure 3-1 shows a NAND gate and its truth table. Let us consider that the line A has a stuck-at-1 (SA-1) fault. The presence of the SA1 fault is detected by the faulty gate response when lines A and B are driven logic 0 and 1, respectively. A fault is said to be detected when the expected output of the logic gate differs from the actual logical output. For example, the third and fourth columns of the table shown in Figure 3-1 illustrate the expected (fault-free) and the actual (faulty) responses of the NAND gate. For the test vector A = 0, B = 1, the expected and actual responses differ, and thus, the fault is said to be detected.

The SAF model has been widely used for many practical reasons, including the availability of Computer-Aided Design (CAD) tools for test generation, the technology independence of the SAF model, and the many physical defects that cause SAFs. With increasingly complex devices and smaller feature sizes, it has become more likely that more than one SAF can occur at the same time in an IC.

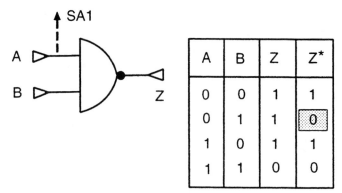

Figure 3-1 The SAF model and its fault-free and faulty Boolean behavior.

However, a large number of possible multiple SAFs force the test generation effort to be impractical. For example, let us assume that a circuit contain n lines. Each line is capable of having three distinct states; SA0, SA1, and fault-free. Therefore, there are $3^n - 1$ multiple (affecting two lines) SAFs possible. A typical IC may contain a hundred thousand or more lines that may result in an enormous number of possible faults. Therefore, it is common practice to assume the occurrence of only Single-Stuck-Line (SSL) faults in the IC. Hence, for an IC with n lines, only $2n$ SSL are possible.

3.3.1.2 Fault Equivalence, Dominance, and Collapsing. A fault f1 is considered to be equivalent to fault f2, if their family behaviors for all possible input stimuli are indistinguishable from each other at the primary IC outputs. Therefore, a stimulus detecting f1 will also detect f2 and conversely. Such faults can be bunched into an equivalent fault class. This point can be further illustrated with Figure 3-1. Consider once again a two-input NAND gate. A SA0 fault at B (B/0) causes a SA1 fault at Z (Z/1) for all input conditions of the NAND gate. Therefore, both of these faults are indistinguishable from each other at the primary IC outputs. A priori knowledge of the fault classes in a network is useful in fault diagnosis. Furthermore, fault detection is simplified by using equivalent fault classes to reduce or collapse the set of faults into fault classes that need to be considered for test generation [87, 88]. The fault resolution of a network depends on how widely equivalent faults are separated. Hence, knowledge of the equivalent fault classes is helpful in problems such as test point placement and logic partitioning to increase the fault resolution [92].

Poage [12] was the first to describe the concept of fault dominance. He believed that if all tests for a fault f1 also detected f2 but only a subset of tests for f2 detect f1, then f2 could be said to dominate f1. However, Abraham [11] argued that it was safer to consider a test for f1 which would ensure that both faults were detected. Hence, f1 is said to dominate f2. For example, in Table 3-1, A/1 does not cause Z/0 behavior for all input conditions of the NAND gate. A test for A/1 will also test for Z/0. The converse is not true because Z/0 is also tested by the test for B/1. Hence, according to Abraham, the detection of A/1 will ensure the detection

3.3 ■ Levels of Fault Modeling

TABLE 3-1 The SAF classes in a two-input NAND gate.

A	B	Z	Fault Classes
1	1	0	A/0, B/0, Z/1
1	0	1	B/1, Z/0
0	1	1	A/1, Z/0
0	0	1	Redundant Test

of Z/0 as well. Therefore, A/1 is said to dominate Z/0. However, according to Poage, Z/0 dominates A/1. The concepts of fault equivalence and fault dominance allow us to collapse different SAFs into a single fault class. In general, for an n-input gate there will be $n + 1$ fault classes [11]. Table 3-1 shows different fault classes and test vectors needed for their detection for a two-input NAND gate.

These concepts can be applied to larger circuits [87–94], but in general it is a complex computational problem. It is shown [91, 92] that the problem of identifying fault equivalence in arbitrary networks belongs to the class of computationally difficult problems called NP complete. Nevertheless, significant attention has been devoted to finding equivalent fault classes in circuits bigger than single-logic gate [87–90, 92–94].

Schertz and Metze [88] described fault collapsing as the process of combining faults by means of implication relationships derived from a study of the network concerned. They defined three separate stages of fault collapsing corresponding to three different types of implication relationships. These stages can be explained as follows: Consider a fault table $[T_{ij}]$ in which each test is represented by a row and each fault is represented by a column. The entry t_{ij} is 1 if and only if test i detects fault j. Column j dominates column k if for every 1 in column k there is also 1 in column j (same as the definition by Poage). If two columns of the table are identical, then each dominates the other. The first stage of the fault collapsing corresponds to identical columns in the fault table. The second stage of collapsing corresponds to the unequal columns, that is, the situations where one fault is more readily detected than the other. The third stage corresponds to concerns with the relationship between single and multiple faults. For example, let us consider the two-input NAND gate of Figure 3-1 and its fault classes in Table 3-1 once again. The faults A/0, B/0, and Z/1 are indistinguishable from each other and, hence, represent the first-stage collapsing procedure (first fault class in Table 3-1). The next two rows of the table represent the second stage of fault collapsing. The third stage of fault collapsing is concerned with the multiple faults. For example, Z/0 is indistinguishable from all input lines (A/1, B/1) having multiple SA1 faults at the same time.

In the case of larger circuits that have nonreconverging fanouts, application of the preceding analysis is rather straightforward. In the case of reconverging fanouts, however, the reconverging branch may interfere with the propagation of the fault to a primary output. Goundan and Hayes [92] further simplified the fault-class identification problem for certain cases. Furthermore, they introduced the concepts of intrinsic and extrinsic fault equivalence. These concepts were utilized to derive some general conditions for fault equivalence and nonequivalence for a given network topology. They argued that every switching function can be realized by a two-level network. Therefore, they exploited these concepts in order to identify equivalent fault classes in a two-level network. For a given two-level network, the computational complexity is reduced to 19% of the original value.

The concept of fault collapsing is independent of technology and the fault abstraction level. The fault collapsing techniques have been extended to NMOS and CMOS circuits not only for stuck-at faults but also for transistor stuck-open (SOP) and stuck-on (SON) faults [93, 94].

> **Exercise 1:** *Tabulate all SAF Classes for a three-input NOR gate, collapse them into fault classes, and determine the test vectors.*

3.3.1.3 Mapping of CMOS Defects at Stuck-at Faults.

In fully CMOS technology an m-input logic gate is realized with m p-channel transistors and an equal number of n-channel transistors. The output of the gate is taken where p- and n-channel transistors join each other. Depending on the Boolean state of the inputs, the output is driven either by a p-channel transistor(s) to logic 1 or by an n-channel transistor(s) to logic 0. The output is never driven by both types of transistors at the same time.

For example, Figure 3-2 shows a transistor-level schematic of a two-input NAND gate in the CMOS technology. Two p-channel and two n-channel transistors are needed to realize a two-input NAND gate. Figure 3-2 also illustrates some common defects in the NAND gate and presents a table showing how they are detected by SAF test vectors. Defect d1 causes a short between output Z and power supply VDD, and therefore, results in the Z/1 stuck-at fault. Needless to say, the A = 1. B = 1 test vector will detect it. Similarly, defect d3 causes the Z/0 stuck-at fault which can be detected by A = 0, B = X, or A = X, B = 0 test vectors.

Defect d2 cannot be modeled by the SAF, but it can be detected by SAF test vectors under the assumption that resistance of the defect and the on-resistance of the n1 transistor are substantially smaller than the on-resistance of the p2 transistor.

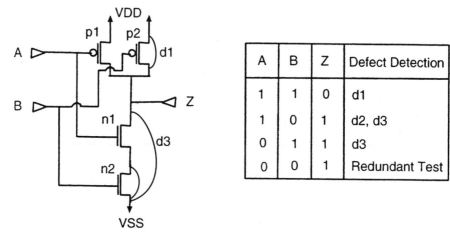

Figure 3-2 Defects in a two-input CMOS NAND gate and their detection.

Test vector A = 1, B = 0 causes transistors p2 and n1 to conduct. Therefore, in the presence of d2, output Z is driven from VDD and VSS simultaneously and has an intermediate nodal voltage. This intermediate voltage is not the same as Z/1 or Z/0. However, this defect is detected, provided that the following logic gate interprets this output level as logic 0 instead of logic 1.

3.3.1.4 Shortcomings of the Stuck-At Fault Model. In spite of its simplicity and universal applicability, the SAF model has some serious drawbacks in representing defects in CMOS technology. It can represent only a subset of all defects. A large number of defects that are not detected by a SA test set cause bad ICs to pass the test. Woodhall et al. [26] reported that the open defects lead to an escape rate of 1210 PPM when CUT was tested with a 100% SA test set.

Next, we illustrate some of the representative examples that are not modeled by SAF.

(i) OPEN DEFECTS. CMOS circuits can be made to memorize their previous values if they are left in the high-impedance state. Such a property has found numerous applications in the data storage and time discrete signal processing area. Furthermore, CMOS circuits offer high-input impedance; hence, floating interconnects retain their previous logic value for a significantly long time. This sequential behavior of CMOS logic gates causes many open defects not to be detected by a SA test set [15].

Consider once again the same two-input NAND gate, see Figure 3-3. Some of the open defects affecting the operation of transistor p1 are not detected. The second test vector (A = 1, B = 0; Figure 3-1) drives the output to logic 1 through transistor p2. The third test vector (A = 0, B = 1) instead drives the output to logic 1 through transistor p1. In the presence of the above mentioned defects, the output is not driven logic high in the third test vector but retains its logic high state. Therefore, those defects are not detected by the SA test vector set. However, these defects are detected if the order of test vectors is changed. For example, test vectors

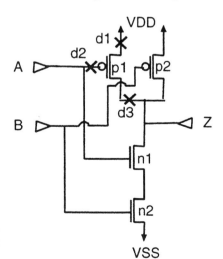

Figure 3-3 Undetected open defects in a two-input CMOS NAND gate by the SA test vectors.

T2, T1, and T3 will detect these faults. A detailed treatment of open defects is presented in the next subsection. However, in general, for open defects two test vectors (T1, T2) are required. The first test vector T1 sets the output to opposite logic value, and T2 attempts to change the output state. For logic gates with higher complexity, the SA test vector set cannot detect all open defects.

> **Exercise 2:** *For the three-input NOR gate of Exercise 1, find all open defects that are not detected by the SA test vector set.*

(ii) SHORT DEFECTS. A short defect is defined as an unintended connection between two or more otherwise unconnected nodes. Often such defects are also referred to as bridging faults or simply as bridges. Shorts are the dominant cause of faults in modern CMOS processes. In CMOS logic gates, shorts cannot be modeled as the wired-OR or wired-AND logic. The circuit level issues (e.g., W/L of driving transistors, defect resistance, logic thresholds of subsequent logic gates, Boolean input logic conditions, etc.) play an important role in their detection. The SAF is not adequate in representing bridging faults, although a large number of shorts lead to SAFs. Hence, the SA test vectors are not suited to detect such defects.

Shorts in ICs can be classified as internal bridges and external bridges. Internal bridges are those that affect the nodes within a logic gate. Shorts shown in Figure 3-2 are examples of this category. The external bridges are those that affect nodes within two or more logic gates. Figure 3-4 illustrates an example of an external bridge and its electrical model. Besides the circuit-level issues, the detection of external bridging faults also depends on exciting node Y and Z to opposite logic values. For a complex circuit, it is a nontrivial task. All potential locations for such bridging defects need to be identified. Techniques such as Inductive Fault Analysis can be useful in finding such locations.

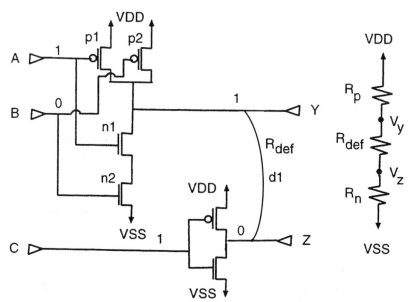

Figure 3-4 External bridging defect not detected by the SA test vectors.

3.3.2 Transistor-Level Fault Modeling

The SAF model has limitations in representing defects in CMOS circuits. In general, many defects in CMOS circuits may be represented by the SAF model and are detected by the SA test set. Other defects are not modeled by the SAF model but are detected by the SA test set. However, still a potentially large number of defects are not modeled by the SAF model and are not detected by the SA test set [11]. Therefore, research was directed at transistor-level fault models that represent faulty behavior with better accuracy. However, transistor-level fault models result in a significantly large number of faults compared to the SAF model for a given circuit. Considerable research effort has been directed toward the better understanding of defects and their influence on circuit behavior [15–26]. This knowledge was expected to result in improved fault models which would lead to efficient, effective tests, and better quality of tested ICs.

As described earlier, a static CMOS logic gate is constructed by a set of p-channel and a set of n-channel enhancement mode transistors between VDD and VSS terminals. An enhancement mode transistor in the absence of a gate voltage does not conduct, and it conducts only when an appropriate gate voltage is applied. For example, a p-channel transistor conducts when its gate terminal is logic 0, and an n-channel transistor conducts when its gate terminal is logic 1. Therefore, a transistor can be treated as a three-terminal ideal switch with a control terminal gate, which controls the flow of data from the source terminal to the drain terminal. In the transistor-level fault modeling, physical defects are mapped onto the functioning of these switches. For a given combination of input logic values, a CMOS logic gate may have one of the following states [18].

1. The output node is driven to VDD via one or more paths provided by conducting p-channel transistors, and no conducting path from output to VSS exists through conducting n-channel transistors.
2. The output node is driven to VSS via one or more paths provided by conducting n-channel transistors, and no conducting path from output to VDD exists through conducting p-channel transistors.
3. The output node is not driven to VDD or VSS via conducting transistors.
4. The output node is driven by both VDD and VSS via conducting transistors.

In the first two cases, the output is logic high and low, respectively. In case 3, the output is in high impedance, and its current logic state is the same as its previous logic state. In case 4, the output logic state is treated as indeterminate, since it depends on resistance ratios of the conducting transistors. However, rarely are logic gates designed to have case 4-type situations.

3.3.2.1 Transistor Stuck-Open Fault Model. The detection of transistor Stuck-Open (SOP) faults has been a difficult problem and has received considerable attention in the past [15–29]. In the presence of a SOP fault, the affected transistor fails to transmit a logic value from its source terminal to its drain terminal. Therefore, the transistor can be treated as a switch that never closes, and it remains open in spite of all possible Boolean input conditions. As is apparent from the name, such faults in enhancement mode transistors are caused primarily by open defects. However, short defects can also cause a transistor to have a SOP fault. Figure 3-5 illustrates some of the defects causing SOP faults in a logic gate. SOP faults are classified by their location as a fault at the source or drain terminal (S/D-line fault) or at the gate terminal (gate-line fault) of a transistor [19]. A S/D line fault creates a break in the data transfer path and clearly causes a SOP fault in the transistor (defects d1 and d3, Figure 3-5). A gate-line fault (defect d2, Figure 3-5) requires a bit of explanation. As explained previously, in the CMOS technology enhancement mode p- and n-channel transistors are utilized. Enhancement transistors have the property that in the absence of any gate voltage the transistor does not conduct. Only when a gate voltage exceeds the threshold voltage, V_T, the transistor starts to conduct. Therefore, a gate-line fault also causes a transistor to be in the nonconduction mode. A short defect between the gate terminal and the source terminal (defect d4, Figure 3-5) of the transistor forces the same voltage on both of these terminals. As a result, the p2 transistor is never in the conduction mode; at the same time, the same defect causes the n2 transistor to be always in the conduction mode (SON fault).

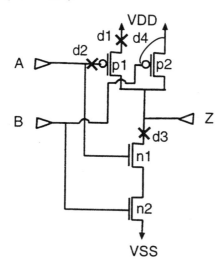

Figure 3-5 Open and bridging defects causing the SOP fault behavior.

Wadsack demonstrated that in the presence of a SOP fault in a CMOS gate, the logic gate shows a memory effect under certain input conditions [15]. Therefore, such faults are not detectable by a SA test set. In general, SOP fault detection requires a two-pattern test sequence ⟨T1, T2⟩ [20]. The first test vector of the sequence, T1, is referred to as the initializing test vector, and the second test vector of the sequence, T2, is referred to as the fault excitation test vector. Test vector T1 initializes the output to logic 0(1), and T2 attempts to set the output to logic 1(0) if the logic gate is fault-free. A failure to set the output to logic 1(0) indicates the presence of a SOP fault.

Some of the SOP faults in static CMOS logic gates require only one test vector, T2, to detect them. If there is only one path from output to VSS (VDD) and the

3.3 ■ Levels of Fault Modeling

SOP fault affects this path, it is not possible to set the output to VSS (VDD). For example, in the case of a two-input NAND gate (Figure 3-3), test vector T2 (A = 1, B = 1) will detect SOP faults in the n-channel transistors. Effectively, such SOP faults disconnect all possible paths from the output to VSS and cause a SA1 fault (Z/1) at the output of the NAND gate which requires only a single test vector to test it [18].

Exercise 3: *For a three-input NOR gate, find out all SOP faults and provide test vectors for their detection. Compare SOP test vectors with SAF test vectors of Exercise 1.*

Exercise 4: *Take the two-input NAND gate of Figure 3-5 with defect d1. Add 0.2 picofarad of capacitance on the output node and Spice simulate it with the SA test set. Does the circuit achieve DC convergence? If not, replace the open defect with a 10-megaohm resistance and resimulate. Is the open defect detected by the SA test set?*

(i) DESIGN FOR SOP FAULT TESTABILITY. A two-pattern test for SOP fault detection can be invalidated (i.e., may not detect the fault it was supposed to detect) by arbitrary circuit delays and glitches if the patterns are not carefully selected. In fact, for some irredundant CMOS complex gates, a robust test (which is not invalidated by arbitrary circuit delays) for SOP faults does not exist [21]. A robust test sequence is a sequence of test vectors in which each successive test vector differs from the previous test vector in only one bit position. As a remedy for robust SOP fault detection, Design for Test (DFT) solutions for complex gates have been proposed [21]. In the first scheme, an addition of two transistors with two independent control lines to each complex gate was suggested. Three test vectors are needed to detect a SOP fault. Figure 3-6 shows the scheme and the test procedure. TN and TP are the fault evaluation vectors (T2). The TN* (TP*) is any value of input I that would have established one or more paths from VSS (VDD) to the output node.

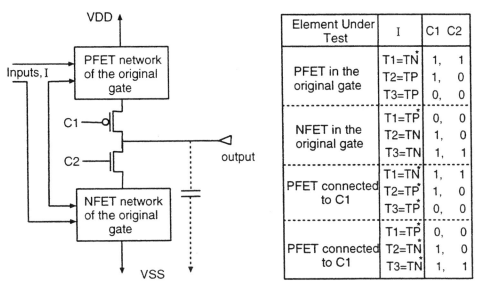

Figure 3-6 Robust SOP fault detection scheme and the three-pattern test procedure [21].

The above-mentioned scheme has a disadvantage in that an SOP fault needs a three-pattern test, which may result in longer test vectors and thus increased test costs. Therefore, Reddy et al. [21] suggested another scheme which would require a two-pattern test, and the test would not be invalidated by arbitrary circuit delays. However, a total of four transistors for each complex gate are needed for implementation. The additional two transistors, PFET2 and NFET2 (P-type Field Effect Transistor and N-type Field Effect Transistor), are added between the output and VDD (VSS) in parallel to the PFET (NFET) network. Implementation of the scheme and the test procedure is shown in Figure 3-7. To test a SOP fault in a PFET, the initialization vector is provided through NFET1 and NFET2. During this time, input I is the evaluation test vector, but its evaluation is blocked by nonconducting PFET1. In the second test vector, T2, since only the control is changed and input I stays the same, the test is not invalidated by the circuit delays. Similarly, other parts of a complex gate are tested for SOP faults. I* in Figure 3-7 signifies any arbitrary input test vector.

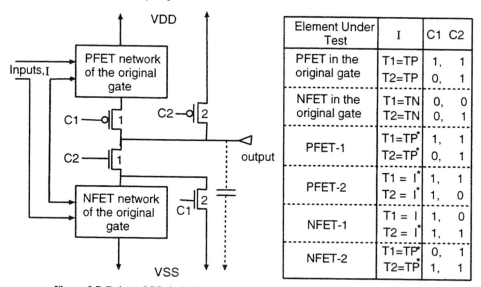

Figure 3-7 Robust SOP fault detection scheme and the two-pattern test procedure [21].

Reddy et al. [22] presented a procedure to find the robust two-pattern test for SOP faults in combinational circuits, if such a procedure existed, which would not be invalidated in the presence of arbitrary circuit delays. They assumed that T2 of the two-pattern test is given, and then they provided a procedure to determine an appropriate initializing input T1. If no appropriate T1 is found, another test T2 is determined and then the procedure is repeated. In case the procedure fails to provide a T1 for all T2s, the implication is that no robust test exists for the fault and the circuit should be redesigned to have the robust SOP test.

Rajsuman et al. presented a test technique for testing of SOP faults with a single test vector [23, 24]. Under this technique, n-channel and p-channel transistors are tested separately. This technique is illustrated in Figure 3-8. Part *a* of the figure

3.3 ■ Levels of Fault Modeling

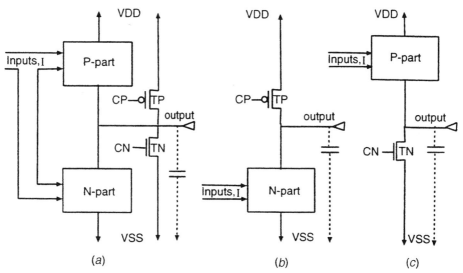

Figure 3-8 Single-pattern SOP fault detection procedure.

shows the addition of two transistors and two control signals to a CMOS logic gate. A full CMOS (FCMOS) logic gate is transformed into a pseudo nMOS (pMOS) gate by adding an extra high-resistive pMOS (nMOS) transistor (Figure 3-8 b,c). The resistance of the transistor should be such that the output is pulled high (low) if none of the nMOS (pMOS) transistors is conducting. In the case when single- or multiple-nMOS (pMOS) transistors are conducting, the output voltage is close to VSS (VDD). Two extra transistors, TP and TN, are needed, which are controlled by two independent signals, CP and CN, respectively. In the normal circuit operation, these transistors are switched off (CP = 1, CN = 0). SOP faults in the nMOS transistors are tested as follows: CP and CN are kept low. Inputs are applied to the nMOS part such that the output is pulled low through each possible Boolean combination of inputs. In the presence of a SOP fault in the nMOS part, the output is not pulled low for one or more input conditions. In fact, these inputs are excitation vectors (T2) of the two-pattern test. Similarly, the pMOS part is tested by keeping CP and CN logic high.

In the presence of a SOP fault, the output of a pseudo nMOS (pMOS) logic gate shows a SA1 (SA0) behavior. Therefore, classical algorithms like D-algorithms, PODEM, and Automatic Test Pattern Generating Programs (ATPGs) can be utilized for test generation. Furthermore, a significant reduction in test generation and test application time is also expected. However, the scheme requires two transistors per logic gate, which is to be tested for SOP faults. In addition, two control lines are required to control the two transistors. These extra transistors themselves are untestable. Furthermore, extra transistors will cause an increase in parasitic capacitance, which will have an impact on circuit performance. In a subsequent article, Jayasumana et al. [25] proposed a DFT solution that required only one transistor and two control lines.

Exercise 5: *For a three-input NOR gate, apply Design for SOP testability schemes (three-, two-, and single-pattern procedures) and find out the test vectors to detect SOP faults. Compare these vectors with SA test vectors of Ex. 1.*

(ii) LAYOUT RULES FOR SOP FAULT DETECTION. In the previous subsection, we saw that a robust two-pattern test generation for SOP fault testability is a difficult problem for a complex IC. Furthermore, SOP DFT schemes have area and performance penalties that restrict their application.

On the other hand, the layout of basic gates has a significant influence on the occurrence of open defects. Layout of a logic gate can be modified so that the probability of a SOP fault is reduced or eliminated. Koeppe presented a set of layout rules to deal with SOP faults [19]. By application of these rules, SOP faults are either reduced or their detection is simplified. He argued that the SOP faults in general are caused by missing contacts, cracks in metal over oxide steps, and dust particles. For the S/D-line faults, faults in only the parallel branches (p-channel transistors in NAND gates) of a basic logic gate (NAND, NOR) require a two-pattern test sequence and are harder to detect by the SA test vector set. To detect such SOP faults, Koeppe suggested a reduction in the contact locations. Figure 3-9 illustrates the conventional and alternative stick diagram for a two-input NAND gate. In the alternative stick diagram, a contact is placed such that its absence affects all parallel branches together. Therefore, such a fault causes a SA (Z/0) fault and is detected by the SA test vector set. Similarly, for gate-line SOP faults, he suggested branchless and fixed-order routing of signals inside the logic gates, so that the chances of an open defect causing a single SOP transistor would be reduced.

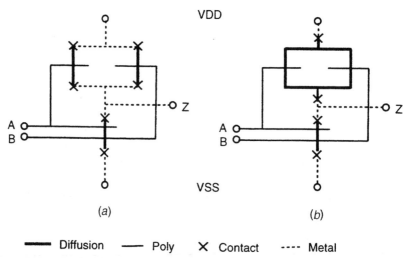

Figure 3-9 Stick diagrams of a two-input NAND gate: (*a*) conventional stick diagram, (*b*) alternative stick diagram to avoid an open S/D line.

To investigate the effectiveness of the rules, Koeppe performed the fault simulation over original and modified layouts. The fault simulation results demonstrated that the SOP fault coverage of the SA test vector set increased substantially for the modified layout. The area overhead for the implementation of the rules was also low.

It was expected to be between 0% and 20%, depending on the application of the rule and the original style of the cell layout. A small performance degradation was also expected since the parasitic capacitance of the transistor drains in the alternative layout was higher than in the original layout.

Exercise 6: *Make the stick diagram for the complex gate shown in Figure 3-10. Modify the stick diagram to avoid S/D-line open defects.*

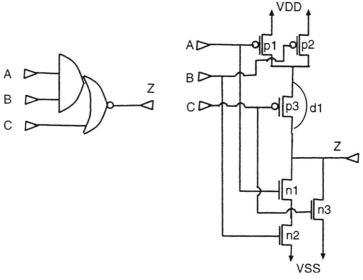

Figure 3-10 A SON fault in a CMOS complex gate.

3.3.2.2 Transistor Stuck-On Fault Model.

A stuck-on (SON) fault forces a transistor to be in the conduction mode regardless of the voltage on its gate terminal. Figure 3-10 illustrates one of the defects causing transistor SON faults. The figure shows a three-input AND-NOR complex gate. A bridging defect between source and drain of the p3 transistor causes a SON fault. The SON fault causes the state-dependent degradation of Boolean output levels. Therefore, their detection depends on circuit-level parameters. In order to detect the SON fault in transistor p3, ABC is chosen among 001, 0X1, or X01. In the fault-free case, the output should be logic low. However, due to the SON fault, a conflict is created between n3 and p-channel transistors which causes a resistive ladder between VDD and VSS. The voltage on the output Z depends on the "on-resistance" of the transistors n3, resistance of the defect d1, and the conducting p-channel transistors. Often, this voltage lies in the ambiguous region between logic 0 and logic 1. Hence, such faults are very difficult to detect through Boolean testing.

A unique property of CMOS circuits can be exploited to test for SON faults. The steady-state current (I_{DDQ} as it is popularly known) consumption in CMOS circuits is extremely low. A million transistor ICs may have an I_{DDQ} value of less than 100 nanoamperes. As we know, a SON fault at appropriate input logic conditions causes a resistive ladder between VDD and VSS nodes. The steady-state current flow through this resistive path will give the indication of a SON fault. For most practical situations, the difference between faulty and fault-free I_{DDQ} is sufficiently large for an unambiguous fault detection [30].

Similarly, some efforts have been made to detect SON faults with delay testing, though with limited success. It can be shown that a SON fault affecting an n-channel transistor in a primary CMOS gate will cause an extra delay in the $0 \rightarrow 1$ output transition under certain input conditions. In a study of CMOS gates, Vierhaus et al. [31] found that in general SON faults cannot be safely tested with delay testing but are effectively tested with I_{DDQ} testing.

> **Exercise 7:** *Spice simulate the complex gate shown in Figure 3-10 with the SA test vectors. Observe the output voltage and I_{DDQ} current. Which test vectors have high I_{DDQ} current? Why?*

3.3.3 Layout-Level Fault Modeling

Layout-level fault modeling is motivated by several factors. First and foremost among them is the inability of logic and transistor-level fault models to represent physical defects. There are many defects (e.g., gate-oxide defects) that will degrade the transistor behavior which cannot be mapped onto a transistor-level fault model. Nor can the bridging faults in the interconnect be mapped to higher level fault models. Higher packing densities coupled with smaller feature sizes have made such defects more likely in current technologies. In other words, a large number of potential defects cannot be modeled by transistor- or logic-level fault models. The rising quality objective of 100 PPM (parts per million) or less necessitates that the layout information is exploited to generate better and effective fault models. All faults are assumed to be equally probable in logic- and transistor-level fault models. However, this is rarely the case in silicon. Some faults are more likely than others. This information should be exploited not only for efficient, effective, and economic test generation but also for creation of defect-insensitive layouts. This subject will be treated at length in the IFA section of this chapter.

3.3.4 Functional-Level Fault Modeling

Over the years, semiconductor technology has matured to such an extent that a common substrate today contains a variety of digital and analog functional blocks. The motivation behind this integration is to offer cheaper, reliable system solutions. Today, a single IC package often includes all functional blocks of an entire microcontroller or Digital Signal Processing (DSP) processor, including the memories and analog interfaces. These functional blocks or system components must now share a common substrate and the manufacturing process. This development has resulted in dramatic changes for testing.

In spite of advances in CAD tools and CPU power, it is no longer possible to fault simulate a complex IC at the transistor level of abstraction. Instead, a complex IC is divided into many functional modules or macros [32]. In many cases, it is possible to model (or map) the impact of defects on the functionality of the macro. Once this mapping is known, efficient and effective test vectors can be generated. However, this mapping must be repeated for each transistor-level implementation.

Testing of semiconductor Random Access Memory (RAM) is a typical example of functional-level fault modeling. However, RAM fault modeling, test algorithm development, and testing are mature disciplines by themselves. A lot of attention has been paid to modeling and testing of faults in RAMs [33, 43, 44, 61–69]. It is beyond the objectives of this chapter to address these aspects here. Furthermore, the IFA section includes two case studies on defect-oriented RAM testing. These studies also provide a brief insight into RAM fault modeling practices. Moreover, for a tutorial overview of RAM fault modeling and testing, an interested reader is referred to [61–64]. In this subsection, we briefly address the functional-level fault modeling, taking RAMs as a vehicle.

Thatte and Abraham [33] suggested a functional test procedure for semiconductor RAM testing. They argued that all RAM decoder and Read/Write logic faults can be mapped on RAM matrix as intercell coupling faults. An address decoder is combinational, selecting a unique RAM cell for a given address. Assuming that under the faulty conditions, the address decoder stays combinational, we observe that it will behave in one of the following ways.

1. The decoder will not access the addressed cell. In addition, it may access a nonaddressed cell.
2. The decoder will access multiple cells, including the addressed cell.

Both of these faulty situations can be viewed as coupling faults involving two or more RAM cells. Similarly, the impact of Read/Write logic faults is viewed as the SAF or coupling fault in the RAM matrix. On the basis of these arguments, they evolved efficient (complexity $n \log n$) compared to complex (n^2) algorithms prevalent in the 1970s. Similarly, other attempts had been made to model RAMs [43, 44, 65–69], Programmable Logic Arrays (PLAs) [34], and microprocessors [35, 36].

3.3.5 Delay Fault Models

In digital circuits the input-output relationship is Boolean. The logic- and transistor-level fault models describe the steady-state malfunctioning of the Boolean relationship but cannot model the faulty delay behavior of a logic element. Timing (or delay) is also an important designed parameter in the input-output relationship. An otherwise good IC may fail to perform correctly in a system if it fails to meet designed timing specifications. With increasing system complexities and higher operational frequencies, timing is becoming a very important aspect of the design. Furthermore, rising quality expectations motivate testing for the correct temporal behavior, commonly known as delay testing [70]. Generally, prefabrication timing is verified at each successive level of design hierarchy. At each level, the objective of the analysis is either to determine the maximum operational frequency at which the circuit will behave correctly or to guarantee that the circuit operates without any malfunction at the prespecified clock rate. Once a chip is fabricated, it still must be tested for the prespecified clock frequency. A circuit is said to have a delay fault if the output(s) of the circuit fails to reach its final value(s) within the specified timing constraint.

A timing or delay fault in an IC could be caused by a number of factors, including subtle manufacturing process defects, transistor threshold voltage shifts, parasitic capacitance, and improper timing design. A substantial research effort has been directed toward delay fault testing [70–80]. Broadly, two fault models have been proposed for delay faults in the literature: the gate delay and the path delay fault model.

3.3.5.1 Gate Delay Fault Model. Each gate in an IC is designed with a specified nominal delay. However, under the gate delay fault model, the faulty gate may assume considerably larger delay. In a digital IC only a few basic gates are used; hence, the test complexity is relatively small. It appears that SA or SOP faults are special (limiting) cases of the gate delay faults. For example, in the case of a SAF, the logic gate output has an infinite delay for a class of input stimuli. Similarly, for a SOP fault the transistor has infinite delay. However, an important distinction can be made between SA, SOP faults, and delay faults. Unlike SA or SOP faults, a gate delay fault does not necessarily cause the circuit to malfunction. In other words, a faulty gate may assume a significantly larger delay than its nominal delay, and still the circuit may work within the timing constraints. Therefore, in general, an evaluation scheme for a delay fault test must not only compute whether or not a delay fault is detected but also calculate the size of the fault. The size was defined as the Fault Detection Size (FDS) [71] or as detection threshold [72] for that fault and a given test T. The FDS for a delay fault and its test T has the property that T is guaranteed to detect any fault at that site of a value greater than the FDS. However, the best FDS achievable for any gate delay fault detecting test is the corresponding slack at the fault site [71, 72]. The slack of a signal is defined as the difference between clock period and the propagation delay of the longest delay path through that signal. The quality of the delay test set depends on how small a delay fault can be tested by a test.

The major drawback of this model is that the interconnect delay is not considered. This deficiency was acceptable when the feature size was relatively large and the gate delay was relatively large compared to the interconnect delay. However, scaling of process dimensions has changed this equilibrium. The transistor switching time has been reduced dramatically due to smaller geometries. As line widths are further scaled into the deep submicron regime, the device switching speed continues to improve. However, delays due to interconnect have not scaled. In fact, the interconnect resistance has increased owing to a smaller cross-sectional area. Furthermore, there is no significant decrease in the interconnect capacitance either. Therefore, interconnect RC product has increased [81–84]. Therefore, the gate delay fault model is restrictive in its application to finer geometries. Furthermore, the gate delay fault model does not take into account the cumulative effect of small delay variations from the primary inputs to the primary outputs.

3.3.5.2 Path Delay Fault Model. The path delay fault model considers the cumulative delay of the path from the primary inputs to the primary outputs. The path delay fault model, in addition to the single isolated failures, also considers distributed delay effects owing to statistical process variations. A faulty situation may arise in spite of the fact that each individual component meets its individual delay specifications. A path delay test will detect both localized and distributed delay defects. Path delay faults may also provide a mechanism for monitoring process variations that may have a significant impact on critical paths. Furthermore, they may provide an ideal vehicle for speed-sorting since they have the most accurate description of the clock speed at which timing failures begin to occur [76].

The path delay fault model has the disadvantage, however, that it is practical to generate tests only for a small number of total paths in a given circuit. Hence, the path delay fault coverage tends to be low [77]. For all practical purposes, the delays in the longest and the shortest paths (critical paths) are considered. If these delays are within the clock cycle, the circuit is considered to be fault-free; otherwise it contains a path delay fault.

3.3.5.3 Robust and Nonrobust Delay Faults, Testing Considerations. Delay fault testing assumes that delay of a gate (or a path) depends on the transition propagated from the input to the output. However, the fault is independent of the vector that induces the given transition. Testing a delay fault requires a two-pattern test, $\langle v_1, v_2 \rangle$ [70]. Similar to the SOP testing, the first test vector, v_1, is called an initializing test vector, and the second vector, v_2, is called a fault exciting test vector.

Regardless of the fault mode (gate or path delay), a two-pattern test may be categorized as robust or nonrobust. A robust test detects the targeted delay faults regardless of the presence of other delay faults in the circuit. Numerous classifications of the robust path delay fault test exist—for example, the hazard-free robust test, single- multiple-input changing tests, and single- multiple-path propagating tests [79]. The important property of such a test is that the test is not to be invalidated by variable delays of the fan-in gates (paths) to the targeted path (path) delay fault. On the other hand, a nonrobust test detects the fault if no other delay faults affect the circuit. These faults are statically sensitizable. A fault is statically sensitizable if there exists at least one input vector which stabilizes all side inputs of the faulty gate (path) at noncontrolling values [79]. In general, it is possible to determine whether or not a given two-pattern test is robust by examining the logic structure of the circuit under test. The actual circuit delays are not important in this determination [80].

3.3.6 Leakage Fault Models

Static CMOS circuits have very low quiescent current (I_{DDQ}). Most of the manufacturing defects in CMOS ICs exhibit state-dependent elevated I_{DDQ}. Therefore, I_{DDQ} testing is a powerful test method in manufacturing process defects detection. A defect-free MOS transistor has nearly infinite impedance; hence, there should not be any current flow between gate and source, gate and drain, or gate and substrate (well). However, some defects such as gate-oxide short will cause leakage current flow between the gate and other nodes of the transistor. In general, a leakage fault may occur between any two given nodes of a MOS transistor. Therefore, Nigh and Maly [85] and Mao et al. [86] independently proposed the leakage fault model for MOS transistors. The fault model is as follows:

1. f_{GS}—the leakage fault between gate and source
2. f_{GD}—the leakage fault between gate and drain
3. f_{SD}—the leakage fault between source and drain
4. f_{BS}—the leakage fault between bulk and source
5. f_{BD}—the leakage fault between bulk and drain
6. f_{BG}—the leakage fault between bulk and gate

These faults include not only the gate-oxide defect causing leakage but also the leakages between various diodes required to realize a MOS transistor. Furthermore, Nigh and Maly [85] suggested that well to substrate diode defects are not necessary to consider explicitly, for leakage or latchup caused by them is easily observable.

Leakage faults such as gate-oxide shorts or p-n junction pinholes occur quite frequently in CMOS processes. Furthermore, reduced geometries cause an increase in electric field in MOS transistors which may cause successive degradation of gate-oxide, and so on. Typically, small leakage faults do not cause a catastrophic failure of an IC. However, they are potential reliability hazards [30].

3.3.7 Temporary Faults

A major portion of digital system malfunctions is caused by temporary faults. They are harder to detect because at the time of testing they are not reproduced. There are two types of temporary faults: transient faults and intermittent faults.

3.3.7.1 Transient Faults. Transient faults are nonrecurring temporary faults. Typically, they are caused by alphaparticle radiation or power supply fluctuations. Transient faults can also be caused by capacitive, inductive coupled disturbances and external electromagnetic fields. They cannot be repaired because they do not cause physical damage to the hardware. Dynamic logic and memories are particularly susceptible to such faults.

In the context of memories, a class of transient faults are popularly known as *soft faults* or *soft errors*. Such failures occur because of alphaparticle hits or electromagnetic noise [95, 97–100]. When an alphaparticle hits a cell node (p-n junction), many electron hole pairs are generated in the substrate. These carriers are collected by the cell node which destroys the "0" or "1" information in the cell. In Dynamic Random Access Memory (DRAM) alphaparticle-related soft errors have been a serious problem. However, high-density Static Random Access Memory (SRAM) are also known to suffer from alphaparticle-induced soft errors [100].

To overcome these failures, many countermeasures have been proposed. Previously, alphaparticles emitted by radioactive contaminants in chip metallurgy and packaging have been the major source of soft errors. For high-reliability applications, care is taken to eliminate the source of radioactive contamination. In the case of DRAMs, Stapper and Klaasen [99] suggested a number of layout measures to reduce soft errors. These measures include the usage of deep-trench capacitors to increase the storage capacitance as well as to reduce the alphaparticle-induced upset probability, building the pass transistor in N-well, and so on. Although the layout and packaging considerations provide memories with a high degree of alpha-particle immunity, Error Correction Code (ECC) techniques [99, 100] are also used in memories to improve soft error immunity.

3.3.7.2 Intermittent Faults. Intermittent faults recur and appear at regular intervals. These faults are caused by the circuit parameter degradations or aging [37]. Such degradations are progressive until a permanent failure occurs. These faults are also the result of design sensitivity to environmental conditions such as ambient temperature, humidity, and vibrations. The frequency of their occurrence depends on how effectively an IC (system) is protected against environmental conditions through cooling, shielding, and filtering.

3.4 INDUCTIVE FAULT ANALYSIS

The circuit layout influences the impact of a defect and thus the faulty circuit behavior to a large extent. This information is often ignored while developing fault models at transistor or logic level. In one of the earliest papers on the subject, Galiay et al. [38] pointed out that the layout information should be utilized while developing the fault model and hence the test. They argued that the assumption, *all failures can be modeled by stuck-at-0, stuck-at-1,* will become less and less sound as IC density increases. In a study on 4-bit microprocessor chips, they found that most of the failures were due to shorts and opens. Furthermore, gate-level fault models do not adequately represent the faulty behaviors. To test a given circuit, they suggested that an analysis of the failure mechanisms at the layout level should be carried out. In another study, Banerjee and Abraham [39] demonstrated that understanding the effects of physical failures on digital systems is essential to design tests for them and to design circuitry to detect and tolerate them. In yet another study, Shen, Maly, and Ferguson [40] proposed a methodology of mapping physical manufacturing process defects to circuit-level faulty behaviors caused by these defects. In this manner, layout and technology specific faults are generated and ranked according to their likelihood. They concluded that manufacturing process defects can give rise to a much broader range of faults that can be modeled using single-line stuck-at fault models. Similarly, other studies have pointed out that a wealth of information can be extracted from layout [41–43].

These early and thought-provoking publications [38–41] formed the basis of what is known today as Inductive Fault Analysis (IFA). IFA differentiates itself from the conventional fault modeling approaches of assuming faulty behaviors on nodes or logic gates. It derives the circuit or logic-level fault model starting from particular physical defects. In other words, a higher-level fault model is formulated by examining defects at a lower level—hence, the word "inductive" which means that the higher-level fault information is induced from lower level defects. Often IFA is mentioned as realistic fault modeling. The term *realistic* signifies that each fault has a physical basis (i.e., a defect). A test approach based on IFA is also referred to as Defect-Oriented Testing.

In order to exploit the potential of IFA fully, it is important to understand the relationship between manufacturing defects and IC faults. For example, many defects that may influence IC performance are controlled before the functional testing. Therefore, they are not included in IFA-based test generation. Similarly, manufacturing defects in an IC and their impact on its performance are strongly governed by the IC design and its layout. IFA can also be exploited to find out, among other things, areas in the design that are difficult to test. Using this information, we can improve the design robustness and yield. The following subsections present an overview of the defect–fault relationship and IC design and layout-related sensitivity.

3.4.1 The Defect–Fault Relationship

The ever increasing quest for higher functionality on a single IC has led to shrinkage of device geometries and increase in chip area. Unfortunately, both of these developments have caused ICs to become more susceptible to various yield loss mechanisms. In final terms, it is the yield of an IC that determines whether or not complex functionality integration is an economically good proposition. Hence, it has become increasingly relevant to know different yield loss mechanisms. The IC manufacturing process involves a sequence of basic processing steps that are performed on a set of wafers. Maly et al. [45] observed that the outcome of a manufacturing operation depends on three major factors: the process controlling parameters or control, the layout of the IC, and some randomly changing environmental factors, called disturbances. The control of a manufacturing operation is the set of parameters that should be manipulated for desired changes in the fabricated IC structure (e.g., setting of equipment). The layout of an IC is the set of masks distinguishing IC areas that need to be processed for each manufacturing step. The disturbances are environmental factors that influence the result of the manufacturing operation. The manufacturing process disturbances have been studied in great detail [45, 46, 47]. These are classified as:

- Human errors and equipment failures
- Instabilities in the process conditions
- Material instabilities
- Substrate inhomogeneities
- Lithography spots

A detailed treatment of manufacturing process disturbances can be found in Chapter 1 of this book. However, for the purpose of realistic (or IFA-based) fault modeling, it is important to know that not all of the disturbances influence the IC performance equally. All these disturbances influence the processed topology of the IC. In other words, these disturbances deform the IC and hence can be grouped according to the class of deformations. A disturbance is the phenomenon that leads to deformation in an IC. For example, a contamination (disturbance) on the wafer causes a break (deformation) in the metal line. In this case, the deformation is geometrical. Similarly, a poor temperature control (disturbance) during the growth of gate-oxide may result in a lower threshold voltage (electrical deformation). In general, all process disturbances can be classified as geometrical and electrical deformations [45].

Figure 3-11 shows this classification and its relationship to different IC failures. The lower half of the figure shows the classification of physical phenomena which cause yield loss and the upper half of the figure shows the fault classification (structural, performance faults). Geometrical and electrical deformations have local as well as global influence on IC functionality/performance. The global influence occurs when a particular parameter, say the transistor threshold, is affected over the complete wafer. The term *local* is used when the influence on the parameter is limited to a region smaller than a wafer. Often these local deformations are called defects such as break and short in conductors. In addition, spot defects that are primarily lithographic form a part of geometrical deformations. In principle, each class of physical deformation is capable of causing a variety of faults; however, some are more likely (solid lines) than others. For example, all global effects are more likely to cause soft performance failures. Similarly, spot defects are more likely to cause structural or hard performance failures. Since global deformations affect the complete or a large part of a wafer, they are quickly detected by test

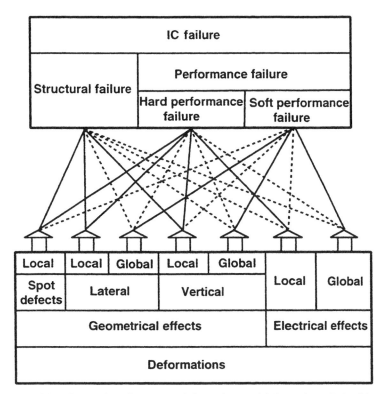

Figure 3-11 IC manufacturing process deformations and deformation relationships with IC faults [45].

structures designed for them (see Chapter 1). Furthermore, in a well-controlled fabrication environment, such problems are kept under control. Therefore, for IFA-based fault modeling and testing, only the local deformations or defects are taken into account.

3.4.2 IC Design and Layout-Related Defect Sensitivity

The design of a modern IC is a complex task. Often designs tend to exploit the very maximum of what a manufacturing process can offer. Issues such as time to market, time to money, and reduced product life cycle further complicate the decision-making process in the design. Faster, smaller, yet complex, low-power dissipation usually are the design objectives. Often these objectives are in conflict with each other, and rarely do they agree with so-called robust design practices. Typically, it takes a lot of product engineering and a couple of design iterations to stabilize the design for reasonably good yield. In spite of all this effort, it has been observed that the yield of certain designs, for a given chip area, is lower than that of others. Design-related sensitivity to yield can be divided into two major classes, which are further divided into subclasses.

- **Defect Sensitive Design** The operational frequency of a digital IC is determined by its critical path. The critical path is the data path in an IC with the largest delay. For good functioning of the chip, the critical path delay should be less than the clock period. However, the actual delay of the critical path is governed by the process. The design margin between the critical path and the clock frequency should be reasonable. Otherwise parametric process variations or spot defects may result in timing (parametric) failures. Furthermore, the amount of physical area the critical path has on the chip is also important because higher area will increase the probability of a defect landing onto the critical path. An otherwise innocuous defect in the critical path may increase its parasitic capacitance or resistance, resulting in a timing failure. Similarly, the logic implementation also has an influence over the timing-related sensitivity of the design. For example, the timing criticality is much higher in the dynamic logic implementation than in the static logic implementation. Furthermore, in the dynamic logic implementations, logic levels are often defined under the dynamic conditions. Many defects influence the dynamic behavior and hence cause failures. On the other hand, the impact of such defects on the static logic is not so severe and causes increased delay. If this delay is not in the critical path, often it does not lead to a failure. Moreover, the type of synchronous logic implementation, random logic, on board memories, and PLA all have yield-related repercussions because they have different inherent sensitivities to the defects.
- **Defect-Sensitive Layout** As mentioned earlier, the IC layout also contains a wealth of yield-related information. Researchers [48, 49] have defined the concept of *critical area*. This area is defined for a defect or radius R, as the area on the die in which the center of a circular defect has to fall for a fault to occur in the circuit (see Figure 3-12). For a given defect density distribution

3.4 ■ Inductive Fault Analysis

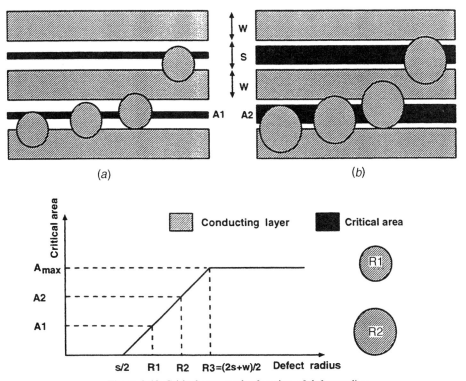

Figure 3-12 Critical area as the function of defect radius.

of a particular defect, we can find the most probable defect radius. For a given defect radius of a particular defect, the critical area can be computed. It can be deduced that the higher the critical area in a given IC, the lower will be the yield. This knowledge can be used to minimize the critical area. Thus, the placement and routing strategies and their implementation have an impact on the yield. IFA can be exploited to find out the layout/design-related yield sensitivity. Moreover, through this methodology better design and layout can be developed. Similarly, it can be argued that clocking strategies and its routing and various possible implementations will have an impact on circuit yield.

3.4.3 Basic Concepts of IFA

IFA is a systematic approach for determining what faults are likely to occur in a VLSI circuit. This approach takes into account the technology of the implementation, the circuit topology, and the defect statistics of the fabrication plant. In a followup publication Maly, Ferguson and Shen [40] formalized concepts of IFA as follows:

IFA is a systematic procedure to predict all the faults that are likely to occur in a MOS integrated circuit or subcircuit. The three major steps of the IFA procedure are:

(1) generation of physical defects using statistical data from the fabrication process, (2) extraction of circuit-level faults caused by these defects, and (3) classification of fault types and ranking of faults based on their likelihood of occurrence.

The major steps of the IFA are shown in Figure 3-13. The circuit layout and the manufacturing defect statistics form inputs for the analysis. The IFA methodology takes into account only local deformations or defects for analysis purposes. The size and the probability of defects are defined by the Defect Density Distribution (DDD). For an IC fabrication plant (fab), DDDs are normally available for each layer and defect types. DDDs define how large the probability of a defect in a

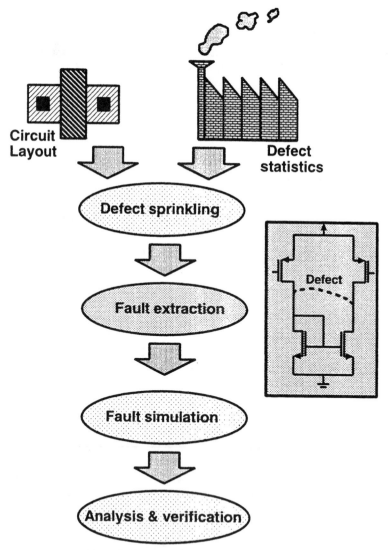

Figure 3-13 A graphical representation of IFA.

certain layer is with respect to other layers and how the probability of defects in a layer depends on the size of the defect. For a typical double-metal single-poly CMOS process, these defects include:

- Extra and missing material defects in conducting and semiconducting layers
- Presence of extra contacts and vias
- Absence of contacts and vias
- Thin- and thick-oxide pinholes
- Junction leakage pinholes

The defects are sprinkled onto the layout in a random manner. For this purpose, CAD tools like VLASIC (VLSI Layout Simulation for Integrated Circuits) [50, 51] and DEFAM (DEfect to FAult Mapper) [Chapter 4] are utilized. The defects are modeled as absence or presence of material on the layout. Only a subset of all defects cause a change in the circuit connectivity. For example, an unintended short is created between two nodes, or an open defect causes a break in the connectivity. All these defects are extracted, and their impact on circuit behavior is modeled at an appropriate level of abstraction for fault simulation. Subsequently, the abstracted defects (i.e., faults) are fault simulated with given test stimuli. The fault simulation information is exploited for providing Design for Testability (DFT) solutions, building fault tolerance into the circuit, and so on. Finally, simulation results are verified by the silicon data.

The IFA technique takes into account only a subset of process defects, namely, local defects (Figure 3.11). The global defects or deviations cause widespead failures or near failures. Since the impact of global defects is present over a relatively large area, these are detected rather quickly. Moreover, special test structures are available to test these erroneous conditions. Hence, such defects are detected early in the manufacturing process before the functional/structural testing. Furthermore, in a well-controlled and monitored process, major process errors causing global defects are detected and solved with relative ease.

3.4.4 Practical Experiences with IFA

In the last decade, a substantial research effort has been devoted to IFA and its exploitation in order to find realistic fault models, test and DFT solutions, and so on. In this subsection, we discuss the salient features of some of the IFA experiments. For a detailed treatment of the subject, the reader is referred to various references.

3.4.4.1 IFA Experiment on a SRAM. In one of the first IFA experiments with silicon results, Dekker et al. applied the technique over an 8k × 8 CMOS SRAM at Philips Research Labs. A realistic fault model of the SRAM was developed, which included the following fault classes [43]:

1. A SA0 (SA1) fault in a SRAM cell if the logic value of the cell cannot be changed by any action.

2. A SOP fault in a SRAM cell if it is not possible to access the cell.
3. A transition fault in a SRAM cell if the cell fails to undergo at least one of the transitions $0 \rightarrow 1$ or $1 \rightarrow 0$.
4. A SRAM cell is state coupled to another cell.
5. A SRAM cell is suffering from a multiple access fault.
6. A SRAM cell is suffering from a data retention fault.

Two SRAM test algorithms of complexity 9N and 13N were developed, which covered this fault model completely. Here, N represents the number of addresses in the SRAM. The 9N algorithm signifies that 9 Read and/or Write operations are carried out on each address location. Hence, the complexity of the algorithm increases with the number of operations in it.

In order to see the effectiveness of IFA-based algorithms, these along with 19 other popular RAM algorithms were applied to a total of 1192 SRAM devices out of 9 wafers. Devices that suffered from total failures, stuck-at faults, or incorrect functioning were discarded because all algorithms failed them. A set of 480 devices were analyzed. For this comparative study, each algorithm was assigned a score number. The maximum achievable score for any algorithm was 2. The higher the score, the better was the fault coverage of the algorithm for realistic faults. IFA-based 13N algorithm with data retention test scored the second highest score of 1.83. The highest score of 1.91 was achieved by the algorithm that had a seven times larger test set. IFA-based 13N and 9N test algorithms scored even better than the 30N test algorithm. This experiment was the first to demonstrate the effectiveness of the IFA technique to evolve simpler, efficient test solutions with silicon data.

Exercise 8: *Take the layout of a CMOS SRAM cell. Conduct the IFA experiment using a small coin or a disc shaped object to determine the fault model. Compare your fault model with that of [43].*

3.4.4.2 IFA Experiment on Standard Cells. In another study, Ferguson and Shen [43] extracted and simulated realistic CMOS faults using the IFA technique. A CAD tool, FXT, was developed, which had capabilities of automatic defect insertion/extraction for a reasonably sized layout. This tool was used to analyze five circuits from a commercial CMOS standard cell library. The five circuits were (1) a single-bit half adder cell, (2) a single-bit full adder cell, (3) a counter, (4) another counter, and (5) a 4 × 4 multiplier. They sprinkled more than 10 million defects in two counters, which caused approximately 500,000 faults in them. Similarly, over 20 million defects were sprinkled in the multiplier, which caused approximately 1 million faults in it. Approximately one-twentieth of defects caused faults; this conforms to the fact that most defects are too small to cause faults. The majority of extracted faults could be modeled as bridging faults, break faults, or a combination of these two. For example in the 4 × 4 multiplier, the bridging and break faults amount to 48% and 42%, respectively.

Almost all the remaining faults were transistor SON faults, which can also be represented as a bridging fault between the source and drain of the transistor. Similarly, transistor SOP fault is equivalent to a break fault. Therefore, almost all the faults could be represented as bridging, break faults, or a combination of the two. The only two categories that were not equivalent to the above-mentioned categories were new transistors and exceptions. A new transistor in the layout is created by a lithographic spot on poly or diffusion mask. Less than 0.7% of faults fall into this category.

The SAF model fared rather poorly in modeling extracted faults. In the case of the 4 × 4 multiplier, only 44% of the bridging faults could be modeled by the SAF. For nonbridging faults, only 50% of the faults could be modeled as SAFs. Hence, for the multiplier, less than 50% of all extracted faults could be modeled as SAFs. A similar comparison is carried out with graph-theoretic (transistor-level) fault models. It is estimated that only 57% of the extracted faults could be modeled as graph-theoretic fault models. Although this is higher than what is modeled by the SAF model, the majority of non-SAF extracted faults could not be modeled, for two reasons. First, many non-SAF faults bridge input nodes together and are not modeled with the graph-theoretic approach either. Second, approximately 70% of the transistors in the analyzed circuits were pass transistors or components of inverters. Pass transistors are not modeled in graph-theoretic fault modeling, and in SAF models most transistor faults occur in inverters that cause change in logical functionality. Only 1% of the extracted faults could be represented by the transistor SOP fault model.

Further analysis was carried out to find out how well SA and exhaustive test sets detect the extracted faults. In order to reduce the simulation time, only extracted bridging faults were simulated in counters. The SA test set could detect between 73% and 89% of the circuit's bridging faults. Even under the unrealistic assumption that all nonbridging faults are detected by the SA test set, the 100% SA test set could detect between 87% and 95% of extracted faults. The fault coverage of extracted bridging faults by the exhaustive test set was relatively higher. The exhaustive test set detected between 89% and 99.9% of the extracted bridging faults. As a solution for better fault coverage, quiescent current monitoring (I_{DDQ}) was suggested. It was implied that I_{DDQ} will provide the best test set for bridging fault detection.

> **Exercise 9:** *Take the layout of a three-input NOR gate and perform an IFA experiment manually. Determine how many fault classes are detected by the SA and SOP test set. Design a test set to detect undetected faults.*

3.4.4.3 The IFA Experiment on an Embedded DRAM.

Sachdev and Verstraelen [44] used the IFA method to develop a fault model and test algorithms for embedded DRAMs. Embedded DRAMs are special in many ways. Not only are they almost analog devices, which must operate in a hostile digital environment, but also they are harder to test owing to system-limited controllability and observability. In addition, they must be designed with layout density reaching the limits of available technology. High packing density, standard manufacturing process implementation, and dynamic nature of operation make embedded DRAMs susceptible to catastrophic as well noncatastrophic defects.

The fault model development activity was divided into two parts: (1) catastrophic defects based, and (2) noncatastrophic defects based. VLASIC was utilized to sprinkle the defects in the layout. The output of VLASIC was a catastrophic defect (defects that modified the circuit netlist) list. These defects were mapped onto the DRAM cell schematic. Figure 3-14a shows a layout of two adjacent DRAM cells with various spot defects. Figure 3-14b shows the schematic of the cell, with defects causing shorts and opens. Subsequently, these defects are classified into various fault categories. In this manner, the contribution of catastrophic defects to the fault model is determined. This exercise resulted in the following categories:

1. A memory cell is stuck-at 0 or stuck-at 1.
2. A memory cell is stuck-open.
3. A memory cell is coupled to another cell.
4. A memory cell has a multiple access fault.
5. A memory cell suffers from a data retention fault in one (or both) of its states.

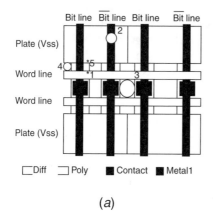

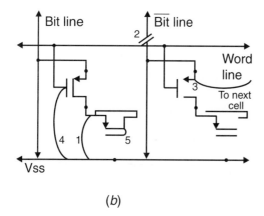

Figure 3-14 Layout of the DRAM with defects (a); and the mapping of defects in the DRAM schematic (b).

The catastrophic fault model was supplemented by the noncatastrophic fault model. The noncatastrophic defects do not modify the circuit's connectivity but typically increase the parasitic capacitive/resistive coupling between circuit nodes.

Owing to the susceptibility of DRAM cells for noncatastrophic defects and the inability of VLASIC to model such defects, a mathematical model of the coupling faults was determined [96]. It was assumed that the two-coupling faults model adequately represents the intercell couplings. This model included 18 types of coupling faults that can affect two arbitrary DRAM cells. Two test algorithms of complexity 8N and 9N were developed which covered the combined fault model. The algorithms were divided into an initialization step and a set of four marches. The 9N algorithm was applied to devices from 34 wafers. Out of the total devices, 579 failed the algorithm. These failures were divided into various categories. A total of 318 devices failed in all four marches, and 201 devices failed in three of the four marches showing all or a large number of cells failing. A set of 21 and 18 devices showed cell SA0 and SA1 faults. A small number of devices had some of the coupling faults. Overall, 96% of the failures could be explained with the faults caused by catastrophic defects. The rest (4%) of the failures could not be explained by the catastrophic defects-based fault model but could be explained by the coupling fault model based on the noncatastrophic defects.

3.4.4.4 IFA Experiment on Analog Circuits. Following the successful IFA applications in digital domain, various experiments were carried out in the analog domain [52, 60]. It is projected as one of the alternatives to analog functional testing. Analog circuits, owing to their nonbinary circuit operation, are influenced by defects in a different manner compared to the digital circuits. This calls for a careful investigation of the occurrence of defects in analog circuits, their modeling-related aspects, and their detection strategies.

The subject of analog IFA is closely related to realistic analog fault model development. In fact, analog fault modeling has been defined as the critical factor in the success of IFA-based analog test methods [52, 53]. The application of popular digital fault models in analog domain is largely unsuccessful in representing faulty behavior. Therefore, Soma [52] and Meixner and Maly [53] proposed fault model development based on commonly found process defects. The evolved realistic fault models form the basis for test generation and fault simulation. Soma applied the IFA technique on several analog building blocks to obtain efficient and effective tests [54, 55].

One of the problems with analog IFA is its application in real-world complex circuits. The circuit-level simulations used to determine faulty behavior are not feasible for complex analog circuits. Harvey et al. [56] utilized high-level models for parts of the circuit. They applied this technique to test a Phase-Locked Loop (PLL) circuit. Kuijstermans et al. [57] applied a similiar approach on a flash A/D converter. Taking ideas from digital macro test concepts [32], they divided an 8-bit flash A/D converter into smaller Spice-level simulatable macros. Circuit simulation is carried out to generate macro-level fault models or the fault signatures. Simulations with higher level models of other cells are used to determine the ability to detect these fault signatures at the circuit boundary. Both of these articles [56, 57] presented the potential of IFA, combined with high-level modeling, to provide analog test solutions. Simulated results presented in these articles demonstrated the high fault coverage of modeled faults by simple test stimuli. In this approach the accuracy of the analysis is limited by the high-level models.

Although the IFA-based technique forms the structured basis of test generation, the validity of this technique for analog circuits is not proven extensively with silicon results. One of the first silicon results was presented by Sachdev [58, 60]. A Class AB amplifier device is analyzed to arrive at realistic fault classes. Simple DC, transient, and AC stimuli in a Spice-like simulation environment detected all the simulated fault classes. A test program is developed using these simple test stimuli and is appended to the existing conventional (specification-based) test program. This exercise is carried out to find the effectiveness of the IFA-based test method with respect to the specification-based test method in a production test environment. Results of the exercise are reproduced in Figure 3-15. A total of 106,784 devices were tested through the conventional test method. The yield of the device was very high, and only 3270 devices were failed by the conventional test method. Only failed devices (3270) were considered for further testing. These devices were tested with the IFA-based test method. Out of this lot, 433 devices passed the test. These passed devices from the IFA-based test method were once again tested with the conventional test method. The results of this test were as follows: 51 devices passed the test, and the rest of the devices (433 − 51 = 382, 0.4% of total tested devices) failed the test again. These failed devices (382) were subjected to a detailed analysis.

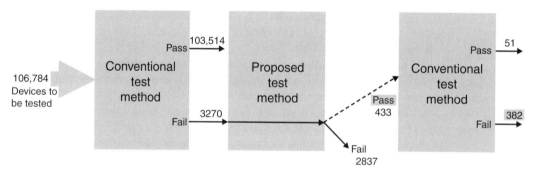

Figure 3-15 The conducted IFA experiment over an amplifier.

The input offset voltage specification contributes to the maximum number of failures (182, 47.6%) which could not be caught by the IFA-based test method. The Total Harmonic Distortion (THD) specification contributed to the second biggest segment of undetected failures (123, 32.2%). Similarly, Signal to Noise Ratio (SNR) measurement failed 20 devices (5.2%). These three categories of failures contributed to the bulk (85%) of the failures that could not be detected by the IFA-based test method. These failures can be attributed to unmodeled faults by the IFA test technique. For example, any differential amplifier has an inherent offset voltage associated with it which is the source of nonlinearity in its operation. Often this offset voltage is minimized by transistor matching, layout, trimming, and compensation techniques. Besides the local process defects, several other factors can manifest themselves as increased offset voltage.

3.4.5 The IFA: Strengths and Weaknesses

The development of realistic fault models is the foremost IFA contribution to the testing world. Previous studies have amply demonstrated that IFA-based tests, in general, are simple, compact, and effective. For example, SRAM algorithms developed by Dekker were found to be more effective than longer algorithms. In the case of analog devices, a 30% reduction in the production test costs for a Class AB amplifier is reported [59, 60]. IFA can also be utilized to design defect-insensitive cells/designs. The cells can be designed such that either the design robustness against the commonly occurring defects is increased or the defect detection is simplified. Similarly, IFA can be utilized for finding out appropriate DFT strategies.

At the same time, the IFA-based test method is limited by the availability of CAD software tools and requires relatively high computer resources in terms of CPU power and data storage. A substantial analysis effort is needed before an IFA-based test method may emerge from the analysis. Furthermore, because of computational and CAD tool-related constraints, only cells and macros can be analyzed. Therefore, ideally this analysis should be carried out in the design environment on a cell-by-cell basis. A bigger design should be partitioned into suitable smaller segments for this analysis.

3.5 SUMMARY

The functional testing for complex digital ICs is prohibitively expensive and does not ensure that the IC will be fault-free. The structural tests that are targeted to detect faulty circuit behaviors provide an alternative. The effectiveness of a structural test is quantifiable in terms of the covered faults. Thus, it allows the user to establish a relationship between test coverage and quality of the tested devices. The test generation for the structural test is considerably simpler owing to the availability of CAD tools.

The structural test, however, requires a fault model that represents likely manufacturing process defects with an acceptable accuracy and provides an objective basis for the structural test generation. Fault models are classified according to the level of abstraction. Gate level (SAF), transistor level (SOP, SON), and functional level are some of the examples of the abstraction levels. The level of abstraction is essentially a compromise between the fault model's ability to represent actual defects and the speed of processing the fault in a fault simulation environment.

The conventional fault modeling approaches do not consider the likely or realistic faults in a given layout of a circuit. The layout of a circuit has a significant impact on the faulty circuit behavior. The IFA takes into account the circuit layout and the defect data from the manufacturing site to generate a list of realistic faults. The word "realistic" signifies that each fault has a physical basis (i.e., defect). In this manner, the circuit layout-dependent fault models are evolved. Reported experiments illustrate the effectiveness of the method in realistic fault model generation with success.

REFERENCES

[1] Eugene R. Hnatek, "IC Quality—Where Are We?" *Proceedings of IEEE International Test Conference,* pp. 430–445, 1987.

[2] Eugene R. Hnatek, "Integrated Circuits Quality and Reliability," New York: Marcel Dekker, 1987.

[3] B. Mustafa Pulat and Lauren M. Streb, "Position of Component Testing in Total Quality Management (TQM)," *Proceedings of International Test Conference,* pp. 362–366, 1992.

[4] E. Takeda et al., "VLSI Reliability Challenges: From Device Physics to Wafer Scale Systems," *Proceedings of IEEE,* vol. 81, no. 5, pp. 653–674, May 1993.

[5] E. J. McCluskey and Fred Buelow, "IC Quality and Test Transparency," *Proceedings of International Test Conference,* pp. 295–301, 1988.

[6] R. D. Eldred, "Test Routines Based on Symbolic Logical Statements," *Journal of ACM-6,* pp. 690–708, Jan. 1959.

[7] S. Funatsu, N. Wakatsuki, and T. Arima, "Test Generation Systems in Japan," *Proceedings of 12th Design Automation Symposium,* pp. 114–122, 1975.

[8] E. B. Eichelberger and T. W. Williams, "A Logic Design Structure for LSI Testability," *J. Design Automation and Fault Tolerant Computing,* vol. 2, no. 2, pp. 165–178, May 1978.

[9] T. W. Williams and K. P. Parker, "Design for Testability—A Survey," *Proceedings of the IEEE,* vol. 71, no. 1, pp. 98–113, Jan. 1983.

[10] M. A. Breuer and A. D. Friedman, "Diagnosis and Reliable Design of Digital Systems," Woodland Hills, CA, Computer Science Press, 1976.

[11] J. A. Abraham, "Fault Modeling in VLSI," *VLSI Testing,* edited by T. W. Williams, Vol. 5, North-Holland: 1986, pp. 1–27.

[12] J. F. Poage, "Derivation of Optimum Tests to Detect Faults in Combinational Circuits," *Proceedings of Symposium on Mathematical Theory of Automata,* pp. 483–528, 1963.

[13] P. Hayes, "Fault Modeling for Digital Integrated Circuits," *IEEE Transactions on Computer-Aided Design of Circuits and Systems,* CAD-3, pp. 200–207, 1984.

[14] J. P. Hayes, "Fault Modeling, IEEE Design & Test of Computers," pp. 88–95, Apr. 1985.

[15] R. L. Wadsack, "Fault Modeling and Logic Simulation of CMOS and MOS Integrated Circuits," *Bell Systems Technical Journal,* vol. 57, no. 5, pp. 1449–1474, May–June 1978.

[16] R. Chandramouli, "On Testing Stuck-Open Faults," *Proceedings of 13th Annual International Symposium on Fault Tolerant Computing Systems,* pp. 258–265, June 1983.

[17] Y. M. El-Ziq and R. J. Cloutier, "Functional-Level Test Generation for Stuck-Open Faults in CMOS VLSI," *Proceedings of International Test Conference,* pp. 536–546, 1981.

[18] S. M. Reddy and Sandeep Kundu, "Fault Detection and Design for Testability of CMOS Logic Circuits," *Testing and Diagnosis of VLSI and ULSI*, edited by F. Lombardi and M. Sami, New York: Kluwer, pp. 69–91, 1989.

[19] S. Koeppe, "Optimum Layout to Avoid CMOS Stuck-Open Fault," *Proceedings of 24th ACM/IEEE Design Automation Conference*, pp. 829–835, 1987.

[20] S. K. Jain and V. D. Agrawal, "Test generation for MOS Circuits Using D-Algorithm," *Proceedings of 20th Design Automation Conference*, pp. 65–70, 1983.

[21] S. M. Reddy, M. K. Reddy, and J. G. Kuhl, "On Testable Design for CMOS Logic Circuits," *Proceedings of International Test Conference*, 435–445, 1983.

[22] S. M. Reddy, M. K. Reddy and V. D. Agrawal, "Robust Test for Stuck-Open Faults in CMOS Combinational Logic Circuits," *Proceedings of 14th International Symposium on Fault Tolerant Computing*, pp. 44–49, 1984.

[23] R. Rajsuman, A. P. Jayasumana, and Y. K. Malaiya, "CMOS Stuck-Open Fault Detection Using Single Test Patterns," *Proceedings of ACM/IEEE Design Automation Conference*, pp. 714–717, 1989.

[24] R. Rajsuman, A. P. Jayasumana, and Y. K. Malaiya, "CMOS Open-Fault Detection in the Presence of Glitches and Timing Skews," *IEEE Journal of Solid-State Circuits*, vol. 24, no. 4, pp. 1061–1129, August 1989.

[25] A. P. Jayasumana, Y. K. Malaiya, and R. Rajsuman, "Design of CMOS Circuits for Stuck-Open Fault Testability," *IEEE Journal of Solid-State Circuits*, vol. 26, no. 1, pp. 58–61, Jan. 1991.

[26] B. W. Woodhall, B. D. Newman, and A. G. Sammuli, "Empirical Results on Undetected CMOS Stuck-Open Failures," *Proceedings of International Test Conference*, pp. 166–170, 1987.

[27] H. Cox and J. Rajaski, "Stuck-Open and Transition Fault Testing in CMOS Complex Gates," *Proceedings of International Test Conference*, pp. 688–694, 1988.

[28] C. Di and J. A. G. Jess, "On Accurate Modeling and Efficient Simulation of CMOS Open Faults," *Proceedings of International Test Conference*, pp. 875–882, 1993.

[29] K. J. Lee and M. A. Breuer, "On the Charge Sharing Problem in CMOS Stuck-Open Fault Testing," *Proceedings of International Test Conference*, pp. 417–425, 1990.

[30] J. M. Soden, C. F. Hawkins, R. K. Gulati, and W. Mao, "I_{DDQ} Testing: A Review," *Journal of Electronic Testing: Theory and Applications (JETTA)*, vol. 3, pp. 291–303, 1992.

[31] H. T. Vierhaus, W. Meyer, and U. Glaser, "CMOS Bridges and Resistive Faults: I_{DDQ} versus Delay Effects," *Proceedings of International Test Conference*, pp. 83–91, 1993.

[32] F. P. M. Beenker, K. J. E. van Eerdewijk, R. B. W. Gerritsen, F. N. Peacock, and M. van der Star, "Macro Testing, Unifying IC and Board Test," *IEEE Design and Test of Computers*, pp. 26–32, Dec. 1986.

[33] S. M. Thatte and J. A. Abraham, "Testing of Semiconductor Random Access Memories," *Proceedings of International Conference on Fault Tolerant Computing,* pp. 81–87, 1977.

[34] J. E. Smith, "Detection of Faults in Programmable Logic Arrays," *IEEE Transactions on Computers,* C-28, pp. 845–853, 1979.

[35] S. M. Thatte and J. A. Abraham, "Test Generation for Microprocessors," *IEEE Transactions on Computers,* C-29, pp. 429–441, 1980.

[36] D. S. Brahme and J. A. Abraham, "Functional Testing of Microprocessors," *IEEE Transactions on Computers,* C-33, pp. 475–485, 1984.

[37] S. Mourad and E. J. McCluskey, "Fault Models," *Testing and Diagnosis of VLSI and ULSI,* edited by F. Lombardi and M. Sami, New York: Kluwer, pp. 49–68, 1989.

[38] J. Galiay, Y. Crouzet, and M. Vergniault, "Physical Versus Logical Fault Models in MOS LSI Circuits: Impact on Their Testability," *IEEE Transactions on Computers,* vol. C-29, no. 6, pp. 527–531, June 1980.

[39] P. Banerjee and J. A. Abraham, "Characterization and Testing of Physical Failures in MOS Logic Circuits," *IEEE Design and Test of Computers,* pp. 76–86, Aug. 1984.

[40] J. P. Shen, W. Maly, and F. J. Ferguson, "Inductive Fault Analysis of MOS Integrated Circuits," *IEEE Design and Test of Computers,* pp. 13–26, Dec. 1985.

[41] W. Maly, F. J. Ferguson, and J. P. Shen, "Systematic Characterization of Physical Defects for Fault Analysis of MOS IC Cells," *Proceedings of International Test Conference,* pp. 390–399, 1984.

[42] F. J. Ferguson and J. P. Shen, "Extraction and Simulation of Realistic CMOS Faults Using Inductive Fault Analysis," *Proceedings of International Test Conference,* pp. 475–484, 1988.

[43] R. Dekker, F. Beenker, and L. Thijssen, "Fault Modeling and Test Algorithm Development for Static Random Access Memories," *Proceedings of International Test Conference,* pp. 343–352, 1988.

[44] M. Sachdev and M. Verstraelen, "Development of a Fault Model and Test Algorithms for Embedded DRAMs," *Proceedings of International Test Conference,* pp. 815–824, 1993.

[45] W. Maly, A. J. Strojwas, and S. W. Director, "VLSI Yield Prediction and Estimation: A Unified Framework," *IEEE Transactions on Computer Aided Design,* vol. CAD-5, No. 1, pp. 114–130, Jan. 1986.

[46] S. K. Gandhi, *VLSI Fabrication Principles,* New York: John Wiley and Sons, 1983.

[47] S. M. Sze, *VLSI Technology,* New York, McGraw-Hill, 1983.

[48] W. Maly, W. R. Moore and A. J. Strojwas, "Yield Loss Mechanisms and Defect Tolerance," SRC-CMU Research Center for Computer Aided Design, Dept. of Electrical and Computer Engineering, Carnegie Mellon University, Pittsburgh, PA 15213.

[49] Albert V. Ferris-Prabhu, *Introduction to Semiconductor Device Yield Modeling,* Artech, 1992.

[50] H. Walker and S. W. Director, "VLASIC: A Catastrophic Fault Yield Simulator for Integrated Circuits," *IEEE Transactions on Computer Aided Design of Integrated Circuits and Systems,* CAD-(5)4, pp. 541–556, Oct. 1986.

[51] H. Walker, *VLASIC System User Manual Release 1.3,* SRC-CMU Research Center for Computer Aided Design, Dept. of Electrical and Computer Engineering, Carnegie Mellon University, Pittsburgh, PA.

[52] M. Soma, "An Experimental Approach to Analog Fault Models," *Proceedings of Custom Integrated Circuits Conference,* pp. 13.6.1–13.6.4, 1991.

[53] A. Meixner and W. Maly, "Fault Modeling for the Testing of Mixed Integrated Circuits," *Proceedings of International Test Conference,* pp. 564–572, 1991.

[54] M. Soma, "A Design for Test Methodology for Active Analog Filters," *Proceedings of International Test Conference,* pp. 183–192, 1990.

[55] M. Soma, "Fault Modeling and Test Generation for Sample and Hold Circuit," *Proceedings of International Symposium on Circuits and Systems,* pp. 2072–2075, 1991.

[56] R. J. A. Harvey, A. M. D. Richardson, E. M. J. Bruls, and K. Baker, "Analogue Fault Simulation Based on Layout Dependent Fault Models," *Proceedings of International Test Conference,* pp. 641–649, 1994.

[57] F. C. M. Kuijstermans, M. Sachdev, and L. Thijssen, "Defect Oriented Test Methodology for Complex Mixed-Signal Circuits," *Proceedings of European Design and Test Conference,* pp. 18–23, Mar. 1995.

[58] M. Sachdev, "Defect Oriented Analog Testing: Strengths and Weaknesses," *Proceedings of 20th European Solid State Circuits Conference,* pp. 224–227, Sept. 1994.

[59] M. Sachdev, "A Defect Oriented Testability Methodology for Analog Circuits," *Journal of Electronic Testing: Theory and Applications (JETTA),* vol. 6, no. 3, pp. 265–276, June 1995.

[60] Manoj Sachdev, and Bert Atzema, "Industrial Relevance of Analog IFA: A Fact or a Fiction," *Proceedings of International Test Conference,* Oct. 1995, pp. 61–70.

[61] M. S. Abadir and H. K. Reghbati, "Functional Testing of Semiconductor Random Access Memories," *ACM Computing Surveys,* vol. 15, no. 3, pp. 175–198, Sept. 1983.

[62] A. J. van de Goor, *Testing Semiconductor Memories: Theory and Practices,* New York: John Wiley and Sons, 1991.

[63] B. Prince, *Semiconductor Memories,* New York: John Wiley and Sons, 1991.

[64] B. F. Cockburn, "Tutorial on Semiconductor Memory Testing," *Journal of Electronic Testing: Theory and Applications (JETTA),* vol. 5, no. 4, pp. 321–336, Nov. 1994.

[65] D. S. Suk, and S. M. Reddy, "A March Test for Functional Faults in Semiconductor Random Access Memories," *IEEE Transactions on Computers,* vol. C-30, no. 12, pp. 982–985, Dec. 1981.

[66] J. P. Hayes, "Detection of Pattern-Sensitive Faults in Random Access Memories," *IEEE Transactions on Computers,* vol. C-24, no. 2, Feb. 1975, pp. 150–157.

[67] J. Savir, W. H. McAnney, and S. R. Vecchio, "Testing for Coupled Cells in Random Access Memories," *Proceedings of International Test Conference,* pp. 439–451, 1989.

[68] C. A. Papachristou and N. B. Sahgal, "An Improved Method for Detecting Functional Faults in Semiconductor Random Access Memories," *IEEE Transactions on Computers,* vol. C-34, no. 2, pp. 110–116, Feb. 1985.

[69] M. Sachdev, "Reducing the CMOS Ram Test Complexity with I_{DDQ} and Voltage Testing," *Journal of Electronic Testing: Theory and Applications (JETTA),* vol. 6, no. 2, pp. 191–202, Apr. 1995.

[70] M. A. Gharaybeh, M. L. Bushnell, and V. D. Agrawal, "Classification and Test Generation for Path-Delay Faults Using Single Stuck-Fault Tests," *Proceedings of International Test Conference,* pp. 139–148, 1995.

[71] A. K. Pramanick, and S. M. Reddy, "On the Computation of the Ranges of Detected Delay Fault Sizes," *IEEE International Conference on CAD,* pp. 126–129, 1989.

[72] V. S. Iyenger et al., "On Computing the Sizes of Detected Delay Faults," *IEEE Transactions on CAD,* vol. 9, no. 3, pp. 299–312, 1990.

[73] G. L. Smith, "Model for Delay Faults Based upon Paths," *Proceedings of International Test Conference,* pp. 342–349, 1985.

[74] E. S. Park, B. Underwood, T. W. Williams, and M. R. Mercer, "Delay Testing Quality in Timing-Optimized Designs," *Proceedings of International Test Conference,* pp. 879–905, 1991.

[75] K. T. Cheng, "Transition Fault Simulation for Sequential Circuits," *Proceedings of International Test Conference,* pp. 723–731, 1992.

[76] B. Underwood, W. O. Law, S. Kang, and H. Konuk, "Fastpath: A Path-delay Test Generator for Standard Scan Designs," *Proceedings of International Test Conference,* pp. 154–163, 1994.

[77] P. Varma, "On Path Delay Testing in a Standard Scan Environment," *Proceedings of International Test Conference,* pp. 164–173, 1994.

[78] I. Pramanick and A. K. Pramanick, "Parallel Delay Fault Coverage and Test Quality Evaluation," *Proceedings of International Test Conference,* pp. 113–122, 1995.

[79] M. Sivaraman and A. J. Strojwas, "Test Vector Generation for Parametric Path Delay Faults," *Proceedings of International Test Conference,* pp. 132–138, 1995.

[80] A. Pierzynska and S. Pilarski, "Non-Robust versus Robust," *Proceedings of International Test Conference,* pp. 123–131, 1995.

[81] K. C. Saraswat and F. Mohammadi, "Effect of Scaling of Interconnections on the Time Delay of VLSI Circuits," *IEEE Transactions on Electron Devices,* vol. ED-29, no. 4, pp. 645–650, 1982.

References

[82] H. B. Bakoglu and J. D. Meindel, "Optimal Interconnection Circuits for VLSI," *IEEE Transactions on Electron Devices,* vol. ED-32, no. 5, pp. 903–909, 1985.

[83] D. S. Gardner, J. D. Meindel, and K. C. Saraswat, "Interconnection and Electromigration Scaling Theory," *IEEE Transactions on Electron Devices,* vol. ED-34, no. 3, pp. 633–643, 1987.

[84] S. Bothra, B. Rogers, M. Kellem, and C. M. Osburn, "Analysis of the Effects of Scaling on Interconnect Delay in ULSI Circuits," *IEEE Transactions on Electron Devices,* vol. ED-40, no. 3, pp. 591–597, 1993.

[85] P. Nigh and W. Maly, "Test Generation for Current Testing," *IEEE Design & Test of Computers,* pp. 26–38, Feb. 1990.

[86] W. Mao, R. Gulati, D. K. Goel, and M. D. Ciletti, "QUIETEST: A Quiescent Current Testing Methodology for Detecting Leakage Faults," *Proceedings of International Conference on CAD,* pp. 280–283, 1990.

[87] E. J. McCluskey, and F. W. Clegg, "Fault Equivalence in Combinational Logic Networks," *IEEE Transactions on Computers,* vol. c-20, no. 11, pp. 1286–1293, Nov. 1971.

[88] D. R. Schertz, and G. Metze, "A New Representation for Faults in Combinational Digital Circuits," *IEEE Transactions on Computers,* vol. c-21, no. 8, pp. 858–866, Aug. 1972.

[89] Kilin To, "Fault Folding for Irredundant and Redundant Combinational Circuits," *IEEE Transactions on Computers,* vol. c-22, no. 11, pp. 1008–1015, Nov. 1973.

[90] B. K. Roy, "Diagnosis and Fault Equivalence in Combinational Circuits," *IEEE Transactions on Computers,* vol. c-23, no. 9, pp. 955–963, Sept. 1974.

[91] O. H. Ibarra, and S. K. Sahni, "Polynomial Complete Fault Detection Problems," *IEEE Transactions on Computers,* vol. c-24, no. 3, pp. 242–249, Mar. 1975.

[92] A. Goundan and J. P. Hayes, "Identification of Equivalent Faults in Logic Networks," *IEEE Transactions on Computers,* vol. c-29, no. 11, pp. 978–985, Nov. 1980.

[93] H. C. Shih and J. A. Abraham, "Fault Collapsing Techniques for MOS VLSI Circuits," *Proceedings of Fault Tolerant Computing Symposium,* pp. 370–375, 1986.

[94] M. L. Flottes, C. Landrault, and S. Pravossoudovitch, "Fault Modelling and Fault Equivalence in CMOS Technology," *Proceedings of European Design Automation Conference,* pp. 407–412, 1990.

[95] S. Voldman et al., "CMOS SRAM Alpha Particle Modelling and Experimental Results," *Proceedings of International Electron Device Meeting,* p. 20.7, 1987.

[96] C. Kuo et al., "Soft-Defect Detection (SDD) Technique for a High Reliability CMOS SRAM," *IEEE Journal of Solid-State Circuits,* vol. 25, no. 1, pp. 61–67, 1990.

[97] K. Takeuchi et al., "Origin and Characteristics of Alpha-Particle-Induced Permanent Junction Leakage," *IEEE Transactions on Electron Devices,* vol. 37, no. 3, pp. 730–736, Mar. 1990.

[98] Z. Hasnain and A. Ditali, "Building-in Reliability: Soft Error–A Case Study," *Proceedings of International Reliability Physics Symposium,* pp. 276–280, 1992.

[99] C. H. Stapper, and W. A. Klaasen, "The Evaluation of 16-Mbit Memory Chips with Built-in Reliability," *Proceedings of International Reliability Physics Symposium,* pp. 3–7, 1992.

[100] M. Fukuma, H. Furuta, and M. Takada, "Memory LSI Reliability," *Proceedings of IEEE,* vol. 81, no. 5, pp. 768–775, May 1993.

4

Functional Yield Modeling

Gary C. Cheek and Geoff O'Donoghue

4.1 INTRODUCTION

The development of models to estimate the functional yield of an integrated circuit (IC) on a manufacturing process technology is fundamental to many aspects of IC manufacturing. A model that results in accurate estimates of manufacturing yield can help predict product cost and profitability, determine the optimum utilization of wafer fabrication equipment, and be used as a metric for the manufacturing organization when evaluated against actual product yields. Yield models are also critical to support decisions regarding new technology developments, the benefits of scaling products to advanced technologies, and the identification of problem products.

Functional yield is defined as those chips on a wafer that meet the nominal performance specification. In some cases, a product is fully functional but does not meet the specification sheet performance for one or more parameters (e.g., speed, accuracy, or power) due to a parametric variability. Therefore, functional yield is often referred to as *hard yield* and parametric yield as *soft yield*. Hard yield loss (functional fail) results from an unintended alteration of the geometric pattern on a wafer. A hard modification of the intended geometric layout of a product may result in catastrophic failure. The modeling and analysis of these types of functional failures are the subject of this chapter.

Functional yield models have historically been used to evaluate memory-intensive digital circuitry. These densely packed circuits, assuming no redundancy, have a high probability of electrical failure if a defect is present. In dense digital applications, the chip area is often used in the yield model calculations. Use of the chip area has worked reasonably well since the chip area approximates the area of the chip that is sensitive to the presence of a defect. The yield modeling methods and assumptions associated with memory-intensive products have been applied to random logic applications. However, for a more generalized modeling approach, it is necessary to account for chips with a mix of memory and digital, as well as for mixed-signal chips that may have a significant amount of the chip area accounted for by passive components such as capacitors and resistors with low-defect sensitivity. Defects on IC layers are caused by many random sources, including contamination from equipment, process or handling, mask defects, and airborne particles. This chapter focuses on random defects and their effect on product functionality rather than on systematic defects. Several terms that are commonly used to define yield loss or defects are as follows.

> **Global** Associated with large areas on a wafer, the whole wafer, or whole batches of wafers and usually affects the yield loss in a similar way. Global losses are caused by gross equipment malfunctions or mask misalignment.
>
> **Local** Associated with a small area or point on a wafer.
>
> **Random** Occurring anywhere, at any time on any wafer at any masking level. Both local and global defects can be random. Causes of random defects include contamination and scratches.
>
> **Systematic** Associated with whole chips on a wafer that can be functional, but owing to a systematic process variable, these chips often are not functional (e.g., the edge exclusion area, the laser scribe area, or a repeating mask defect).
>
> **Correlated** Associated with defects that appear to group or cluster together spatially on a wafer.
>
> **Uncorrelated** Defects located on a wafer without correlation between defect locations.

The second section of this chapter discusses the basic statistics associated with developing random defect yield models and derives some common yield models. The third section discusses four classes of yield models, ranging from the simplest to an advanced yield model. The fourth section defines yield models in terms of the key model components: defect density, area and Y_o. Finally, in the fifth section, applications of yield models are described based on actual examples from digital and mixed-signal products on manufacturing technologies.

4.2 BASIC YIELD STATISTICS: RANDOM DEFECTS

Yield models are usually presented as a function of D_o, the *average* number of defects per unit area and the area of the chip A_d. These parameters are combined to form fault density $\lambda = D_o \times A$. The relationship between the number of faults and the chip area is complicated. It depends on the circuit complexity, the density of photolithographic patterns, the number of photolithographic masks used in the

4.2 ■ Basic Yield Statistics: Random Defects

process technology, the minimum geometry used in the design, and so on. Random defects that contribute to D_o are usually defined as any physical anomaly that causes a circuit to fail. Thus, D_o is comprised of "killer" defects. Physically, these defects include shorts, resistive paths or opens in conducting layers, misalignment, photoresist splatters and flakes, pinholes, scratches, contamination, localized metallurgical anomalies, and crystallographic defects.

Yield models are therefore developed using statistics associated with random events. The assumptions concerning the ability to distinguish defect types from one another, the size distribution of defects, and the degree of random distribution of defects across a wafer all affect the statistical confidence that is achieved by using a specific yield model. Two basic modeling approaches are described in this section and are valid for all four yield model classes [12].

In the case of clearly distinguishable defect types on a single-process layer, Metal 1 for example, the number of unique ways in which three defects, each of a different type, such as Metal Short, Metal Open, and Metal 1 to Metal 2 Short, can be distributed on N chips on a wafer is:

$$N^M \tag{4-1}$$

If there are three chips on this hypothetical wafer, those three defects can be distributed over N_1, N_2, N_3 in

$$N^M = 3^3 = 27 \tag{4-2}$$

different ways. The truth table is given in Table 4.1.

TABLE 4-1 Truth Table of Unique Combinations

	N1	N2	N3		N1	N2	N3
1	M1M2M3			15	M3		M2M1
2		M1M2M3		16		M1M2	M3
3			M1M2M3	17		M1M3	M2
4	M1M2	M3		18		M2M3	M1
5	M1M3	M2		19		M1	M2M3
6	M2M3	M1		20		M2	M1M3
7	M1M2		M3	21		M3	M2M1
8	M1M3		M2	22	M1	M2	M3
9	M2M3		M1	23	M1	M3	M2
10	M1	M2M3		24	M2	M1	M3
11	M2	M1M3		25	M2	M3	M1
12	M3	M2M1		26	M3	M1	M2
13	M1		M2M3	27	M3	M2	M1
14	M2		M1M3				

The statistics in Table 4-1 are often referred to as Maxwell-Boltzmann. Maxwell and Boltzmann visualized the energy among molecules as being randomly distributed, resulting in an exponential behavior. In a similar fashion, defects can be modeled as being randomly distributed among all chips on a wafer, which results in an exponential probability of finding chips without any defects. Maxwell-Boltzmann statistics indicate that the above three defect types that are distinguishable from each other can be placed on the three chips (N_1, N_2, N_3) in

$$Z_1 = N^M \qquad (4\text{-}3)$$

unique different ways if all possible combinations are used. If one chip is removed (i.e., contains no defects), the number of ways to distribute the M defects in the $(N - 1)$ remaining chips is

$$(N - 1)^M \qquad (4\text{-}4)$$

different ways. The *probability* that a chip will contain zero defects of any type can be given as

$$\frac{(N-1)^M}{N^M} = \left(1 - \frac{1}{N}\right)^M \qquad (4\text{-}5)$$

Substituting $M = N A D_o$, the yield is defined as the number of chips having zero killer defects

$$Y = \lim_{N \to \infty} (1 - 1/N)^{NAD_o} = \exp^{(-AD_o)} \qquad (4\text{-}6)$$

if $A D_o$ is held finite. This equation suggests that as the chip size becomes very small (i.e., the number of chips per wafer gets very large), the yield becomes a decreasing exponential function of the fault density. This same result can be arrived at using more conventional Poisson statistics. The Poisson statistics are an approximation of the Maxwell-Boltzmann or "Binomial" distribution when large sample sizes are used. The Poisson probability distribution function is given by

$$P(X = x) = \frac{\exp^{-\lambda} \lambda^x}{x!} \qquad (4\text{-}7)$$

where X is the number of faults per chip and λ is the fault density. X can have integer values of $X = (0, 1, 2, \ldots)$. The yield is defined at $X = 0$ (no faults)

$$Y = P(X = 0) = \exp^{-\lambda} = \exp^{-AD_o} \qquad (4\text{-}8)$$

If it is now assumed that the defects (M_1, M_2, M_3) are completely indistinguishable from each other or are identical (M, M, M) and assuming again that there are three chips on the wafer, the truth table indicates that for all possible combinations of locating the defects on the three chips, there are only 10 combinations that are uniquely identifiable. Thus, the three defects (M, M, M) that are identical may be placed in the N chips in:

$$Z_2 = \frac{(N + M - 1)!}{(M!(N - 1)!)} \qquad (4\text{-}9)$$

unique ways. If one chip does not have any defects present, the probability of finding one of the M defects spread over the $(N - 1)$ chips is given by

$$Z_3 = \frac{(N + M - 2)!}{(M!(N - 2)!)} \qquad (4\text{-}10)$$

4.2 ■ Basic Yield Statistics: Random Defects

The yield can then be derived, similar to the Maxwell-Boltzmann case as:

$$Y = \frac{\frac{(N+M-2)!}{(M!(N-2)!)}}{\frac{(N+M-1)!}{(M!(N-1)!)}} = \left[\frac{(N-1)!}{(N-2)!}\right]\left[\frac{(N+M-2)!}{(N+M-1)!}\right] \quad (4\text{-}11)$$

Using the expression

$$X! = X(X-1)! \quad (4\text{-}12)$$

we can reduce the above equation to

$$Y = \frac{(N-1)}{N+M-1} = \frac{\left(1 - \frac{1}{N}\right)}{\left(1 + \frac{M}{N} + \frac{1}{N}\right)} \quad (4\text{-}13)$$

Substituting $M = N A D_o$ in the limit as N tends to infinity

$$Y = \lim_{N \to \infty} \frac{\left(1 + \frac{1}{N}\right)}{\left(1 + \frac{M}{N} + \frac{1}{N}\right)} = \frac{1}{(1 + A D_o)} \quad (4\text{-}14)$$

if $A D_o$ is held finite. This equation is the basic Seeds yield model [18] derived independently by Price [16] using the above Bose-Einstein statistics. The Bose-Einstein distribution is usually encountered in the statistics of indistinguishable particles which have no constraint on the number that can occupy a given state. This assumption is not so valid for integrated circuit fabrication processes since defects are often visual and can be distinguished from one another. However, metallic inclusions, leading to pipes (i.e., shorts between emitter and collector, or pinholes in the gate-oxide on other insulators) are examples of defects that are indistinguishable from each other [4]. In technologies where indistinguishable defects are the major contributors to yield loss, a negligible error will result if their distribution is described by the Bose-Einstein distribution.

4.2.1 Yield Model Derivations

Yield models establish the relationship between the chip layout and the manufacturing defect density. This relationship is depicted in Figure 4-1.

This section provides a derivation of commonly used functional yield models. In manufacturing practice, defect and particulate densities can vary widely from wafer to wafer and lot to lot. As a result of this variability, the Poisson model was found to underestimate the yield. An accurate yield model needs to incorporate a variable defect density or defect distribution. Murphy [9] reasoned that the value of the defect density, D, must be summed over all chips and wafers using a normalized

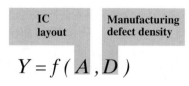

$$Y = f(A, D)$$

Figure 4-1 The yield model defines the relationship between the product circuit layout and the manufacturing defect density.

probability distribution function of defect densities $F(D)$. The incorporation of $F(D)$ into the yield integral is given by

$$Y = \int_0^\infty F(D) \exp^{(-AD_o)} dD \quad \text{where} \int_0^\infty F(D) dD = 1 \quad (4\text{-}15)$$

The various forms of the probability distribution function form the basis for the subtle differences between many of the analytical models.

4.2.1.1 Poisson Model. The Poisson model requires that defects are perfect points and are uniformly and spatially uncorrelated across a wafer with an average defect density of D_o and that each point defect will result in a fault. The Poisson model assumes that the probability distribution function is a delta function, that is,

$$F(D) = \delta(D - D_o) \quad (4\text{-}16)$$

normalized to D_o, the average defect density as given in Figure 4-2a.

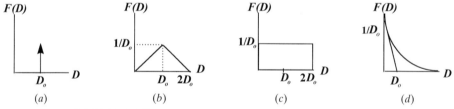

Figure 4-2 (a) Probability distribution function (delta function) for the Poisson model; (b) probability distribution function (triangular) for the Murphy model; (c) probability distribution function (rectangular) for the Murphy model; (d) probability distribution function (exponential) for the Seeds model.

Since the average defect density, D_0, can be found anywhere on the wafer, the yield can be determined from

$$Y_{\text{Poisson}} = \int_0^\infty F(D) \exp^{(-AD_o)} dD = \exp^{(-AD_o)} \quad (4\text{-}17)$$

The Poisson model is simple and easy to derive, and the estimates of yields for small-area chips using this model were reasonably good in the earlier days of IC manufacture [1]. However, if a D_o value from the Poisson model is calculated based on small-area chips, using this same D_o value for large-area chip yield calculations has always resulted in a pessimistic yield prediction compared to actual measured data.

4.2 ■ Basic Yield Statistics: Random Defects

4.2.1.2 Murphy Model. Murphy, who first proposed the concept of a variable defect density, believed that a well-shaped Gaussian distribution would be a reasonable estimate for $F(D)$. However, since he was unable to integrate the expression, he approximated it using a triangular form as indicated in Figure 4.2b. Use of a Gaussian probability distribution has been shown not only to be solvable, but also to have applications in industrial manufacturing [26]. In this case, the probability distribution function is normalized to D_o, the average defect density, and ranges from 0 to 2 D_o. As a result, there are two values of $F(D)$, notably:

$$F(D) = \frac{D}{D_o} \quad \text{for } 0 \leq D \leq D_o \tag{4-18}$$

$$F(D) = \left(\frac{2}{D_o} - \frac{D}{D_o^2}\right) \quad \text{for } D_o \leq D \leq 2D_o \tag{4-19}$$

Murphy also used a rectangular probability distribution function which, (see Figure 4.2c) indicates a constant value of $1/D_o$ for all values of D, that is,

$$F(D) = \frac{1}{D_o} \quad \text{for } 0 \leq D \leq 2D_o \tag{4-20}$$

The Murphy yield models for the two distributions above are given as:

$$Y_{\text{Murphy}_{\text{Triangular}}} = \int_0^\infty F(D)\exp^{(-AD_o)} dD = \left[\frac{1 - \exp^{(-AD_o)}}{A_c D_o}\right]^2 \tag{4-21}$$

$$Y_{\text{Murphy}_{\text{Rectangular}}} = \int_0^\infty F(D)\exp^{(-AD_o)} dD = \left[\frac{1 - \exp^{(-2AD_o)}}{2A_c D_o}\right] \tag{4-22}$$

The Murphy (triangular) yield model is widely used today in industry to determine the effective manufacturing process defect density. Figure 4.3 indicates the

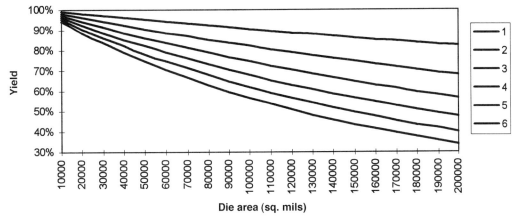

Figure 4-3 Murphy model with yield plotted as a function of defect density (in defects/cm^2).

calculated yield as a function of chip area for a number of different defect density values.

As the fault density becomes very large, $A\,D_o \gg 1$, the above equations simplify to:

$$Y_{\text{Murphy}_{\text{Triangular}}} = \frac{1}{(2A\,D_o)} \qquad (4\text{-}23)$$

$$Y_{\text{Murphy}_{\text{Rectangular}}} = \frac{1}{(AD_o)^2} \qquad (4\text{-}24)$$

These equations oversimplify actual manufacturing environments and are not widely used.

4.2.1.3 Seeds Model. Seeds was the first to verify Murphy's predictions by evaluating chip multiples using a technique called windowing. Seeds surmised that the high yields for the chip multiples he had observed were caused by a large population of low-defect densities and a small proportion of high-defect densities. He therefore proposed using an exponential defect density distribution given by

$$F(D) = \frac{1}{D_o}\exp\left(\frac{-D}{D_o}\right) \qquad (4\text{-}25)$$

and indicated in Figure 4.2d. In other words, the probability of observing a low-defect density was significantly higher than the probability of observing a high-defect density. The resulting yield model, after integration, is given by

$$Y_{\text{Seeds}} = \int_0^\infty F(D)\exp^{(-A\,D_o)}\,dD = \frac{1}{(1+AD_o)} \qquad (4\text{-}26)$$

Although the Seeds model is straightforward, its yield predictions for large-area chips are rather optimistic; as a result, the model has not been widely used in the VLSI era [25]. As mentioned earlier, Price derived the Seeds yield equation by using Bose-Einstein statistics. The Price yield model can be written as

$$Y = \prod_{i=1}^{n} \frac{1}{(1+A\,D_i)} \qquad (4\text{-}27)$$

where D_i is the defect density of each defect-producing mechanism or process step as the wafers pass through the line. This method of line partitioning forms the basis for current yield management practices.

4.2.1.4 Gamma Model. Okabe, Nagata, and Shimada [1] recognized the physical nature of defect distributions and sources of variability and proposed the use of Erlang and gamma defect probability functions. Stapper has written several papers concerning the development and applications of the gamma distribution for use in IC modeling. The gamma model is often referred to as the Stapper model. The gamma distribution is given as

$$F(D) = \frac{1}{\Gamma(\alpha)\beta^\alpha} D^{\alpha-1}\exp(-D/\beta) \qquad (4\text{-}28)$$

where α and β are the two distribution parameters, $\Gamma(\alpha)$ is the gamma function, and D, α and β are all greater than zero. In this distribution, the average density of defects is given by $D_o = \alpha\beta$, the variance of D is given by $\text{var}(D) = \alpha\beta^2$, and the coefficient of variation is given by

$$\frac{\sqrt{\text{var}(D)}}{D_o} = \frac{1}{\sqrt{\alpha}} \qquad (4\text{-}29)$$

The yield model is derived by substituting the gamma distribution function into the Murphy yield equation and is given by

$$Y_{\text{gamma}} = \left(1 + \frac{AD}{\alpha}\right)^{-\alpha} \qquad (4\text{-}30)$$

The statistics associated with the above yield model are known as the negative binomial distribution. Strictly speaking, the negative binomial distribution is only defined for integer values of $\alpha \geq 1$. However, the yield model has no such restrictions and is well defined for all real values of $\alpha > 0$. This model has therefore come to be known as the generalized negative binomial distribution [23]. In the above yield model, α is usually referred to as a "cluster" parameter and increases with decreasing variance in the distribution of defects. One key strength of the gamma distribution function is its capability to emulate other distribution functions by the selection of the α parameter value. When the variability of defects across a wafer is low, that is, spatially uncorrelated with little or no clustering, α is high and the gamma distribution shape approaches the delta function

$$F(D) = \delta(D - D_o) \qquad (4\text{-}31)$$

which will result in the Poisson yield model given as

$$Y = \lim_{\alpha \to \infty} \left(1 + \frac{AD_o}{\alpha}\right)^{-\alpha} = \exp^{(-AD_o)} \qquad (4\text{-}32)$$

If the variability of defects across a wafer is significant (i.e., exhibiting spatial extent), α becomes low and the resulting distribution shape tends toward an exponential, which suggests many areas with no defects and a few areas with many defects, or the Seeds model given as

$$Y = \lim_{\alpha \to \infty} \left(1 + \frac{AD_o}{\alpha}\right)^{-\alpha} = \frac{1}{1 + AD_o} \qquad (4\text{-}33)$$

Thus, by selecting values of α in the gamma model, yield equations previously discussed can closely be emulated as indicated in Table 4.2.

The value of α needs to be determined empirically. Cunningham [3] details several methods to determine α from laser reflectometry (particle counting) and

TABLE 4-2 Emulation of Yield Models by Changing Alpha Value in the Gamma Model

Amount of Clustering	Value of Alpha	Yield Model
None	10–1000	Poisson
Some	4.2	Murphy
Much	1	Seeds

monitor wafer analysis. Some authors have found that values of $\alpha = 2$ provide a good approximation for a variety of logic and memory circuits [5, 21]. Thus, if the critical area of the circuit and the defect density of the manufacturing area are independently measured, as explained later in this chapter, the gamma model can be used to accurately model most any circuit type (RAM cells, logic, drivers) in digital or analog process technology, in most any fab area, by proper selection of the values of α. The key yield models and the associated distribution functions are summarized in Table 4-3.

TABLE 4-3 Summary of Yield Models and Distribution Functions

Yield Model	Distribution Function
$Y_{\text{Poisson}} = \exp^{(-AD_o)}$	$F(D) = \delta(D - D_o)$
$Y_{\text{triangular}} = \left[\dfrac{1 - \exp^{(-AD_o)}}{AD_o}\right]^2$	$F(D) = \dfrac{D}{D_o}$ for $0 \leq D \leq D_o$
	$F(D) = \left(\dfrac{2}{D_o} - \dfrac{D}{D_o^2}\right)$ for $D_o \leq D \leq 2D_o$
$Y_{\text{Seeds}} = \dfrac{1}{(1 + AD_o)}$	$F(D) = \dfrac{1}{D_o}\exp\left(\dfrac{-D}{D_o}\right)$
$Y_{\text{gamma}} = \left(1 + \dfrac{AD_o}{\alpha}\right)^{-\alpha}$	$F(D) = \dfrac{1}{\Gamma(\alpha)\beta^\alpha} D^{\alpha-1} \exp(-D/\beta)$

4.3 CLASSES OF YIELD MODELS

Many yield models have been developed. Some models are empirical and apply to a specific technology in a specific manufacturing area. Other models have limited applicability to modern manufacturing equipment, techniques, or practices. The four models described earlier are commonly applied to real-world manufacturing data and are divided into four classes based on implementation methodology [7].

4.3.1 Class I Yield Models

The most straightforward application of the yield models is as a simple analytic tool or performance metric. In this way, a yield model is used to estimate the yield of one chip in terms of other chips that have already been manufactured using the same process technology. To estimate the yield of a chip with an area A_d, the measured yields of several chips with different areas are plotted as a function of chip area. Solving a yield equation, such as the Poisson equation for the average defect density D_o, gives

$$D_o = \frac{1}{M} \sum_{j=1}^{M} \frac{-\ln(Y_j)}{A_{dj}} \quad (4\text{-}34)$$

where M is the number of products used to obtain the average. Thus, the average defect density D_o that is characteristic of a specific process technology can be determined as either the best fit of the parameter D_o to the yield versus chip area curve or the average defect density as defined in Equation 4-34. This general *back calculation* method to determine average defect density is very easy to implement, quick to use, and reasonably accurate for smaller chip sizes. This method is widely used in the semiconductor industry. The average defect density D_o is very useful as a process control monitor and as an improvement metric for product yields.

There are some limitations, however, to the use of such a *global* defect density. The data used in the global defect density calculation contain yield loss due to all mechanisms, including parametric variability, edge chip loss, prober or tester malfunction, and any other manufacturing loss. Thus, nondefect yield loss mechanisms are rolled into the global defect density.

There is a risk of substantial error in predicting yield for chip sizes outside the range of chip areas used to calibrate the model and for chips that are not manufactured using the same technology. The risk of error is also increased as device and interconnect feature sizes decrease, and if nondefect yield loss mechanisms have a significant contribution to yield loss.

4.3.2 Class II Yield Models

The second class of yield model involves increasing the level of granularity into the defect density term used in the simple analytical yield models. The increased granularity is derived from attempting to identify the actual number of defects present on key layers on the integrated circuit. The use of measured defect density in a simple yield model significantly increases the model accuracy.

Knowledge of the key level by level defect density data enables assignment of yield improvement responsibility to particular steps in the manufacturing process (e.g., metal deposition, contact etch, polysilicon photolithography). Knowledge of the level-specific defect density may also be used to separate the hard physical components of yield loss from the parametric variability component of yield loss.

The identification of level-by-level defect density can be obtained by evaluating long-loop monitors, which include products or defect monitors that can be used for delayering, or short-loop electrical monitors and equipment particle monitoring. Long loop is defined as processing material through all the steps in the technology, whereas short loop is defined as processing material through a selected set of steps in the technology.

Use of long-loop monitors such as a product or a static random access memory (SRAM) allows the measurement of physical defects when actual underlying circuit topography is present. However, these techniques are tedious and time consuming.

Short-loop monitors such as snake and comb structures enable rapid determination of an electrical D_o for an individual level. The defects are easy to locate, and it is possible to correlate these defects with electrical failures. The drawback of the short-loop monitors is that topography is not necessarily present and the photo effects are not the same. An alternative view is that the short loop represents the best topography possible for the specific defect monitoring step. If the yield loss at that step is significantly high, then a problem exists at that particular process step.

To estimate the yield of a product, the level-by-level defect density measured on a monitor needs to be evaluated [20]. A cleverly designed electrical monitor will allow the determination of the average number of killer defects or faults for each key level. A defect density may be determined by the equation:

$$D_{\text{level}_i} = \frac{N_{\text{level}_i}}{A_{\text{monitor}}} \tag{4-35}$$

where the subscript, i, represents the number of each key level, N is the number of killer defects, and A is the area of the monitor. This level-by-level defect density can then be used in a simple analytic yield model rewritten to accommodate the extra terms, that is,

$$Y = \exp^{-(D_{\text{level}_1} A_{\text{Chip}} + D_{\text{level}_2} A_{\text{Chip}} + \ldots + D_{\text{level}_n} A_{\text{Chip}})} \tag{4-36}$$

$$= \prod_{i=1}^{n} \exp^{(-D_{\text{level}_i} A_{\text{Chip}})} \tag{4-37}$$

The key points associated with this second class of yield models include (1) the use of monitors on a level-by-level basis to measure the killer defect density, (2) the ability to partition a manufacturing process into small units, each represented by a short-loop monitor to identify key yield-detracting processes, and (3) the development of component yield models. Because the defect density is measured and not derived as a function of chip area or product yield, the yield model can be used with increased confidence to predict yield for larger chip sizes and as a metric.

4.3.3 Class III Yield Models

The third class of yield models accounts for the interaction between the physical layout and the distribution of defects on a wafer and requires a stricter definition of the D_o and A terms. The defect density term is defined in terms of the size distribution of defects that are likely to occur on a wafer (a physical parameter), and the area term is defined in terms of a measure of the defect sensitivity of the circuit layout called *critical area*.

The defect size distribution is important in defining defect density and critical area, because this size distribution defines the expected number of defects by size that may be present on a wafer. The critical area is a measure of the area on each level of a circuit layout that is sensitive to a defect, such that if a defect is present, a circuit fault will occur. The most dense layers with the smallest feature sizes (e.g., polysilicon, metal) will be more sensitive to defects, having a larger critical area compared with layers (e.g., nwell, passivation) that are not as dense and have larger feature sizes.

A yield model such as the gamma model can now be expressed in terms of the measured critical area and the measured defect density for each level i:

$$Y = \prod_{i=1}^{n} \left(1 + \frac{A_i \times D_{oi}}{\alpha_i}\right)^{-\alpha_i} \quad (4\text{-}38)$$

In this class of yield model, the A_i and D_{oi} values will be significantly different from the A_d and D_o values derived from the chip area or global defect density. The Poisson probability distribution function is defined when the number of faults per chip is zero ($X = 0$) in Equation 4-7. However, the possibility can exist for multiple faults to be present, for example, $X = 1, 2, 3 \ldots$ in Equation 4-7, and continue to have fully functional chips. In this case, there would be redundant chip areas that can be linked into the circuit should one section of the chip be faulty. Redundancy is broadly used in memory circuitry where the area penalty for the extra (potentially unused circuits) outweighs the benefit of having extra circuitry available that could enable the chip to fully function [2]. The benefits and cost of redundancy can be evaluated when using Class III and IV models; however, this calculation is beyond the scope of this work.

In essentially every case, the defect density in Class III yield models will be higher than the value calculated in Class I or II models. The reason is that in most cases the critical area is less than the chip area, and for the same yield the defect density must be higher. It is difficult and inadvisable to compare the defect density of the Class I and Class III models. The defect density in the Class I model assumes that the defects are point defects, whereas the defect density in the Class III model is based on the size distribution of defects. The defect size distribution, the defect density, and the critical area are all measurable quantities and must be determined *independently* in the Class III models. An example of the estimation of yield using the measured level-by-level defect densities and the critical area is given in Table 4-4.

TABLE 4-4 Level-by-Level Fault Density Analysis

	Critical Area	Defect Density	Fault Density	Yield
	mm * mm	#/cm²	#	%
Metal opens	1.145	3.000	0.034	96.62%
Metal shorts	2.326	1.500	0.035	96.57%
Poly opens	0.245	3.500	0.009	99.15%
Poly shorts	2.356	4.000	0.094	91.01%
Sum of fault densities			0.172	84.19%

The measured data for defect density and critical area are entered for some of the key manufacturing levels for the product. The fault density ($\lambda_i = D_{oi} \times A_i$) for each level is determined. The yield for each level can also be calculated, in order to identify the levels that contribute to the greatest yield loss. The overall product yield can also be estimated by summing all the λ_i using the Poisson model:

$$Y = \exp\left(-\sum_{i=1}^{n} \lambda_i\right) \quad (4\text{-}39)$$

By using the gamma model, the defect "clustering factor," α, can also be incorporated into the calculation by determining the yield for each key level as in Table 4-5 but for various values of α.

TABLE 4-5 Level-by-Level Yield Using Poisson and Gamma Models

Level Critical Area	Defect Density	Fault Density	Poisson Model	Gamma Model			
				Alpha 0.01	Alpha 1	Alpha 4.2	Alpha 100
mm * mm	1/cm * cm	#	EXP(-DA)				
15.57	3.5	0.545	58.0%	96.1%	64.7%	59.9%	58.1%
13.38	3.5	0.468	62.6%	96.2%	68.1%	64.1%	62.7%
15.65	2.8	0.438	64.5%	96.3%	69.5%	65.9%	64.6%
2.04	2.8	0.057	94.4%	98.1%	94.6%	94.5%	94.4%
4.88	4.5	0.220	80.3%	96.9%	82.0%	80.7%	80.3%
28.96	2.0	0.579	56.0%	96.0%	63.3%	58.1%	56.1%

An α value can be associated with each level by evaluating the measured yield values with the calculated values for each α value. The cluster factor can then be determined (empirically) and can be used as an equipment performance metric or as an indicator of process control. The α value for each level is like a signature for the process and can give the process engineer important information about the production equipment. A changing value of α for any masking level may indicate a need for equipment cleaning or other maintenance. These techniques are equally applicable to monitors or product wafers. Initially, the average defect density on a key level using short-loop monitors or wafers needs to be determined, that is

$$D_m = \frac{-\ln Y_m}{A_{c_m}} \quad (4\text{-}40)$$

It is possible to estimate the yield for a product at the same photo level on the same process technology by using the short-loop monitor defect density in a product yield equation:

$$Y_p = \exp^{-(A_{c_p} D_m)} \quad (4\text{-}41)$$

and by substituting

$$Y_p = \exp\left(\left(\frac{A_{c_p}}{A_{c_m}}\right) \ln Y_m\right) \quad (4\text{-}42)$$

$$Y_p = Y_m^{\left(\frac{A_{c_p}}{A_{c_m}}\right)} \quad (4\text{-}43)$$

Thus, the monitor yield can be scaled to the product yield with an exponent given by the ratio of the CRITICAL AREAS of the product and the monitor [23]. This scaling is possible only when the ratios of critical area are used, not the product or the monitor chip areas.

4.3.4 Class IV Yield Models

The preceding yield model derivations have focused on local defects. While often the primary yield loss mechanism in most manufacturing processes, global yield loss mechanisms, which may have no dependence on chip size or layout density, are usually present to some extent. Usually, global yield loss is spatially correlated. Global yield loss often manifests itself as yield loss due to variability in key electrical parameters, such as transistor gain or threshold voltages, that could have spatial extent caused by process variabilities, such as temperature or thicknesses being out of control. Thus, global yield loss is often identifiable as a pattern on a wafer, (e.g., annular or an identifiable group of failing chips). Class IV models encompass the global yield loss by including a multiplier term called Y_o into the yield model:

$$Y = \prod_{j=1}^{n} Y_{o_j} \prod_{i=1}^{m} \exp^{-(D_i A_i)} \tag{4-44}$$

where Y_{oj} represents the yield loss for different global yield loss mechanisms. The Y_o term is not related to physical defect density or critical area as defined in Class III or IV models:

$$Y_{o_j} \neq f(D_o, A) \tag{4-45}$$

Table 4-6 summarizes the four yield model classes in terms of defect density, area, and nondefect model components.

TABLE 4-6 Yield Model Classes

Yield Model Class	Nondefect	Defect Density	Area
I $Y = f(D_o, A)$	—	Curve fit D_o	Chip area
II $Y = f(\Pi D_{oi}, A)$	—	Measured D_o	Chip area
III $Y = f(\Pi D_{oi}, \Pi A)$	—	Measured D_o	Critical area
IV $Y = \Pi Y_o * f(\Pi D_{oi}, \Pi A)$	Spatial	Measured D_o	Critical area

4.4 YIELD MODEL COMPONENTS

Functional yield models have been defined in terms of three independent parameters, D_o, A, and Y_o. These parameters are statistically independent and can be directly measured. The following sections detail each parameter and how they are used in the four classes of yield models.

4.4.1 Defect Density Term in Yield Models

Defect density is most often referred to as a manufacturing performance metric, derived for Class I models. Once the defect density is calculated, it is used in product yield planning. The defect density, when calculated in this manner, is a mathematical approximation or a curve fit parameter and may not be representative of the physical defect density. Calculated in this manner, all defects comprising the defect density are killer defects.

Alternatively, a physical interpretation of defect density incorporates the measured size distribution of defects. Using a physical definition of defect density, not all defects are killers; that is, a defect composed of additional material that forms between two conducting tracks of the same material cannot cause a failure if that defect is significantly smaller than the spacing between the tracks. Physical defect density must also be defined relative to a minimum defect size, typically, the minimum dimension (width or spacing) of the particular level in the technology.

4.4.1.1 Defect Density for Class I and II Yield Models. A defect density can easily be extracted as a dependent variable by rearranging the yield equation. This can be done for several chips or for a single chip. Examples are given for a single chip using Poisson and gamma models:

$$D = \frac{-\ln Y}{A} = D(A) \text{ (Poisson)} \tag{4-46}$$

$$D = \frac{\alpha(\sqrt[\alpha]{1/Y} - 1)}{A} = D(A) \text{ (gamma)} \tag{4-47}$$

Class I models work best for similar products on the same mature technology, and the difference in chip area is not greater than a factor of about 2 or 3. Because there can be so many unknown yield loss mechanisms, especially on new technologies, this method should be applied carefully. The key usefulness of this definition of D_o is as a course metric of wafer fabrication process performance.

4.4.1.2 Defect Density for Class III and IV Yield Models. The physical definition of average defect density can be made very explicit. The patterns for successive levels in the manufacturing process are defined by tooling, which allows patterns to be printed on a wafer surface. The desired pattern is the actual circuit layout and is highly specific as to its intent. A "defect" is an unintended pattern in the form of extra material or missing material. In defining the physical defect density, some terms need to be defined.

4.4.1.3 Definition of Commonly Used Yield Model Terms.

Contamination Foreign material on a wafer surface or embedded in a thin film of material. Contamination can be from human skin, dirt, and dust from airborne sources, an oxidized gas, residual chemicals, or some piece of equipment fixturing possibly sputtered during a deposition process. Contamination can also be the same material chemically as the material on the wafer, but nonhomogeneously incorporated (e.g., flakes of SiO$_2$ from the fixturing of a deposition system either embedded into the film or located on the surface). Particles are a form of contamination.

Defect An alteration in the desired physical pattern intended to be printed. A defect can include extra material on a layer such as metal or polysilicon in an area where no material was intended to exist or missing material such as an oxide pinhole or a break in a metal line where that specific material was intended. Some typical defects include metal stringers, opens, shorts, notches, extra material in the form of splotches, bridges, or hillocks. Defects can also result from surface scratches, contaminated photoresist, or surface contamination that interact with the intended process.

Fault A circuit electrical failure caused by a defect in a pattern. Faults are addressed in Class I and II yield models. Faults can result from either point defects (local) or defects that have spatial extent (global). Point defects can include oxide pinholes or crystallographic defects and metallic inclusions or bipolar "pipes."

From these definitions, a hierarchy can be developed. Contamination is a physical, random event that may or may not result in a defect. A defect may or may not result in a fault. The relative magnitude of occurrence is given in Table 4-7.

TABLE 4-7 Relative Magnitude of Contamination, Defects, and Faults on a Wafer

Contamination ≫ Defects ≫ Faults

The correlation that exists between contamination, defects, and faults is totally empirical and very weak, with little analytical basis. Attempting to map contamination to defects or defects to faults is time consuming and difficult. Control and measurement of contamination is important as a wafer fab metric and as an equipment performance indicator. The measurement of contamination by light scattering techniques, scan electron microscopy (SEM) evaluations, and holography are discussed elsewhere in this book.

4.4.1.4 Defect Size Distribution. In a mature manufacturing line, the defect density has been experimentally determined to follow an inverse power law relationship with respect to the size distribution (x) as indicated in this section [24].

$$D_o(x) \propto \frac{N}{x^p} \qquad (4\text{-}48)$$

where x is defect size and N is a technology parameter, specific to a level in a technology. This inverse cube power law is natural to all manufacturing areas, and the defects that constitute it are assumed to be located randomly on a wafer surface. The actual defect size distribution is fundamental to Class III and IV models. Neither the critical area nor the physical defect density can be determined without the defect size distribution. Figure 4-4 indicates the relationship between the defect size and the product layout.

The defect density term in a Class III and Class IV yield model has a specific definition based on the following concept. If a defect of a particular size causes a fault, then a larger defect at the same location will also cause a fault. Alternatively, a defect at the same location smaller than the minimum geometry for that level (e.g., Metal 1 spacing) will not result in a fault. An example of the effect of defects of different sizes at the same location is given in Figure 4-4. The two adjacent metal lines will be shorted together by a defect greater in size than the spacing(s) between them and will not be shorted by defects smaller than the spacing. In a circuit layout, the spacing between tracks will vary from the minimum spacing allowed by the technology to multiples of the minimum spacing. The example is for the spacing between intralevel structures; it is also valid for opens in structures.

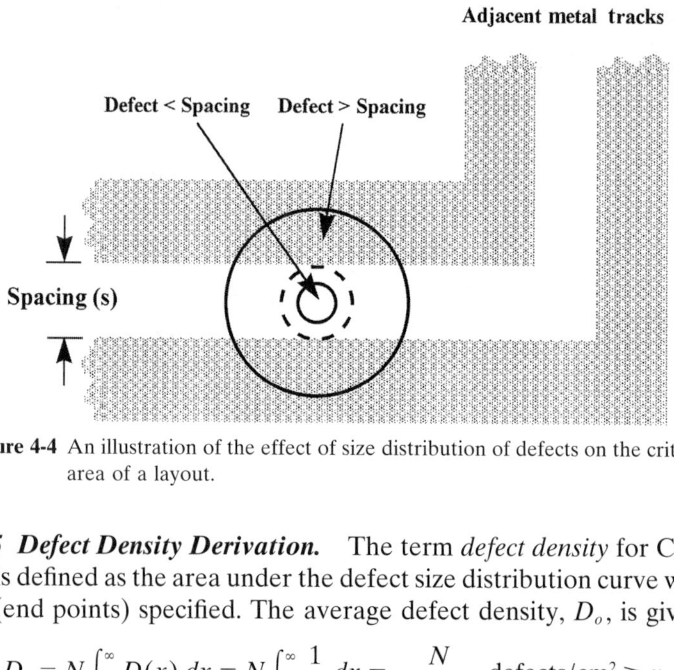

Figure 4-4 An illustration of the effect of size distribution of defects on the critical area of a layout.

4.4.1.5 Defect Density Derivation.

The term *defect density* for Class III and IV models is defined as the area under the defect size distribution curve with specific size limits (end points) specified. The average defect density, D_o, is given by

$$D_o = N \int_{x_0}^{\infty} D(x)\, dx = N \int_{x_o}^{\infty} \frac{1}{x^p}\, dx = \frac{N}{2 \times x_o^{p-1}} \text{ defects/cm}^2 \geq x_o \qquad (4\text{-}49)$$

A value for physical defect density must always be stated relative to the lower integration limit or the size reference, x_o. The reference defect size, x_o should have some empirical relationship to the minimum design rules. For example, if x_o is *larger* than the minimum technology design rule, the calculated defect density will be *low* and the observed yield will be inconsistent with the model. If the x_o value is a factor of 3 to 10 times less than the minimum design rule, the defect density will be *high*, but the Class III and IV models will be highly accurate. Typically, defect density is quoted for a specific manufacturing area and for a specific process technology (often the most aggressive in the manufacturing area) using an x_o of about 50% of the smallest geometrical feature size allowed.

The defect density as defined here is very different from that defect density used in Class I and Class II yield models. When the defect size distribution is not explicitly quantified, all defects are assumed to be point defects, and all defects are assumed to be killer defects. This strict definition of defect density allows the comparison of different technologies, as the defect densities can be normalized and compared. This comparison allows the effects of scaling products between technologies to be evaluated. The power that the denominator is raised to, usually considered to be 3, is an interesting number. It allows the random defect limited yield of chips to be reduced exactly in proportion to the minimum design rules for a technology. In other words, the inverse cube law is yield neutral. An exponent greater than 3 would suggest that scaling a product would result in a decrease in yield, while an exponent less than 3 would suggest that yield may actually increase as the product is scaled.

4.4.1.6 Spatial Distribution of Defects. The distribution of random defects over the surface of a wafer, or more specifically, over the surface of a wafer at a given processing step, may be either spatially correlated or uncorrelated. Rogers [17] developed a rigorous theory predicting the occurrence of an event (defect) in a particular region that depends on whether the occurrence probability of an additional event is related to or independent of the prior occurrence of other events in that region. Rogers indicated that different statistics need to be applied to the problem, depending on the probability for "another" defect to occur in a region already occupied by a defect. If the probability of occurrence of a defect to occur in a region *decreases* with the number of defects that have already occurred in that region, the binomial model results. If the occurrence of a defect is *independent* of the occurrence of previous defects in the region, then the Poisson model results. Finally if the probability of occurrence of a defect *increases linearly* with the number of defects that have already occurred in that region, the negative binomial model applies.

The spatial defect distribution assumed for most yield models is a random uncorrelated distribution. In Class IV models, defects have been modeled when they are allowed to cluster together. Measured defect distributions on different wafers have various degrees of clustering or spatial correlation. Using the gamma model, we can empirically determine the clustering factor α from the probe data/ wafer maps and some nonlinear curve fitting algorithms. The cluster factor can be determined by evaluating the yield for different groupings of chips, that is, $n = 1$, 2, 3, 4 . . . and evaluating the nonlinear relationship between Y and n (group size) using α and λ as independent variables in the gamma yield model equation. Defect distributions that use the above models are referred to as "clustered-random." While the defects do cluster, the locations and size of these clusters remain random events.

4.4.2 Area Term in Yield Models

Two circuit area terms are used in yield models: *chip area*, used in Class I and Class II models, and *critical area*, used in the Class III and Class IV models. The difference between use of chip and critical area depends on the level of complexity required of the yield model.

Chip area is typically defined as the area of the chip defined by the product of the orthogonal distance in X and Y between the same point on adjacent chips on a wafer. The chip X and Y dimensions are often referred to as the probing dimensions. The X and Y probing dimensions are composed of concentric squares. The scribe or kerf forms a ring outside the active area. The scribe is removed during the sawing operation, when the individual chips on a wafer are separated. Within the scribe is the pad ring, containing the interconnect pads that connect the active circuitry contained within the pads to the package, board, or multichip module. The dimensions of the scribe and pad rings are typically in the 50 μ to 200 μ range. Since the scribe is eventually removed, and the pads are relatively large structures, it is unlikely that many fault-causing defects will occur in these areas. Thus, the chip area used for yield modeling can be reduced from the probing dimensions to the pad or active dimensions. An organization commonly standardizes on one definition for the chip dimensions in order to ensure consistency.

Critical area has been used to describe the concept that not all parts of a chip layout have an equal likelihood of failing due to the presence of a defect [29]. The critical area concept allows a level of granularity and accuracy when calculating the defect sensitivity of a chip layout. The concept is illustrated qualitatively in Figure 4-5.

The dark filled areas are the Metal 1 pattern for a circuit cell. The shaded area is all of the sensitive spacing at the minimum spacing for that technology. The critical area is a quantitative measure of the spacing sensitivity at all dimensions for a cell or for an entire chip. Thus, the critical area is highly dependent on the defect size distribution for the intended manufacturing technology. Although the critical area concept is straightforward, care needs to be taken in the measurement approaches.

4.4.3 Probability of Fail

The important concept behind the measurement of critical area and the electrical verification of a fault is probability of fail [23]. Several calculation methods have been reported that determine critical area, and all methods calculate the probability of fail. The Monte Carlo or dot-throwing method simulates the effect of defects by combining many dots (defects) with the physical layout patterns and recording the number of modified patterns that result in a fault [28, 29]. This procedure is repeated for several defect sizes. Figure 4-6 indicates the Probability of Failure (POF) for a product that has both analog and digital functionality.

Because of the density of the Metal 1 and Metal 2 layers in the digital section of this chip, the POF is significantly higher for smaller defect sizes. This result means that the critical area is larger and that a consequent increase in the defect-related yield loss will be observed. The virtual artwork method searches the physical layout pattern database for equal spacings between patterns [6]. The sum of equal spacings and widths is used to generate a histogram or virtual artwork from which the probability of fail can be calculated. The union method uses an algorithmic geometric test for a range of defect sizes to each pattern in the physical layout database [13].

4.4 ■ Yield Model Components

Figure 4-5 Subcircuit metal layer graphically indicating critical area (shaded).

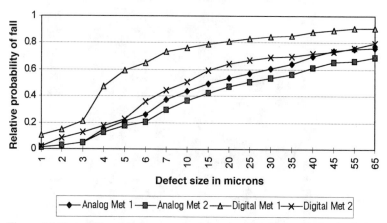

Figure 4-6 Probability of Failure curves determined by Monte Carlo simulation for a chip layout including both analog and digital circuitry.

The probability of fail is a strong function of defect size. This function typically has the form as indicated in Figure 4-7 with $POF(x)$ plotted versus defect size, normalized to a minimum dimension x_o which is the same x_o in Equation 4-49, the lower integration limit or defect size reference. The POF also functions as a global design rule checker, as there should be no fails at less than the minimum design rule.

4.4.4 Computation of Critical Area

The critical area is defined as follows:

$$A_c = A \int_{x_o}^{\infty} POF(X) \cdot D(X) \, dX \qquad (4\text{-}50)$$

where A is the chip area, x_o is the defect size reference, and $POF(x)$ and $D(x)$ are the probability of fail and the defect size distribution, respectively. A graphical

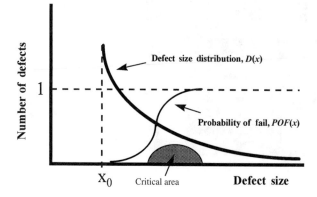

Figure 4-7 Graphical representation of the Critical Area Integral.

4.4 ■ Yield Model Components

interpretation of Equation 4-50 is given in Figure 4-7. The preceding formula contains a design specific parameter, $POF(x)$ and a technology-specific parameter $D(x)$. Use of the critical area can be reduced to the calculation of yield entitlement, or the $POF(x)$ term can be optimized during the design process to minimize critical area.

4.4.5 The Y_0 Term in Yield Models

Modeling local yield loss is well understood, but point defects are often not the only yield loss mechanism. Although it is difficult to model global yield loss analytically, use of spatial analysis tools can help verify if the measured yield loss is consistent with a point defect model.

Spatial analysis typically requires wafer maps that are typically the output of automatic probe test equipment. The failing chips on a wafer are often categorized into similar failure bins (e.g., functional fail, speed fail). Good chips are usually designated as 1 and failing chips as 0.

One type of spatial analysis is to partition the wafer map into different geometric groupings such as segment or radial partitioning. The distribution of failing chips should generally be equal in all areas of the wafer if the primary yield loss mechanisms is random defects. Nonuniformities may indicate a global type of yield loss. This technique can be applied to single wafers or to stack wafer maps. A stack map consists of a counter for each chip location, tracking the incidence of a particular bin or bin combination. A special case of a stack map is a reticle map. A reticle map consists of a counter for each chip location within the reticle array. The reticle stack is used to identify repeating defects. For examples, see Figure 4-8.

4.4.5.1 Windowing Technique.
An estimate of the value of Y_o can be derived from a windowing technique whereby chips are grouped into windows of increasing size, creating superchips. The effective yield of each superchip is calculated. The yield of each window size is graphed against the effective chip size; the Y intercept

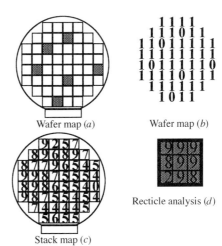

Figure 4-8 Examples of wafer maps, stack map, and reticle analysis.

is a measure of the effective yield with an area of zero (hence the term Y_o). If the intercept is at 100% yield, there is no global yield loss content. If the intercept is less than 100%, then Y_o is defined as

$$Y_o = 1 - Y_{\text{intercept}} \qquad (4\text{-}51)$$

4.4.5.2 Randomness Testing.
Techniques have also been developed to evaluate a wafer map and characterize the degree to which the failing chips are spatially correlated to each other [15]. If the location of defects on a wafer is random (i.e., no spatial correlation), then the location of failing chips on that wafer should also be random. Should there exist any spatial correlation between the location of adjacent failing chips, a systematic defect source may exist. A graphical description of local (spatially uncorrelated) versus global (spatially correlated) yield loss on a wafer map is given in Figure 4-9.

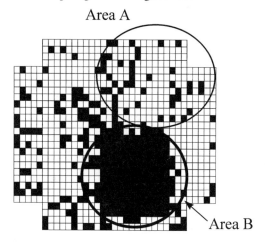

Figure 4-9 Wafer map indicating areas that have local (Area A) and global (Area B) yield loss.

A pattern recognition algorithm has been developed that treats randomly located good chips as noise and the nonrandomly located failing chips as a pattern. The analysis is divided into two steps. The first is to apply a randomness test to identify the presence of a pattern. In the second, the algorithm applies a nearest neighborhood method to quantify the spatial extent of the pattern. The measured yield is defined as

$$Y_M = Y_o \times Y_r \Rightarrow Y_r = \frac{Y_M}{Y_o} \qquad (4\text{-}52)$$

where Y_m is measured yield, Y_o is the nonrandom loss component, and Y_r is the defect limited yield. The separation of yield loss components is useful in the analysis of Class IV models. Advanced techniques have been developed to determine the Y_o content [10].

4.5 APPLICATIONS OF FUNCTIONAL YIELD MODELS

The following examples demonstrate the usage of functional yield models in an industrial environment. These examples demonstrate how models can help optimize

4.5 ■ Applications of Functional Yield Models

circuit layout and product cost, minimize reliability hazards, and improve product yields.

4.5.1 Low-Yield Cutoff and Chip Costing

The normalized wafer yield distribution for more than 2000 six-inch wafers of a mixed-signal product is shown in Figure 4-10. A low-yield cutoff value is often negotiated between manufacturing and product development organizations because of the increased cost of the chips on low-yielding wafers, and the attendant increased reliability hazards present in wafers that are outside of a normal yield distribution. A low-yield cutoff means that wafers with yield less than the cutoff are rejected and are not considered to be acceptable material. The low-yield cutoff effectively defines the acceptable defect density for a manufacturing process.

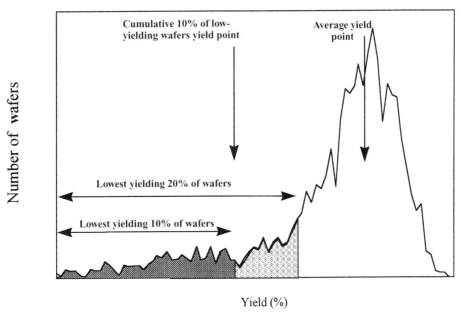

Figure 4-10 Wafer yield distribution indicating two low-yield cutoff points, 10% and 20% of the lowest yield wafers.

In this example, the lowest yielding 10% of all wafers in the wafer yield distribution contribute 4% of the total chips from all the wafers in the distribution, whereas the lowest yielding 20% of the wafers contribute 12% of the total chips. Thus, if the cutoff limit is set so that the lowest 10% of the wafers are rejected, there is a 4% shortfall to the planned manufacturing volume for end use.

Several methods are applied to the estimation of the low-yield cutoff limit, such as setting a fixed value based on previous experience, applying normal statistics with a 3 or 4 standard deviation limit, or using a Class III or IV model. Because the cost of a chip is proportional to the inverse of yield, the cost of chips from low-yielding wafers can be significantly higher than that of chips produced from wafers within 1 or 2 standard deviations of the average yield. In this example, the cost of a chip from a wafer at the lowest 10% cumulative point in Figure 4-10 is approximately 2.5 times the cost of a chip from a wafer with average yield from this distribution. If the low-yield cutoff were established and the lowest 20% yielding wafers were rejected, the average chip cost for over 1 million chips would have decreased by 10%. There is also an increased reliability risk associated with low-yielding wafers. Wafers or lots that have yields outside the normal distribution often have time-dependent failure mechanisms that could result in field fails.

4.5.2 Spatial Yield Distributions: Y_o

The components of yield loss should be separated early in the product life cycle. The presence of a yield-limiting mechanism has consequences for planning, cost, manufacturing efficiency, and reliability. The wafer maps shown in Figure 4-11 represent two circular sections from the first wafers of a new product on 8-inch submicron technology.

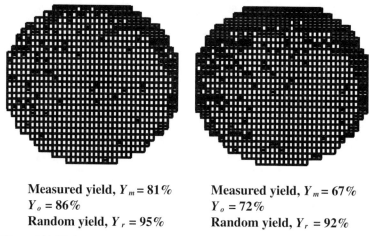

Measured yield, $Y_m = 81\%$ Measured yield, $Y_m = 67\%$
$Y_o = 86\%$ $Y_o = 72\%$
Random yield, $Y_r = 95\%$ Random yield, $Y_r = 92\%$

Figure 4-11 Wafer maps for two 8-inch wafers with different pattern yield loss (Y_o) and similar DLY loss (Y_r).

The average measured yield was lower than expected. However, after analysis of the spatial distribution of failing chips, a pattern of yield loss was identified on all wafers. The underlying Defect Limited Yield (DLY) was higher than the initial data suggested. Although it was possible to visually identify a systematic problem on the top parts of the wafer maps, use of an analytical method enabled an accurate estimate of the underlying DLY to be calculated and a focused process improvement activity to take place. In volume production, average yield was close to the DLY estimate, or about 20% higher than the initial measured yield. As a result, the DLY estimates were used for planning and cost structure development.

4.5.3 Yield Distributions

The shape of the measured yield distribution provides information regarding global yield loss. Normalized yield distribution for a large-volume IC product is shown in Figure 4-12. The **average** yield in Figure 4-12 is approximately 10% lower

4.5 ■ Applications of Functional Yield Models

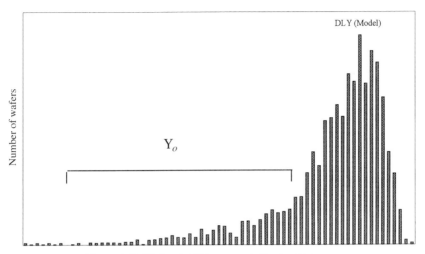

Figure 4-12 Wafer yield histogram indicating Y_o losses.

than the Class III model DLY estimation. The yield variance with respect to the model is associated with some Y_o loss on the left side of the distribution that shifts the average yield down by 10%. The cause of the Y_o loss is due to handling, and systematic losses are due to the product having high-performance mixed-signal circuitry.

The Y_o yield loss was found to be due to one faulty piece of equipment that caused the formation of a large number of defects. The systematic yield loss was identified by evaluating the difference between the predicted and measured DLY and analyzing the shape of the wafer yield distribution. Y_o losses are often present in analog circuitry. These losses occur because functions such as analog-to-digital conversion or digital-to-analog conversion are inherently sensitive to more variations in the manufacturing process than to random defects.

4.5.4 Critical Area in Product Design: SRAM Example

Class III and IV yield models have been used in the development of custom and standard cell chip designs. The effect of increasing the widths and spacings in parts of the chip layout can be evaluated against potential changes in chip size and yield. Evaluating the consequence of changing width and spacing dimensions on yield can also be applied to regular repeated structures, such as memory cells. The example in Figure 4-13 shows the Metal 1 pattern for a six-transistor memory cell. One section of the Metal 1 pattern has been chosen for modification, designated as spacing dimension x. The objective is to decrease the defect sensitivity of the cell, while maintaining the *same* cell size and not violating the technology layout design rules. The spacing dimension, X, was increased from 0.8 μm to 1 μm. The resultant effect on yield is summarized in the table in Figure 4-13.

Increasing the spacing dimension X by 0.2 μm decreases the Metal 1 shorts contribution to yield loss by 1% for 10,000 cells. This method can also be applied to other dimensions on different levels. The example shows that it is possible to evaluate the effect on yield of desensitizing the defect sensitivity of part of a chip

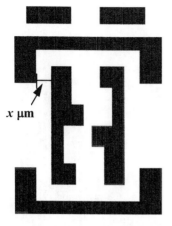

Cell	Critical Area (10K cells) (Metal 1 shorts)	Yield Estimate (Poisson model)
SRAM ($x = 0.8\mu m$)	0.0418 cm²	93.9%
SRAM ($x = 1.0\mu m$)	0.0356 cm²	94.8%

SRAM cell critical area and yield estimate with spacing (x) values of 0.8 μm and 1.0 μm on Metal 1 pattern for 10,000 cells.

SRAM cell, Metal 1 pattern

Figure 4-13 Metal level of single SRAM cell and the yield improvement for the Metal 1 layer by increasing the spacing dimension X for 10,000 cells.

layout by increasing a spacing or width dimension, while not increasing the overall chip size or violating the technology layout design rules.

Another example indicates the layout of a metal level ordered as a histogram of the total length of the minimum spacing 2 λ in Figure 4-14. For this product layout, there are a small number of very long tracks that constitute almost 3% of the total length of the 2 λ spacing. It is necessary to directly measure the critical area to evaluate the yield improvement if the spacing between the longest tracks were to be increased. Critical area can also be useful in evaluating memory redundancy, by calculating the amount of redundant cells required for a certain yield to

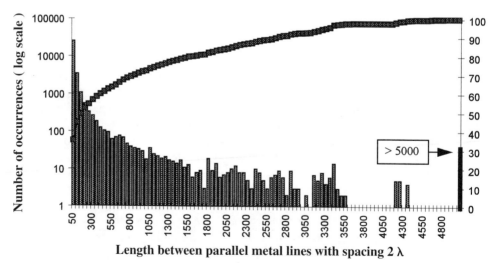

Figure 4-14 Histogram of the sum of lengths with equal spacing of 2 λ (λ is the minimum spacing) for a metal level of a chip layout.

4.5 ■ Applications of Functional Yield Models

be achieved and the cost of those extra cells in terms of increased chip size. Recently, critical area analysis has also been applied to analog circuits, although this is more difficult and will be more ad hoc and informal [19].

4.5.5 Critical Area in Yield Calculation

The graphs in Figure 4-15 indicate the measured yield for almost 50 products as a function of chip area (*a*) and effective critical area (*b*), based on data from IBM [27]. The data in Figure 4-15*a* show a good correlation of the model and the actual yields up to a chip area of 50 mm^2, but a weak correlation beyond that size. The data in Figure 4-15*b* show a strong correlation between the model and the actual yield for the same products in Figure 4.15*a* when plotted as a function of effective critical area (normalized to unit logic circuits). These data show the validity of the Class I model for small chip areas and the difficulty in predicting yield for much larger chip sizes.

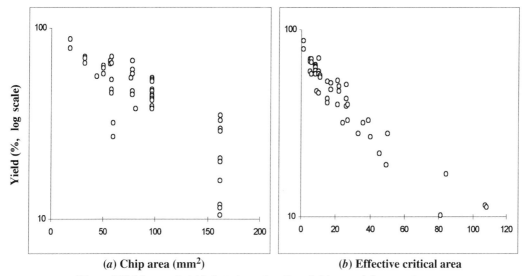

Figure 4-15 Measured yield plotted as a function of chip area (*a*) versus the same data plotted as a function of effective critical area (*b*) for approximately 50 products.

4.5.6 Use of Yield Models for Scaling Applications

Models are often used to provide an indicator of yield on future technologies or circuit redesigns. For a given manufacturing area, all the products are subject to the same defect density. Products will have a different DLY based on (1) chip size, (2) circuit type and density, (3) design rules, and (4) layout methodology being applied (i.e., auto-route versus custom). There could be nonrandom components of yield loss, such as described by Class IV yield models; however, this type of yield loss is not scalable. Random yield losses can be scaled from one product to another if the critical area and the defect density and defect size distribution are well understood.

Chip area can be used to estimate the yield for one size chip based on another size chip. Using chip area to scale is valid only if the circuit type is the same and the chips are to be manufactured in the same fab location. The accuracy of using Class I models for scaling increases if the chip size is less than the yield baseline chip size and becomes somewhat inaccurate as chip size increases. If the overall fault density of one size chip is $\lambda_1 = D \times A_1$, then the fault density for a chip with a different area can be expressed as $\lambda_2 = \lambda_1 \times A_2/A_1$ assuming the same defect density. A simple example is given in Table 4-8.

TABLE 4-8 Scaling Yield Using Chip Area

Chip Area (cm^2)	0.48	1.0
Fault density	0.96	2.0
Poisson model yield (%)	38.3	13.5

Excluding Y_o losses, the yield for the two chip sizes scales with λ. The above methodology can be used for "simple" linear shrinks. That is, every circuit feature is scaled down by exactly the same linear ratio (i.e., from 0.8 μm technology to 0.6 μm technology). This type of linear shrink is often done for digital logic circuitry. However, it is usually not possible to linearly scale mixed-signal circuits since the device performance for analog components does not necessarily scale in a linear fashion. One interesting feature of a true linear shrink is that the *yield* on the scaled product will be the same as the yield for the original product. This result is a consequence of the $1/X^3$ defect distribution discussed earlier in this chapter [14]. The key purpose of scale, therefore, is not to improve yield but to manufacture more chips on the wafer with the same yield, thereby lowering the cost per good chip. Table 4-9 presents a simple example of a linear shrink.

TABLE 4-9 Effects of a Linear Shrink on Yield

Feature Size (μm)	0.8	0.6
Chip area (cm^2)	1.0	0.56
Defect density	2.0	3.56
Fault density	2.0	2.0
Poisson model yield (%)	13.6	13.6
Good chips/wafer	N	1.77 N

Thus, a linearly scaled chip will have the same yield, and the number of chips will increase by the ratio of the areas. This example is valid only for random defects that have a $1/X^3$ distribution; it does not hold for systematic or nonrandom yield loss components.

When using Class III models, the yield can be calculated directly without introducing chip area ratios. Product scaling to new technologies when the scaling factors are often nonlinear or different for each level of the product requires the use of critical area. In mixed-signal technology, there is not a single factor to scale from as in the chip area scaling example because each level of the product may scale with a different factor and the level that has the "largest" feature size or pitch dimension may not be the area limiting factor. Use of critical area and defect size distribution and density allows the costs associated with new technology scaling to be accounted for and optimized a priori.

Critical area relies on the availability of measured defect density and defect size distribution data. Critical area, with a common reference for calibration, enables the defect density on products manufactured on different technologies in different manufacturing locations (i.e., Foundry A, Foundry B, ...) to be directly compared. Use of critical area yield models is interesting since this type of comparison cannot be done using Class I or II yield models.

4.6 SUMMARY

Use of functional yield models in an industrial environment has been discussed. The origin of several of the commonly used yield models is described through the development of the basic yield model mathematics. Four classes of yield models are defined, and their applicability to the industrial environment is detailed. The key yield model parameters are clearly defined, and their use is described for each of the various classes of yield models. For a comprehensive reference list on many of the topics covered in this chapter, see [8].

ACKNOWLEDGMENTS

The authors acknowledge the contributions of Paul Koralishn and Scott Munroe. The authors also acknowledge Analog Devices Inc. for supporting this development effort.

4.7 EXERCISES AND SOLUTIONS

4.1. Given the yield/chip area data in the following table, calculate the average defect density D_o as defined in [20]. Calculate the expected yield for a chip two times and five times the size of the largest chip in the table. Assume that the yield for the 2x and 5x chips is based on the yield of a chip with area $A = 0.2$ cm and yield 70%, but the actual defect yield is 80%. There is 10% of yield loss not due to defects. Calculate the error or risk associated with this estimate.

Yield (%)	Chip Area (cm^2)
84	0.1
82	0.13
72	0.2
73	0.22
75	0.17

Answer.

$$D_o = \frac{1}{5} \times \left(\frac{-\ln(0.84)}{0.1} + \frac{-\ln(0.82)}{0.13} + \frac{-\ln(0.72)}{0.2} + \frac{-\ln(0.73)}{0.22} + \frac{-\ln(0.75)}{0.17} \right)$$

$$= 1.6 \text{ Defects/cm}^2$$

$A_{2x} = 0.44 \text{ cm}^2 \hspace{3em} A_{5x} = 1.1 \text{ cm}^2$

$Y_{2x} = \exp(-1.6 \times 0.44) = 49\% \hspace{2em} Y_{5x} = \exp(-1.6 \times 1.1) = 17\%$

$D_o(0.7) = \dfrac{-\ln(0.7)}{0.2} = 1.78 \text{ defects/cm}^2 \hspace{2em} D_o(0.8) = \dfrac{-\ln(0.8)}{0.2} = 1.11 \text{ defects/cm}^2$

$Y_{2x}(0.7) = \exp(-1.78 \times 0.4) = 49\% \hspace{2em} Y_{2x}(0.8) = \exp(-1.11 \times 0.4) = 64\%$

$Y_{5x}(0.7) = \exp(-1.78 \times 1.0) = 17\% \hspace{2em} Y_{5x}(0.8) = \exp(-1.11 \times 1.0) = 33\%$

The risk associated with this estimate is ($64\% - 49\% = 15\%$) for the 2x and ($33\% - 17\% = 16\%$) for the 5x.

4.2. Product X is presently manufactured on 1 μm technology. There is the opportunity to shrink the product X design to 0.8 μm or 0.6 μm technology. Assume the following: a linear shrink, the same wafer size and cost for all technologies, an increase in the gross chips per wafer proportional to the area, a cubic defect size distribution, and the Poisson yield model. The 1 micron D_o is 2 defects/cm² > 1 μm, the 0.8 μm D_o is 4 defects/cm² > 0.8 μm, and the 0.6 μm D_o is 8 defects/cm² > 0.6 μm.

Answer: Normalize the defect densities to 1 μm, based on the cubic defect size distribution

0.8 μm D_o, 4 Defects/cm² > 0.8 μm = 4/1.56 = 2.56 Defects/cm² > 1 μm

0.6 μm D_o, 8 Defects/cm² > 0.6 μm = 8/2.78 = 2.87 Defects/cm² > 1 μm.

The increase in the number of chips is ×1.56 for the 0.8 μm and ×2.78 for the 0.6 μm. The differences between the yield * gross chips product for the technologies are then:

Chips for 0.8 μm is $\exp(-0.56) * 1.56 = 0.89$

Chips for 0.6 μm is $\exp(-0.87) * 2.78 = 1.16$

The 0.8 μm technology will give 16% improvement in the total number of chips and is the technology to choose.

4.3. 10,000 units of a product with area = 0.5 cm² and 200 chips per wafer are to be produced in three manufacturing areas, with a D_o of 0.9, 1.1 and 1.3 defects/cm², respectively. How many wafers need to be ordered? Use the negative binomial model with $a = 2$ and a combined fab, assembly, and test yield of 95%.

Answer.

$$Y_A = \frac{1}{\left(1 + \dfrac{0.5 \times 0.9}{2}\right)^2} = 72\% \hspace{2em} Y_B = \frac{1}{\left(1 + \dfrac{0.5 \times 1.1}{2}\right)^2} = 62\%$$

$$Y_C = \frac{1}{\left(1 + \dfrac{0.5 \times 1.3}{2}\right)^2} = 57\%$$

4.7 ■ Exercises and Solutions

$$\text{No. of wafers} = \frac{3 \times 10,000}{200} \times \frac{1}{0.95} \times \left(\frac{1}{0.72} + \frac{1}{0.62} + \frac{1}{0.57}\right) = 751 \text{ wafers}$$

4.4. Develop the truth table for the case of three indistinguishable defects placed on three chips on a wafer. Make sure the answer is consistent with Equation 4-9.

	N1	N2	N3
1	MMM		
2		MMM	
3			MMM
4	M	M	M
5	MM	M	
6	MM		M
7	M	MM	
8		MM	M
9	M		MM
10		M	MM

4.5. Prove the yield neutrality of the inverse cube law. Develop a numerical example to show yield increases/decreases as a function of the exact shape of the defect size distribution curve (value of p in Equation 4-48).

4.6. Assume two chips of sizes 20,000 square mils and 250,000 square mils. Each is currently running on a 0.6 micron technology. Use a gamma yield model with a range of alpha values to evaluate the benefits of scaling either of these chips to a 0.5 micron technology. Explain how the inclusion of defect clustering is important as the chip size increases.

4.7. A new product with a critical area of 0.45 cm² is to be produced on a technology with a defect density of 0.5 defects/cm². Three similar products are already being produced on the technology, with critical area and yield data contained in the following table. Analyze the data and calculate the short-term and long-term yield expectations using the Poisson model.

Critical Area (cm²)	Measured Yield (%)
0.1	81
0.2	78
0.4	70

Answer: The expected defect limited yield and the measured yield are compared, and a Y_o estimate is calculated.

Critical Area (cm²)	Expected Yield (%)	Y_o (Ym/Yd)
0.1	95	0.85
0.2	90	0.86
0.4	82	0.85

Based on the measured data, there is a fixed percentage of each product yield that is not defect related. The average Y_o is 0.85; therefore, in the short term, the yield expectation for the new product should incorporate this Y_o. The short yield expectation is $(0.85 \times 79\%) = 68\%$, while the long-term yield expectation, assuming that the cause of the Y_o is identified and the problem fixed is 79% yield.

REFERENCES

[1] W. J. Bertrem, *Yield and Reliability in VLSI Technology,* 2nd ed., ed. by S. M. Sze, New York: McGraw-Hill, 1988.

[2] R. S. Collica et al., "A Yield Enhancement Methodology for Custom VLSI Manufacturing," *Digital Technical Journal* vol. 4, no. 2, pp. 83–99, 1992.

[3] J. A. Cunningham, "The Use and Evaluation of Yield Models in Integrated Circuit Manufacturing," *IEEE Transactions on Semiconductor Manufacturing,* vol. 3, no. 2, pp. 60–71, May 1990.

[4] A. V. Ferris-Prabnu, "On the Assumptions Contained in Semiconductor Yield Models," *IEEE Transactions on Computer Aided Design,* Vol. CAD-11 (8), pp. 966–975, Aug. 1992.

[5] R. S. Hemmert, "Poisson Process and Integrated Circuit Yield Predictions," *Solid State Electronics,* vol. 24, no. 6, pp. 511–515, 1981.

[6] W. Maly, "Modeling of Lithography Related Yield Losses for CAD of VLSI," *IEEE Transactions on Computer Aided Design,* vol. CAD-4, no. 3, pp. 166–177, July 1985.

[7] W. Maly, "Yield Models: A Comparative Study," in *Defects and Faults Tolerance in VLSI Systems,* vol. 2, ed. by C. H. Stapper, New York: Plenum Press, pp. 15–31, 1990.

[8] W. Maly, "Computer-Aided Design for VLSI Circuit Manufacturability," *Proceedings of the IEEE,* vol. 78, no. 2, pp. 356–392, Feb. 1992.

[9] B. T. Murphy, "Cost-Size Optima of Monolithic Integrated Circuits," *Proceedings of the IEEE,* vol. 52, no. 12, pp. 1537–1545, Dec. 1964.

[10] G. O'Donoghue and G. Cheek, Bitstream Defect Analysis Method for Integrated Circuits, U.S. Patent No. 5,497,381, Mar. 1996.

[11] T. Okabe, Nagata, and Shimada, "Analysis of Yield of Integrated Circuits and a New Expression for the Yield," *Electrical Engineering in Japan,* vol. 92, pp. 135–141, Dec. 1972.

[12] H. G. Parks, Personal Communication, 1992.

[13] J. Pineda de Gyvez, J. Jess, "Systematic Extraction of Critical Areas from I.C. Layouts," *Defect and Fault Tolerance in VLSI Systems,* ed. by C. Stapper, vol. 2, New York: Plenum Press, pp. 47–61, 1990.

[14] I. Koren and M. A. Breuer, "On Area and Yield Considerations for Fault Tolerant VLSI Processor Arrays," *Proceedings of the Conference on Advanced Res.,* 1984.

[15] A. Mirza, G. O'Donoghue, A. Drake, and S. Graves, "Spatial Yield Modeling for Semiconductor Wafers," *Proceedings of the Advanced Semiconductor Manufacturing Conference and Workshop,* Boston, pp. 276–281, 1995.

[16] J. E. Price, "A New Look at Yield of Integrated Circuits," *Proceedings of the IEEE* (Lett.), vol. 58, pp. 1290–1291, Aug. 1970.

[17] A. Rogers, *Statistical Analysis of Spatial Dispersion, the Quadratic Method,* London, U.K.: Pion LTD, 1974, Chapter 2, pp. 12–18.

[18] R. B. Seeds, "Yield and Cost Analysis of Bipolar LSI," IEEE International Electron Devices Meeting, Washington, DC, Oct. 1967.

[19] M. Soma, "Challenges in Analog and Mixed-Signal Fault Models," *IEEE Circuits and Devices,* vol. 12, no. 1, pp. 16–19, Jan. 1996.

[20] C. H. Stapper, "Defect Density Distribution for LSI Yield Calculations," *IEEE Transactions on Electron Devices,* vol. ED-20, pp. 655–657, July 1973.

[21] C. H. Stapper and R. J. Rossner, "A Simplified Method for Modeling VLSI Yields," *Solid State Electronics,* vol. 25, no. 6, pp. 487–489, 1982.

[22] C. H. Stapper, "Integrated Circuit Yield Statistics", *Proceedings of the IEEE,* vol. 71, no. 4, pp. 453–470, Apr. 1983.

[23] C. H. Stapper, "Modeling of Integrated Circuit Defect Sensitivities," *IBM Journal of Research Development* vol. 27, no. 6, pp. 549–557, Nov. 1983.

[24] C. H. Stapper, "Modeling of Defects in Integrated Circuit Photolithographic Patterns," *IBM Journal of Research Development,* vol. 28, no. 4, pp. 461–475, July 1984.

[25] C. H. Stapper, "Fact and Fiction in Yield Modeling," *Microelectronic Journal,* vol. 210, no. 1–2, pp. 129–151, May 1989.

[26] C. H. Stapper, "On Murphy's Yield Integral," *IEEE Transactions on Semiconductor Manufacturing,* vol. 4, no. 4, pp. 294–297, Nov. 1991.

[27] C. H. Stapper and R. J. Rossner, "Integrated Circuit Yield Management and Yield Analysis: Development and Implementation," *IEEE Transactions on Semiconductor Manufacturing,* vol. 7, no. 4, Nov. 1994.

[28] D.H.M. Walker, "Yield Simulation for Integrated Circuits," Hingham, MA: Kluwer Academic Press, 1987.

[29] D.H.M. Walker and S. Director, *IEEE Transactions on CAD of IC and Systems,* CAD vol. 5, no. 4, pp. 541–556, Oct. 1986.

5
Critical Area and Fault Probability Prediction

D. M. H. Walker

5.1 INTRODUCTION

Competitive pressures require lower costs and shorter schedules in new product development. This in turn reduces the number of iterations of the design, manufacturing, and test cycle. Fewer iterations can be achieved only if engineers can make more accurate predictions, which in turn is possible only through the availability of a *virtual factory*. There is a particular need for accurate predictions of the yield loss resulting from manufacturing disturbances, as well as predictions of the failure modes that result from these disturbances (see Figure 5-1). Based on the terminology of [1], a *disturbance* can be defined as a deviation from the nominal IC manufacturing environment. A *local disturbance* is a typically large deviation that affects a small region of an integrated circuit. Examples of local disturbances include particulates, liquid droplets, and crystal impurities. Since a local disturbance is usually caused by foreign matter, the term *contamination* is often used. A *global disturbance* is one that causes a typically small deviation over large areas of chips and wafers. An example would be a variation in temperature across the wafer in an oxidation furnace. These disturbances may cause *deformations* to the nominal IC structure, usually termed *defects*. Deformations due to local disturbances are usually termed *spot defects*. Examples include extra or missing conducting, semiconducting, or insulating material. Extra and missing material spot defects are usually modeled as circular disks on the affected IC layer, whereas junction leakage and pinholes in insulating layers are normally modeled as dimensionless points [2]. Spot defects caused by embedded particles are three-dimensional, and evidence indicates that they cause a significant fraction of observed faults [3]. Three-dimensional defects have been modeled as spheres, cylinders, and cubes [4–6].

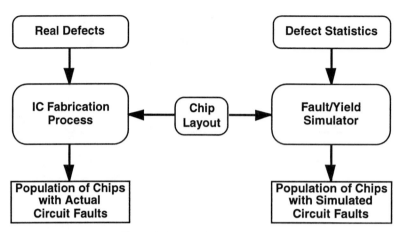

Figure 5-1 Virtual fabrication line.

Defects may in turn cause *circuit performance faults,* often termed *circuit faults* or *faults.* A *hard performance fault,* often called a *catastrophic* or *functional fault,* is one that causes the circuit to have a gross deviation in its behavior; for example, a digital circuit computes a different Boolean function. We define a catastrophic fault as one that makes a unique change to the circuit topology under worst-case bias conditions. (The bias condition accounts for junction breakdown and device punchthrough. The uniqueness permits faults to be distinguished.) Nearly all catastrophic faults can be categorized as *short-circuit faults,* also called *shorts* or *bridging faults,* or *open-circuit faults,* also called *opens* or *breaks.* A *soft performance fault,* often termed a *parametric fault,* or a *soft fault* [7, 8], is a change in parametric behavior, such as speed, power consumption, noise margin, and gain, sufficient to violate performance specifications but not able to affect functional operation. It may also cause a reliability hazard. In the case of catastrophic faults, it is often desirable to determine the change in circuit behavior, or *failure*, that results from it. For example, a short circuit might cause a bit stuck-at-0 in a memory.

Catastrophic faults are caused primarily by spot defects, while parametric faults are caused mainly by global disturbances. These two paths are shown in Figure 5-2. Spot defects can also cause parametric faults, such as delay faults in digital circuits [9] and loss of gain-bandwidth or increased offset voltage in analog circuits [10]. Conversely, global disturbances can cause catastrophic faults; for example, under-etching can cause an open contact between two interconnect layers, or polyactive misalignment can cause MOSFET source-drain shorts. An interaction also takes place between the two disturbances, in that global disturbances can alter the probability that a local disturbance causes a fault. For example, under-etching or lithography overexposure can increase interconnect line-widths, increasing the probability that an extra material defect will cause a short [11].

In this chapter we focus on local disturbances that cause catastrophic faults and functional yield loss. The simulation procedure is often referred to as yield simulation, defect simulation, catastrophic fault simulation, or dot throwing. The inputs and outputs of this procedure are shown in Figure 5-3. The approach used

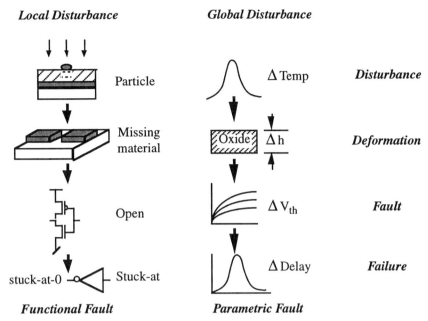

Figure 5-2 Disturbance-deformation-fault-failure relationship.

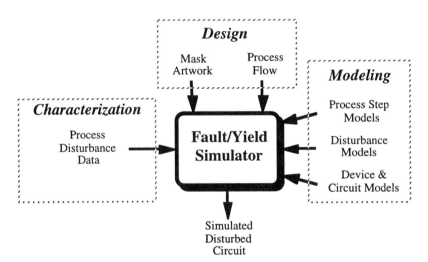

Figure 5-3 Yield simulation.

for catastrophic or hard faults is easily extended to soft faults [8]. A discussion of parametric faults caused by global process disturbances can be found in Chapter 4. We will restrict our discussion to digital circuits, although the same concepts can be applied with little modification to analog circuits.

Applications such as yield prediction and design rule development [12] require the critical area or probability for *any* fault caused by a specific defect type. Applications such as manufacturing fault tolerance [13], test generation and test quality analysis [14], process control [15], and process monitoring and diagnosis [16, 17] require the critical area or fault probability for a specific fault caused by a specific defect type. We will discuss techniques for computing critical areas and fault probabilities for both cases. We also describe algorithms for exploiting the design hierarchy to handle large designs. Applications that need to know specific fault probabilities often need to map these to specific system failure probabilities. For example, manufacturing fault tolerance is usually implemented with spare modules that can replace faulty modules, so faults must be described in terms of module failures. In the case of a redundant memory, failures will be described in terms of bits, rows, columns, and other modules. Since a single fault can cause multiple modules to fail, joint failure probabilities must be computed. For example, a bridging fault between two memory bit lines will cause two columns to fail. We will discuss the mapping from faults to failures in common applications.

In this chapter we introduce the theory of critical area and fault probabiity computation, and we describe how contamination can be mapped to defects, how defects can be mapped to faults, and how faults can be mapped to failures. We then discuss how critical area and fault probabilities are used in design, manufacturing, and test applications.

5.2 THEORETICAL BACKGROUND

Analytical yield models, as described in Chapter 3, compute the yield as a function of the expected number of catastrophic faults and the fault spatial distribution. In addition, these models assume that faults are caused by a number of independent defect sources. The expected number of faults caused by defect type i is computed as $\lambda_i = D_i A_{\text{crit},i}$ where D_i is the defect density and $A_{\text{crit},i}$ is the critical area for defect type i. The critical area is the area on the chip where a defect must occur to cause a catastrophic circuit fault [18–20]. Figure 5-4 shows the critical area for a pair of parallel metal lines and a fixed defect diameter. A defect center must occur in the critical area for it to result in a short-circuit fault. We define critical area $Cr_i(x)$ for a defect of type i and diameter x to be that area where if the defect occurs, a fault

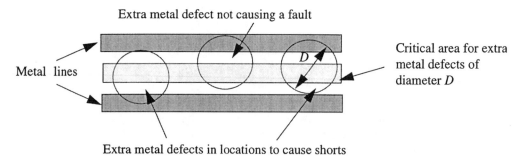

Figure 5-4 Critical area for extra metal defects for defect diameter D for two parallel metal lines.

results. If the defect size distribution $h_i(x)$ for the defect type i is known, the effective critical area $A_{\text{crit},i}$ can be expressed as

$$A_{\text{crit},i} = \int_0^\infty Cr_i(x) \cdot h_i(x) dx \tag{5-1}$$

A typical form of $h(x)$ is [21]

$$h(x) = \frac{(p-1)X_0^{p-1}}{x^p}, \qquad x \geq X_0, \qquad p > 1 \tag{5-2}$$

The X_0 constant is usually determined by the resolution limit of the lithography process. The exponent p is determined by the lithography and particle size distributions and is often assumed to be 3 [22], but recent data indicate that it can be as high as 5 [23].

In many applications, the critical area for each unique fault is needed. $A_{\text{crit},i,f}$ is the critical area in which a defect of type i must occur in order to cause unique fault f. This can be computed by using $Cr_{i,f}(x)$, the critical area for a defect of type i and diameter x to cause fault f. Since

$$Cr_i(x) = \sum_f Cr_{i,f}(x) \tag{5-3}$$

a single method can be used for computing both types of critical areas.

In many applications, it is not the critical area that is needed, but the probability of a fault or failure (POF). The POF and critical area are directly related since

$$POF_{i,f} = \frac{A_{\text{crit},i,f}}{A_{\text{chip}}} \tag{5-4}$$

where A_{chip} is the chip area. In practice, some methods compute critical areas directly, and some compute fault probabilities directly, and these are converted to the appropriate representation as needed by the application. Implicit in Equation 5-1 is the assumption that $h_i(x)$ is uniform across the chip. Implicit in Equation 5-4 is the assumption that D_i is uniform across the chip. Section 5.4.5 describes how these assumptions can be relaxed.

5.3 CONTAMINATION TO DEFECT MAPPING

The most accurate calculation of fault probabilities and critical areas requires knowledge of contamination statistics and a model of how contamination causes defects. For example, an opaque particle can be deposited on a photoresist surface prior to exposure. This particle acts as a mask, preventing the underlying resist from being exposed. The result is that the photoresist will be removed in this area, so the underlying material (e.g., metal) will be etched, resulting in a missing material defect. This sequence is shown in Figure 5-2. Accurate mapping of this disturbance to a defect requires use of three-dimensional lithography and topography simulators [24, 25]. Such simulations are expensive, permitting their use on only a few contamination examples. It is possible to develop efficient approximation models [23, 26], but even these efficient models are relatively slow. An additional drawback is that the required simulation inputs may not be readily available. For example, the particle size distribution or chemical composition may not be known.

For most applications, it is not necessary to know the contamination to defect mapping. Characterization of the manufacturing process in terms of defects is sufficient. The primary application for a contamination to defect mapping is in process diagnosis, where the goal is to isolate the source of yield loss back to the individual unit process step. For example, a bridging fault due to an extra metal defect could be caused by contamination prior to metal deposition, after deposition, after photoresist deposition, during photoresist patterning, or prior to metal etch.

Contamination to defect mapping is useful when adequate defect characterization data are not available or are too expensive to collect. Figure 5-24 shows how this could be done in the DEFAM (Defect to Fault Mappers) simulator. Contamination could be injected into the simulated manufacturing process flow at all the likely points and in all the likely configurations of nearby mask geometry, in order to compute the likelihood that a defect in these locations would cause a fault.

5.4 DEFECT TO FAULT MAPPING

Defect to fault mapping can be performed using two basic approaches: geometrical and Monte Carlo. Geometrical methods compute critical area, while Monte Carlo methods compute fault probabilities. As was shown in Equation 5-4, probabilities can be computed from critical areas, and vice versa. We now examine each method in detail, as well as approaches that try to combine both methods.

5.4.1 Geometrical Methods

Geometrical methods apply a sequence of Boolean polygon operations to the mask artwork to compute the critical area [27–29]. Figure 5-5 shows how $Cr_i(d)$ is computed for an extra material defect with diameter d causing a short-circuit fault.

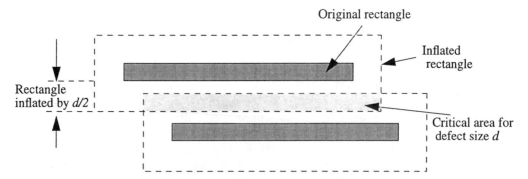

Figure 5-5 Critical area calculation based on polygon operations. The operations shown are for computing the area for an extra material defect to cause a short circuit.

Geometry corresponding to each of the nets is inflated by $d/2$. The intersection region defined by the bloated rectangles of two or more distinct nets then corresponds to the critical area for that defect diameter.

Geometrical methods for computing the critical area for opens due to extra insulating material (missing conducting material) can be viewed as the complement of computing the critical area for shorts [2]. The "space" polygons are inflated and intersected, or, conversely, the wire polygons are shrunk and self-intersected. The drawback of these approaches is that the space is often connected, so there are not separate polygons to be intersected, and Boolean polygon operations will usually cause a polygon to disappear if it shrinks past zero. Representations such as corner-stitched rectangles [30] already divide the space into polygons.

The critical area for shorts caused by pinhole defects is some combination of the overlap area between the two conducting layers separated by the insulating layer, and possibly the edge length at different layer combinations. For example, under certain manufacturing process conditions, MOSFET gate oxide pinholes are a function of the source/drain edge length.

To exactly compute the critical area for faults due to a circular defect would require the use of Euclidean expansion and contraction of polygons, as opposed to the orthogonal expansion shown in Figure 5-5. This computation requires the handling of polygons with arcs. This capability increases analysis time by up to an order of magnitude [31] but has been demonstrated for small cells [32]. As a result, most geometrical analysis uses the orthogonal expansion. This effectively approximates the defect as a square circumscribing the circle, and since the square is larger than the circle, it leads to an overestimate of critical area at layout corners. This can lead to an 8% overestimate in critical area [2], but in practice the error is quite small except for large defect sizes [33]. Since these defects are relatively rare, only a small error is introduced into Equation 5-1. To improve accuracy, several algorithms approximate the arcs with diagonal edges or analytical formulas [34, 35].

Most geometrical methods restrict themselves to rectilinear mask artwork. This improves performance, but more importantly it simplifies the algorithms. A corner-stitched rectangle database can be used to automatically eliminate overlaps in the intersection areas [6, 36], and a scanline approach can be used to limit the search neighborhood [27]. As shown in Figure 5-5 for bridging fault, the critical area analysis can be approximated by considering the layout as a series of parallel wire segments, reducing the problem to only one dimension [33–35, 37–42].

One drawback in the use of Boolean polygon operations to compute critical areas is that we are typically not interested in $Cr_i(x)$, but in $A_{crit,i}$. As shown in Equation 5-1, this requires integrating over x. This integration is usually performed using a trapezoidal approximation, requiring that $Cr_i(x)$ be evaluated for many defect diameters. Performing the polygon operations for these many diameters is expensive and unnecessary. As can be seen in Figure 5-5, increasing the defect diameter slightly will only increase the critical area slightly. It will not change the shape or location of the critical area. This observation led to the concept of the *susceptible site* [43], which is a location in the layout where a critical area will occur. These susceptible sites can be expanded and contracted without the need to recompute the Boolean operations on the entire mask artwork.

An example of critical area computation via susceptible sites in the XLASER system [34] is shown in Figures 5-6 to 5-16. Figure 5-6 shows the layout of cells from a cellular neural network chip [44]. Figures 5-7 to 5-11 show the critical areas for short-circuit faults caused by extra polysilicon defects of different diameters. Figures 5-12 to 5-16 show the critical areas for open-circuit faults caused by missing polysilicon defects of different diameters. Figures 5-17 and 5-18 show the *sensitivity* or probability of failure for extra and missing polysilicon defects of different diameters. As expected, the sensitivity is zero for very small defects and approaches one for large defects.

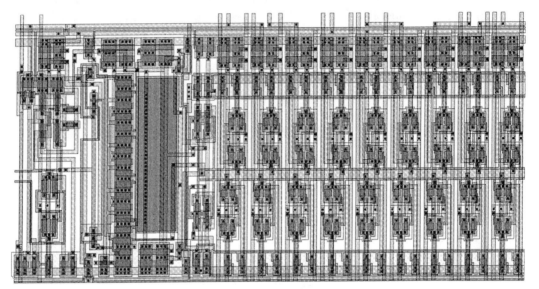

Figure 5-6 Layout of neural network cell.

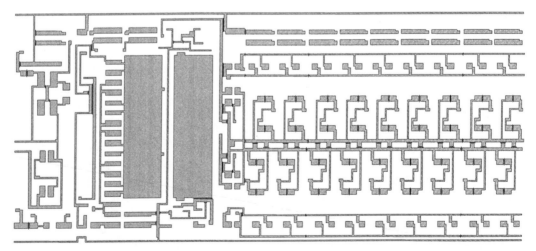

Figure 5-7 Critical area for short circuits caused by a 1 μ extra polysilicon defect.

5.4 ■ Defect to Fault Mapping

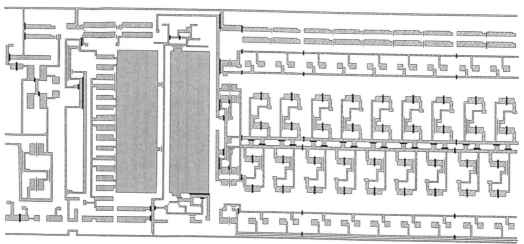

Figure 5-8 Critical area for short circuits caused by a 2 μ extra polysilicon defect.

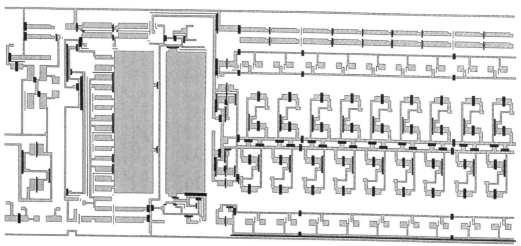

Figure 5-9 Critical area for short circuits caused by a 3 μ extra polysilicon defect.

For most applications, the location and shape of the critical area are not important, only the area. This observation led to the concept of the *virtual artwork* [11]. A virtual artwork is an artificial layout that has the same statistics in terms of wire lengths, widths, and spacings as the original mask artwork [45], so that it will have approximately the same $Cr_i(x)$ value. The virtual artwork is normally chosen to simplify the critical area computation. For example, the mask artwork for one layer is usually converted to a set of parallel lines with various widths, spacings, and lengths. This layout will provide a piecewise linear approximation to the $Cr_i(x)$ function. The virtual artwork also has the advantage that it is normally a much more compact representation of the layout since many wires can be lumped into a single wire. Virtual artworks have not been used for fault-oriented critical area

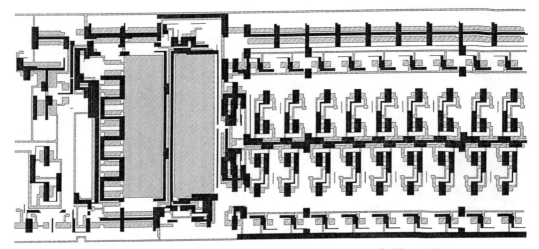

Figure 5-10 Critical area for short circuits caused by a 6 μ extra polysilicon defect.

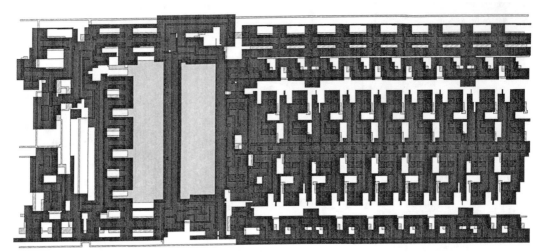

Figure 5-11 Critical area for short circuits caused by a 12 μ extra polysilicon defect.

and fault probability computations. A virtual artwork for that purpose could be constructed, using circuit knowledge in extracting layout statistics, for example, the length that two particular nets run in parallel at a fixed spacing. However, the virtual artwork might be complicated enough that it would provide little benefit over analyzing the original layout.

Geometrical methods can be extended to compute critical areas for individual circuit faults, as shown in Figure 5-19. These approaches typically start with mask artwork that has been extracted and labeled with transistors, net numbers, and so on, and then the critical area analysis is done for each combination of faults [46].

5.4 ■ Defect to Fault Mapping

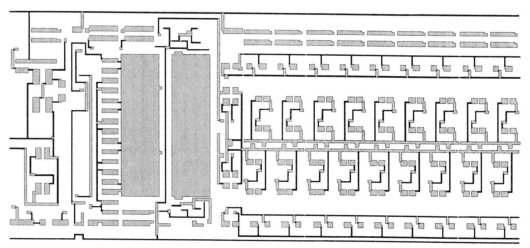

Figure 5-12 Critical area for open circuits caused by a 1 μ missing polysilicon defect.

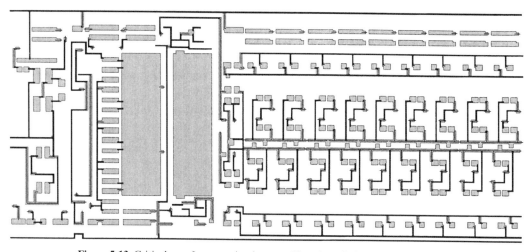

Figure 5-13 Critical area for open circuits caused by a 2 μ missing polysilicon defect.

The critical area analysis can also combine global linewidth variation into its analysis [32].

Since many geometrical methods for critical area computation use polygon operations on the mask artwork, several implementations have been developed based on university or commercial Boolean polygon packages associated with mask design rule checkers. For example, one system [47] was built on top of the Cadence *Edge* system. It could compute $Cr_i(x)$ for bridging faults between two nets due to extra material defects. The system was implemented in only 1000 lines of code, but finding the critical area for bridging faults required $O(n^2)$ time for n nets since the database does not provide a spatial ordering of the nets.

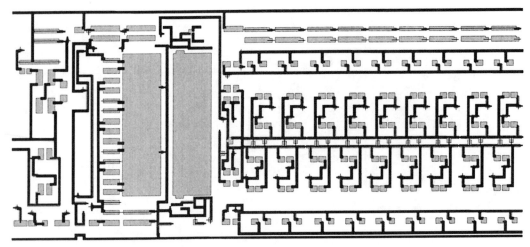

Figure 5-14 Critical area for open circuits caused by a 3 μ missing polysilicon defect.

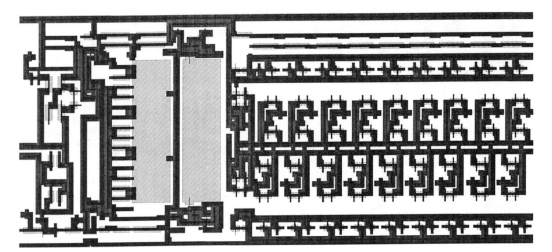

Figure 5-15 Critical area for open circuits caused by a 6 μ missing polysilicon defect.

5.4.2 Monte Carlo Methods

Monte Carlo methods estimate critical area by sampling the layout with defects, using the procedure shown in Figure 5-20. Defects are uniformly introduced on the layout. After each defect is introduced, a circuit extraction is performed and compared to the nominal circuit to determine if the defect caused a fault. The fraction of defects that cause a fault is used to estimate the critical area. Suppose we introduce N defects of type i and diameter d. If we define function $E(i, x, y, d)$ as

$$E(i, x, y, d) = \begin{cases} 1 & \text{Defect of type } i \text{ at } x,y \text{ and size } d \text{ causes a fault} \\ 0 & \text{Defect of type } i \text{ at } x,y \text{ and size } d \text{ causes no fault} \end{cases} \quad (5\text{-}5)$$

5.4 ■ Defect to Fault Mapping

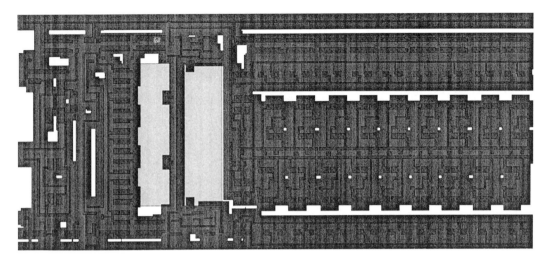

Figure 5-16 Critical area for open circuits caused by a 12 μ missing polysilicon defect.

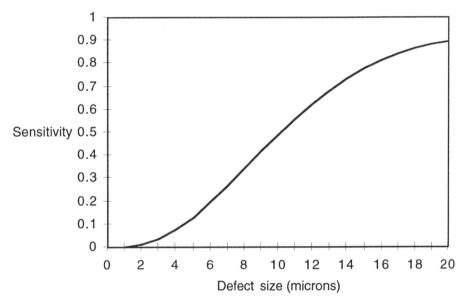

Figure 5-17 Neural network cell sensitivity to short circuits versus extra polysilicon defect size.

then, $Cr_i(d)$ can be approximated by

$$Cr_i(d) = \left(\frac{1}{N}\sum_{j=1}^{N} E(i, x_j, y_j, d)\right) \cdot A_{\text{chip}} \quad (5\text{-}6)$$

Since Monte Carlo sampling is a set of Bernoulli trials, the confidence interval of the critical area estimate can be approximated using the t-distribution, converging at a rate of $1/\sqrt{N}$. If a round defect is used in the circuit extraction, the critical

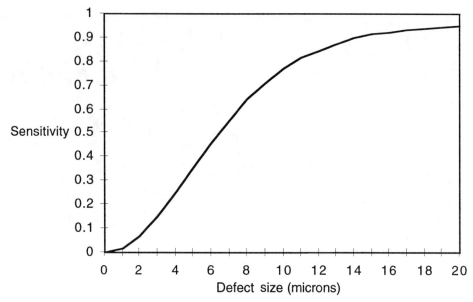

Figure 5-18 Neural network cell sensitivity to open circuits versus missing polysilicon defect size.

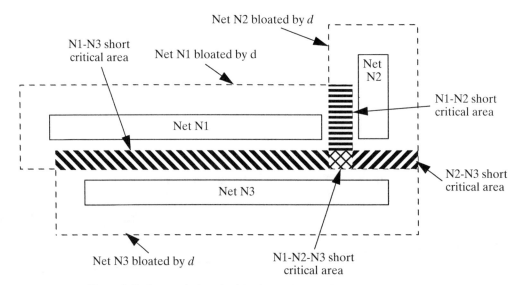

Figure 5-19 Geometrical method for fault-oriented critical area extraction.

area estimate will converge to the "exact" value computed using Euclidean expansion. The primary drawback of Monte Carlo methods is the need for a large sample size to give reasonable accuracy. Their primary advantage is their ability to more readily handle fault types that are more complex than bridging faults.

Equation 5-6 is used to compute $Cr_i(x)$ for different values of x, and the results

5.4 ■ Defect to Fault Mapping

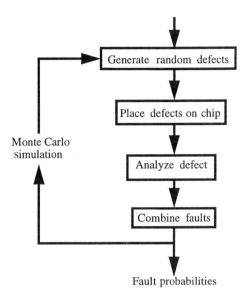

Figure 5-20 Monte Carlo analysis loop.

are then combined with the defect size distribution $h_i(x)$ using Equation 5-1 to compute $A_{\text{crit},i}$. The Monte Carlo approach can be modified so that defect sizes are drawn from the $h_i(x)$ size distribution. In this approach, we do not explicitly evaluate $Cr_i(x)$ but instead compute $A_{\text{crit},i}$ directly as

$$A_{\text{crit},i} = \left(\frac{1}{N} \sum_{j=1}^{N} E(i, x_j, y_j, d_i) \right) \cdot A_{\text{chip}} \tag{5-7}$$

Critical area computed using Equations 5-5 to 5-7 does not specify the critical area for each fault type, such as those shown in Figure 5-19. The methods employed in computing the total critical area for a design can be extended to compute critical area for individual faults too. The analysis of each defect is extended to a local circuit extraction, using either a general-purpose extractor and circuit comparison [48, 49] or a series of single-purpose extractor/comparators specialized to the fault types and process technology [50]. The advantage of the former approach is that process technology, defect, and fault types can be changed without major effort. In the latter approach, new extractors must be written for new fault types, and potentially all extractors must be modified for a technology change. An intermediate approach is to handle most faults with a few special-purpose extractors and to pass rare, complicated faults to general-purpose extractors in a post-processing step [4]. The extraction algorithms can be limited to rectangles [4], or more general-purpose polygon operations can be used [50].

5.4.3 Combined Methods

The DEFAM system, shown in Figure 5-21, is structured so that it can use either a Monte Carlo or a geometrical method for computing fault probabilities [51]. It can even use both methods in the same analysis, since critical areas and fault probabilities can be translated using Equation 5-4.

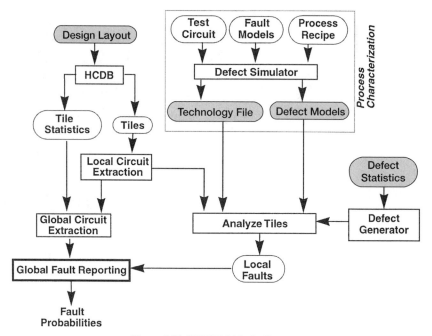

Figure 5-21 DEFAM block diagram.

It was observed that the defect size distribution is usually described analytically, as in Equation 5-2, so the integration in Equation 5-1 can be performed by sampling the layout with a Monte Carlo method and by integrating over the size distribution analytically [28, 52]. This can be done since once a defect is large enough to cause a fault, all larger defects will also cause a fault. So the minimum size for a fault is determined, and then $h(x)$ is integrated analytically from that point. The result is significantly faster than Monte Carlo methods with an accuracy greater than geometrical methods. The primary limitation of this approach is that if larger defects cause different faults than smaller defects, the approach cannot be used to compute $A_{\text{crit},i,f}$.

5.4.4 Three-Dimensional Defects

Implicit in the preceding analysis of critical area is the assumption that only two-dimensional or *footprint* defects are being considered. These are defects that affect only a single layer of the IC structure. To properly analyze three-dimensional defects, *critical volumes* must be computed. Similar to critical areas, fault probabilities can be computed by dividing the critical volumes by the total volume. In the case of Monte Carlo methods, the probability is computed directly by using three-dimensional defects. The problem can often be treated as 2.5-dimensional, using a series of two-dimensional analyses on each layer. In general, as discussed in Section

5.3, a full three-dimensional analysis with sophisticated process models is required to account for problems like step coverage and depth-of-focus.

In the 2.5-dimensional approach, three-dimensional defects are modeled as cylinders, with a height as well as a diameter distribution. The cylinder height and vertical location determine the layers the cylinder intersects. On each of these layers the cylinder looks like a disk. The analysis can be performed using existing two-dimensional algorithms on each affected layer, with the additional knowledge that the disks will be electrically connected if the defect is conducting. A more general approximation is to model the defect as a cube, polyhedron, or sphere, and apply the three-dimensional analogy of Boolean polygon operations to determine critical volumes. An approach extending the idea of susceptible sites to three-dimensions was described in [5].

5.4.5 Spatial Clustering Within Chips

The analysis methods described here assume that the defect density and defect size distribution are uniform within a chip. This permits $Cr_i(x)$ for the entire chip to be computed as the sum of all individual critical areas. If the defect density is not uniform, then these combined areas cannot be used to compute yield loss and fault probabilities for the entire chip. An analogy is with clustering between chips on a wafer, where the chip fault probabilities must be computed separately and combined according to the defect clustering statistics. Evidence indicates that chips are now large enough that intra-die defect clustering does occur, and analytical yield models have been developed for it [53, 54]. These models take a module fault probability as one of their inputs, and this can be computed using existing defect to fault mapping techniques. Hierarchical defect to fault mapping techniques [55, 56] can readily incorporate the clustering statistics when combining module fault probabilities to compute global fault probabilities. Geometrical methods that use design rule checking packages may not be able to do so since the package may only report total area, not the area and location of individual critical areas [57].

5.4.6 Circuit Model Issues

A fault was previously defined as a unique change to the DC circuit topology under worst-case bias conditions. This rules out changes in parasitic values as faults. This creates difficulties when resistors, capacitors, and inductors are used as circuit elements. For example, many SRAM processes use second-level polysilicon for distribution of power and as pullup resistors, the difference being determined by an implant mask. Similarly, analog circuits use polysilicon resistors, parallel-plate capacitors, and the like. The solution is that the fault analysis routines and circuit representation must be capable of treating passive elements both as parasitics and as design elements, with the user specifying the criteria to determine if a significant deviation in element values is a catastrophic fault.

5.5 HIERARCHICAL DEFECT TO FAULT MAPPING

Most defect to fault mapping algorithms operate on a fully instantiated (flat) layout. The largest critical area and fault probability computation reported using a flat analysis was for a 16K-bit CMOS SRAM [58, 59]. Larger designs would require an infeasibly large amount of memory and time.

The standard solution for coping with design complexity is to exploit the high degree of regularity present in the design hierarchy [60–62]. Similar layout patterns have similar fault mechanisms and critical areas. This implies that having evaluated the critical areas and fault probabilities for every unique neighborhood a defect is likely to find itself in, the results can be combined to find the critical areas and fault probabilities for the entire design.

For footprint defects, the approaches used in hierarchical design rule checking and circuit extraction [60, 62–72] can be applied to hierarchical critical area estimation. These algorithms use the general approach of checking each symbol definition, and then checking interactions between symbol instances and between symbol instances and mask geometry out to some interaction distance, normally determined by the maximum design rule, coupling capacitance, and the like. Some algorithms restrict the design so that a symbol instance cannot be overlapped by geometry or other symbol instances. The problem is then reduced to checking symbol definitions and combinations of nearby symbol instances. In general, whenever a symbol instance is overlapped by geometry or another symbol instance, the combination is checked together. Most systems attempt to track combinations that have been seen previously so they only need to be checked once.

Assuming that defect size can be limited to some maximum value, hierarchical versions of the polygon operations shown in Figure 5-5 can be used for critical area analysis [73–75]. One drawback of existing approaches is that only a few [71] guarantee that only unique layout areas will be analyzed. In cases of significant symbol instance and geometry overlap, most algorithms perform a large amount of duplicate analysis. We have found that many "regular" designs, including most of the designs in Table 5-1, have a large amount of overlap. Existing design rule checkers and circuit extractors can handle these designs because they only need to perform simple geometry operations. However, when applying the multiple

TABLE 5-1 Result of Running DEFAM on Some Test Layouts

Design	Trans.	POF Metal 1	POF Metal 2	Savings
16-bit SRAM	182	0.0455	0.0015	1.46
Adder	447	0.166	0.102	8
Multiplier	8,400	0.018	0.00026	10.2
Router	3.5K	0.0221	0.0132	7
Array	85K	0.0423	0.052	9.59
Coprocessor	28.5K	0.0229	0.00062	6.38
Retina	29K	0.11	0.019	29.5
24-port Reg.	64K	0.0991	0.0471	2.611
Crossbar	164K	0.07	0.0031	1143

5.5 ■ Hierarchical Defect to Fault Mapping

geometry operations of critical area analysis, these hierarchical approaches behave poorly.

An additional complexity to using standard hierarchical analysis algorithms is that the critical area must be clipped so that it is not double-counted [76]. For example, if instances of symbols A and B abut, the critical area is really the critical area for A clipped to its bounding box, the critical area for B clipped to its bounding box, and the critical area for the A–B interaction that is not already counting in A or B. This can be counted by first taking the union of the A, B, and A–B critical area polygons and then computing their area. If the A, B, and A–B critical areas are summed, the result will be an overestimate of $Cr(x)$. This overestimate of $Cr(x)$ becomes quite large for large x. However, since this is usually weighted by $h(x)$ as in Equation 5-1 and $h(x)$ falls sharply with x, the typical error in Equation 5-1 will be about 3% [77]. This error can be avoided by using a hierarchical approach.

5.5.1 Identification of Nonoverlapping Layout Areas

One means of identifying unique nonoverlapping layout areas is the hierarchical chip database (HCDB) [73, 74, 78, 79]. The data structure used by HCDB is shown in Figure 5-22. The design is stored as several levels of rectangular units, each of which is called a *tile*. Each tile is a corner-stitched rectangle database [30] with each rectangle referencing tiles at lower levels of the hierarchy. Tiles at the lowermost level of the hierarchy, referred to as *primitive tiles*, represent pieces of fully instantiated layout. Each reference to a tile, referred to as a cell, consists of a transformation and a bounding box to specify the part of the tile used. The bounding box is specified in order to allow portions of a tile to be instantiated, instead of constructing a new tile every time we encounter only a section of the tile. This is similar to the stepping boundary available in many layout systems.

In order to create nonoverlapping tiles, HCDB may modify the designer-specified hierarchy. Symbol definitions and symbol instances are mapped into HCDB tiles and cells. Whenever symbol instances overlap, HCDB creates more

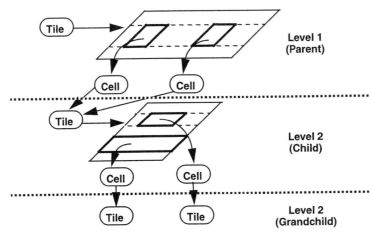

Figure 5-22 HCDB data structure.

tiles, corresponding to the overlap region of the instances. These tiles are referred to as *merge tiles*. New merge tiles are formed for every unique instance overlap. Since geometry overlapping a symbol instance may cause additional fault types, combinations of instances and overlapping geometry form new tiles that are the union of the instance and the geometry. Figure 5-23 shows a design with the unique nonoverlapping tiles identified by HCDB. Note that new tiles were generated owing to both symbol instance overlap and to overlapping geometry. Also note the different bounding boxes for which each tile can be instantiated, shown as dotted lines.

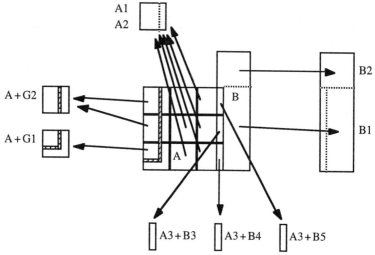

Figure 5-23 Identification of unique nonoverlapping tiles in HCDB. The nine instances of A and one instance of B are divided into three A + B overlaps, two A + geometry overlaps, and two subsections each of A and B. The subsections are identified with a bounding box on the original tiles.

Since defects can span cell boundaries, all the cell combinations that might potentially interact must be identified. This is typically done by identifying all the instances and geometry that fall within a *halo* around each instance [61, 62] or by identifying all instance pairs closer than a minimum distance [64]. A canonical representation can be used to identify unique interactions, in order to minimize fault analysis [51].

5.5.2 Hierarchical Circuit Extraction of Nonoverlapping Layout

Applications requiring $POF_{i,f}$ or $Cr_{i,f}(x)$ must have the layout extracted to label all geometry with net numbers. The same approaches described earlier for hierarchical design rule checking can also be used for circuit extraction. If a nonoverlapping hierarchical representation such as HCDB is used, the problem is simplified. First, each primitive tile is extracted. Geometry may receive multiple net numbers, since the different bounding boxes that a tile can be instantiated with can modify the net topology. After extracting all primitive tiles, nonprimitive tiles are extracted. Since tile instances do not overlap, the extraction is performed by identifying conducting interfaces between neighboring instances and the primitive tile nets that correspond to them.

5.5.3 Hierarchical Defect to Fault Mapping of Nonoverlapping Layout

Once all the unique layout areas and interaction areas are identified, a defect to fault mapping is performed on them using either a geometrical or Monte Carlo method. Computation is performed on $Cr_{i,k_l,f}(x)$ or $POF_{i,k_l,f}$ for defect of type i causing a fault of type f in the lth instance of tile k, and $Cr_{i,k_{l_m},f}(x)$ and $IPOF_{i,k_{l_m},f}$, the critical area and POF in the mth interaction of the lth instance of tile k. These results are combined with the tile statistics in order to compute $Cr_{i,f}(x)$ and $POF_{i,f}$ for the entire chip.

The unique area of the chip determines the fault analysis time. Analyzing only unique tiles greatly reduces the sample size. The last column in Table 5-1 is the area reduction factor due to a hierarchical analysis using HCDB, effectively a measure of the chip regularity.

5.5.4 Global Fault Reporting

The previous section described the computation of critical areas and fault probabilities within a tile and its interactions. Consequently, the information about faults is only local to the tile. This is sufficient for applications such as yield prediction that do not need to know fault details, or applications that only need to know the details of faults within a cell. However, applications such as test generation need *global fault* information, that is, fault descriptions in terms of the entire design.

There are three aspects of global fault reporting. First, the fault must be reported in terms of global net names. This is straightforward in that a local net name can always be converted to a unique global net name by prefixing the path through the hierarchy. So net A in instance 2 of cell B in cell C might be specified as $C.B\_2.A$. Second, a locally reported fault might not exist as a global fault, as shown in Figure 5-24. This can be discovered through net equivalences farther up in the hierarchy. Finally, critical areas and fault probabilities for local faults must be adjusted in terms of the entire chip. For example, in order to determine the

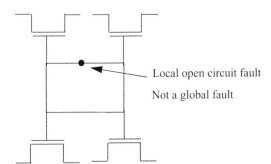

Figure 5-24 Local versus global faults. A local fault is not necessarily a global fault.

critical area for a global *Vdd-Gnd* short, the critical areas for *Vdd-Gnd* shorts in all tiles must be summed.

As an example of probability scaling, suppose nets *Bit* and *BitB* in cell *A* have a 0.1 probability of shorting. If the chip consists of three instances of *A*, then the probability of an *A_1.Bit-A_1.BitB* short is 0.1/3, since a defect has only a one-third chance of occurring in instance 1 of *A*. A detailed formalism for the critical area and fault probability computation can be found in [77]. Note that the probability computation assumes that defects occur uniformly within the die. If this is not the case, a more complex joint probability computation can be substituted.

5.6 FAULT TO FAILURE MAPPING

Many applications require that faults be mapped to chip failures, either in terms of the circuit structure, such as a module failure, or in terms of behavior, such as a faulty response to input stimulus. If the defect to fault mapping was done in a hierarchical fashion, module failures may already be provided. In more complex cases, application-specific rules can be used to map faults to failures. In the case of memory arrays, these rules are relatively simple [2, 13, 29, 80]. For inductive fault analysis, rule-based methods can map many faults to stuck-at failures [81].

In the general case, fault to failure mapping must be done using behavioral models with accuracy sufficient for the application. In some cases, the failure must be in terms of a high-level behavior model. For example, in a neural network, faults should be reported in terms of changes to neuron behavior [82]. Applications such as process diagnosis [83] and current testing [84] require failures in tems of voltage and current outputs for a given test set. This is usually obtained by general or special-purpose circuit or timing simulation [85, 86]. These fault to failure simulations often are much more expensive than the defect to fault mapping.

5.7 APPLICATIONS

Applications can be divided into those that require critical area and fault probability listed separately for each unique fault type, and those that do not. In general, applications that do not use design circuit knowledge do not need to know individual fault probabilities. Examples include yield prediction and design rule development. At the other extreme, applications such as test generation and process monitoring need to know precisely the faulty circuit topology created by each defect. In between, design for defect tolerance needs to map faults to module failures but usually does not need to know precisely how a module fails.

5.7.1 Yield Prediction

As described in Chapter 4, analytical yield models use the chip critical area A_{crit}, the critical area for each defect type i $A_{crit,i}$, or the critical area as a function of defect size $Cr_i(x)$. These are either computed directly using polygon operations or Monte Carlo sampling, or derived from the more detailed fault probability information. For example,

$$A_{\text{crit}} = A_{\text{chip}} \sum_i \left(\frac{D_i}{D} \sum_f POF_{i,f} \right) \tag{5-8}$$

where $D = \sum_i D_i$. The approach used in (5–8) generates much more detail than necessary but is more accurate since each fault can be verified to cause faulty circuit behavior. Note that if the defect density is not uniform within the chip, the critical area can be reported for subsections of the chip. This is straightforward in the case of a hierarchical analysis but can also be provided in a flat analysis by dividing the critical area polygons among regions and computng their area separately. The same technique can be used if the defect size distribution also varies with location. The assumption is that defects can be approximated as random with a single-size distribution per defect type within each region.

To illustrate the result of computing $POF_{i,f}$, the ranked fault probabilities for an 8-bit adder are shown in Figure 5-25. As can be seen, extra first-level metal defects are much more likely to cause faults than second-level metal defects.

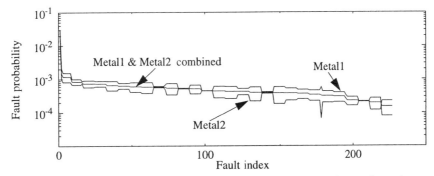

Figure 5-25 Probability of stuck-at faults caused by extra first and second metal defects in an 8-bit CMOS adder. First and second metal defects are assumed to be equally likely. First metal defects are more than twice as likely as second metal defects to cause a fault for most fault types.

5.7.2 Redundancy Analysis

Redundancy analysis requires knowledge of the probability of failure for each module or combination of modules, and the defect spatial distribution, which determines module failure correlation. An example of module failure probabilities is shown in Figure 5-26. The failure modes due to extra first-level metal defects as a function of defect size are given for a CMOS SRAM array. These curves can be convolved with $h(x)$ to compute the critical area and probability of failure. The sparse dotted line shows the simplistic projection of critical area for maximum metal density, which is clearly pessimistic.

If an analytical defect spatial distribution is known (e.g., Poisson, negative binomial) and a simple spare-swapping strategy is used, these failure probabilities

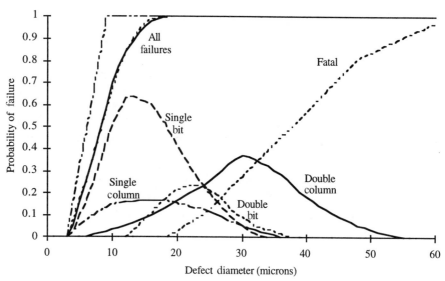

Figure 5-26 SRAM array failure probabilities versus defect size for extra first metal defects.

can be analytically combined to compute the defect tolerant yield. This approach has been applied to memories [13, 29, 80, 87], processor arrays [88, 89], neural networks [90], crossbar switches [91], and underutilized arrays [92]. In designs with a more heterogeneous module set and more complicated reconfiguration algorithm, a Monte Carlo analysis can be performed by placing defects with spatial correlations [93, 94]. Most such analyses make the assumption that only one fault will occur within a module, with the same spatial distribution as the defects, so the defect to failure mapping step can be skipped in the Monte Carlo loop.

5.7.3 Test Generation

As described in Chapter 3, the Inductive Fault Analysis (IFA) approach to functional test generation requires a list of possible circuit faults. Ideally, this list includes fault probabilities or is at least rank-ordered, so that the most likely faults will be tested first, minimizing average test time, and the true fault coverage can be estimated. Faults must be specified in full detail, in terms of how they modify the circuit graph, and then mapped to failures acceptable to the test generator.

A closely related problem is predicting test escape rates. Traditional models assume that faults are equally likely. By using fault probabilities, more accurate estimates of escape rates can be made [95, 96].

5.7.4 Process Diagnosis and Monitoring

Failure analysis is used to help determine the disturbance that caused a fault [9, 15–17, 83, 85, 97–103]. This permits electrical testing of test structures (e.g., memories) or product die to identify or greatly narrow the possible causes of the fault. Electrical testing is much faster and cheaper than traditional failure analysis approaches such as delayering and optical inspection. This in turn speeds up the rate of yield learning, increasing the slope of the yield ramp, and increases the rate of return on manufacturing investment.

Failure analysis essentially requires an inverse mapping from an electrical test failure to a fault to a defect. It has been found that simultaneous current and voltage testing can greatly narrow the number of possible faults [83]. The probability of each defect causing the fault can then be used to narrow the list of possible defects. Ideally, a single defect type is pinpointed. If this is not possible, it is often very useful to localize the region on the chip where the defect occurred, reducing optical inspection time. As is the case with test generation, full details of the fault are necessary to translate them into a voltage/current test failure.

Normal production line testing is used to classify a chip as good or bad. If the chip is bad, it is discarded. But simply discarding bad chips loses information that might increase our understanding of defects and defect mechanisms. To extract this defect information, we need to add special-purpose test vectors and current measurements to the normal production test program. The test results can be put in the form of special bitmaps from which defect "signatures" can be extracted. By simulating testing of all common defects beforehand, we can build a library of defects and the signatures they cause. During production testing, the library is used to match signatures to defects. This was done for a double-poly CMOS SRAM [100]. As part of this work, fault probabilities are computed to determine all of the common memory array circuit faults. Circuit simulation was used to translate these faults into test pattern failures, and from test-pattern failures to the defect signature.

5.7.5 Design for Manufacturability

Design for manufacturability can be thought of as a passive form of defect tolerance. This can take the form of design rule development, design-specific layout selection, and yield-driven layout synthesis. Increasing widths and spacings in design rules will reduce the probability that a spot defect will cause a fault, but at the cost of increased die area. An optimum value maximizes the number of good die on a wafer [11, 104, 105]. For design rule development, only the critical area as a function of the design rules is needed. However, if the design incorporates redundancy, then the failure modes are of interest, since only some of them may be correctable. An example of SRAM failure modes as a function of the scale factor is shown in Figure 5-27. Ideally, the design and design rules should be developed to minimize critical area for the specific process and specific design. As shown in Figure 5-28, reducing design rules may reduce cell area faster than critical area rises, improving overall yield and die cost.

Yield can also be incorporated as a metric into layout synthesis tools. Several yield-driven routers have been developed [36, 106–117]. To maintain speed, these systems typically use a very simple geometrical method of computing critical area based on parallel wires and wire overlaps.

A problem related to design for manufacturability is design for testability. The layout can be modified to minimize the probability of faults that have low detectabil-

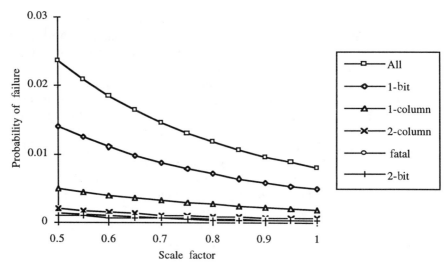

Figure 5-27 SRAM array failure probabilities due to extra first metal defects versus design rule scaling.

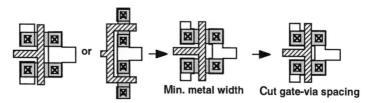

Figure 5-28 Defect sensitivity minimization.

ity [118–120]. For example, if a bridging fault is hard to detect, the two jets can be spaced farther apart, possibly at the expense of increased area. Given a metric such as $\Sigma_f (POF_f/POD_f)$ where POD_f is the probability of fault f being detected by the test set, the goal is to minimize the metric within the chip area constraints. Since fault detectability is computed in terms of a test set, the same fault details as required for test generation are used in design for testabiilty. However, in order to maintain speed during the synthesis process, simple fault critical area calculations are performed similar to those used in yield-driven routing.

5.8 SUMMARY AND RESEARCH DIRECTIONS

In this chapter, we have described techniques for computation of critical areas and probabilities of catastrophic faults based on knowledge of the product mask artwork, manufacturing process flow, and statistics of the process disturbances. The software that performs the fault and critical area analysis acts as a virtual fabrication line in that it simulates the production of chips in the presence of local manufacturing process disturbances. The results can then be used in design, manufacturing, and test applications, such as defect tolerance, manufacturing yield diagnosis, and test generation. We described the existing geometrical and Monte Carlo critical area analysis and fault probability algorithms which cover a spectrum of speed versus accuracy, and algorithm simplicity versus application generality.

The major open problems in critical area and fault probability computation are development of efficient contamination to defect and fault to failure mappings. In particular, applications such as test generation and process diagnosis are limited primarily by the expense of current fault to failure mapping techniques such as circuit-level fault simulation.

5.9 EXERCISES AND SOLUTIONS

5.1. A critical area computation for a pair of parallel wires would show the critical area for a short circuit due to an extra material defect increasing without bound as the defect diameter increases. Does this make sense? If not, why not? What sensible limits would you place on the critical area?

Solution: The critical area cannot increase without bound, since the chip has a finite size. An upper bound on critical area is the chip area. An alternative would be to compute an upper bound based on a maximum defect diameter.

5.2. All widely used geometrical critical area extraction tools use an approximation when dealing with ends and corners of wires. One such approximation is to use orthogonal polygon expansion, which is the same as assuming a square defect. Does this underestimate or overestimate the critical area? Draw an example of a structure that will have a large error in the critical area estimate when considering shorts caused by extra material defects.

Solution: This overestimates critical area since a circular defect is circumscribed by the square. The worst case is a structure that has a high corner to parallel wire ratio, since the error occurs at the corners and ends of wires. Two rows of squares offset by half a pitch, as shown in Figure 5-29, will have a high error. Note that a circular defect also overestimates critical area, since the low-pass properties of layer patterning cause wire corners to be rounded off.

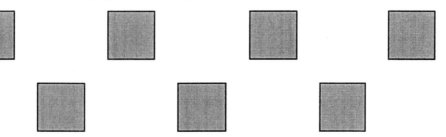

Figure 5-29 Structure with large critical area error for extra material defects.

5.3. One of the key problems in critical area analysis for opens caused by missing conducting material (extra insulating material) defects is reporting where the open occurred in the net. This information is required for applications such as test generation, defect tolerance, and process diagnosis. One method of reporting opens is to report the set of circuit element terminals that remain connected on each side of the break. What are some of the serious drawbacks of reporting opens in terms of element connectivity? What are some of the advantages of using this approach?

Solution: Opens on nets with many elements require large connectivity descriptions. For example, a clock network might connect two clock driver transistors with hundreds of thousands of latch transistors. Similarly, each memory row select or bit line typically connects thousands of transistors. Each open would require at a minimum listing all transistors on each side of the open. Since many unique partitions might occur, the memory required to describe all of them would be infeasibly large. In addition, the large networks will have a large critical area, so Monte Carlo sampling will encounter them often, at a high cost for each defect sample.

An advantage of reporting partitions is that the faulty circuit can be fed directly into a simulator, since the element terminals are easily relabeled with the nodes formed by the break. Another advantage of this approach is that no layout preprocessing needs to be done before analysis of an open. The net on each side of the break can simply be traversed, listing circuit element terminals as they are found.

5.4. Consider a simple design rule development problem. You are to come up with minimum metal spacing rules such that overall chip manufacturing costs will be minimized. Assume that the only defects that occur are extra metal defects that may cause short circuits. Also assume the defect size distribution in Equations 5-2 with $p = 5$ and $X_0 = 0.2$ micron, and that the minimum metal width is one micron. Assume that defects occur randomly (i.e., no spatial clustering).

Solution: Consider a "chip" that consists of parallel metal lines. The goal is to select a spacing that will result in the maximum number of good chips per wafer. For wafer area A_w and chip area A_{chip}, the goal is to maximize $F = (A_w/A_{chip})e^{-DA_{crit}}$ where D is the defect density and A_{crit} is the critical area weighted by the defect size distribution. If we assume a chip made up of N parallel wires with width w, spacing s, and length L, then $A_{chip} = NL(s + w)$. Ignoring end effects, then $A_{crit} = \int_0^\infty Cr(x)h(x)dx$, $Cr(x) = 0$ for $X_0 \leq x \leq s$, $Cr(x) = NL(x - s)$ for $s \leq x \leq 2s + w$, and $Cr(x) = A_{chip}$ for $x \geq 2s + w$. From Equation 5-5, the defect size distribution is $h(x) = (4X_0^4)/(x^5)$, $x \geq X_0$. Note that we assume that $s > X_0$, which will be the case for all real IC fabrication processes. The problem can be solved by substituting for A_{chip} and A_{crit} in the good chip equation, and finding the maximum of F with respect to s. As can be seen, increasing values of p will result in smaller optimum values of s. In practice, it turns out that global process disturbances limit s before spot defects do.

REFERENCES

[1] W. Maly, A. J. Strojwas, and S. W. Director, "VLSI Yield Prediction and Estimation: A Unified Framework," *IEEE Transactions on Computer-Aided Design of Integrated Circuits and Systems,* vol. CAD-5, no. 1, pp. 114–130, 1986.

[2] D. M. H. Walker, *Yield Simulation for Integrated Circuits,* Boston: Kluwer Academic Publishers, 1987.

References

[3] W. Maly, "Computer-Aided Design for VLSI Circuit Manufacturability," *Proceedings of the IEEE,* vol. 78, no. 2, pp. 356–392, 1990.

[4] J. Khare, "A New Methodology for Yield Simulation of Integrated Circuits," M.S. Thesis, Carnegie Mellon University, Department of Electrical and Computer Engineering, 1989.

[5] J. Pineda de Gyvez and S. M. Dani, "Modeling of 3-Dimensional Defects in Integrated Circuits," *IEEE International Workshop on Defect and Fault Tolerance in VLSI Systems,* Dallas, TX, pp. 197–206, 1992.

[6] H. Xue, C. Di, and J. A. G. Jess, "Fast Multi-Layer Critical Area Computation," *IEEE International Workshop on Defect and Fault Tolerance in VLSI Systems,* Venice, Italy, pp. 117–124, 1993.

[7] E. Bruls, "Quality and Reliability Impact of Defect Data Analysis," *IEEE Transactions on Semiconductor Manufacturing,* vol. 8, no. 2, pp. 121–129, 1995.

[8] G. A. Allan and A. J. Walton, "Critical Area Extraction of Extra Material Soft Faults," *IEEE International Workshop on Defect and Fault Tolerance in VLSI Systems,* Lafayette, LA, pp. 55–62, 1995.

[9] W. Maly, P. K. Nag, and P. Nigh, "Testing Oriented Analysis of CMOS ICs with Opens," *IEEE International Conference on Computer-Aided Design,* Santa Clara, CA, pp. 344–347, 1988.

[10] A. Meixner and W. Maly, "Fault Modeling for the Testing of Mixed Integrated Circuits," *IEEE International Test Conference,* pp. 564–572, 1991.

[11] W. Maly, "Modeling of Lithography Related Yield Losses for CAD of VLSI Circuits," *IEEE Transactions on Computer-Aided Design of Integrated Circuits and Systems,* vol. CAD-4, no. 3, pp. 166–177, 1985.

[12] R. Razdan and A. J. Strojwas, "Statistical Design Rule Developer," *IEEE International Conference on Computer-Aided Design,* Santa Clara, CA, pp. 315–317, 1985.

[13] C. H. Stapper, A. N. McLaren, and M. Dreckmann, "Yield Model for Productivity Optimization of VLSI Memory Chips with Redundancy and Partially Good Product," *IBM Journal of Research and Development,* vol. 24, no. 3, pp. 398–409, 1980.

[14] T. Storey, W. Maly, J. Andrews, and M. Miske, "Assessing CMOS Test Quality: Theory vs. Practice," SRC Technical Conference, San Jose, CA, pp. 387–390, 1990.

[15] Y. J. Kwon and D. M. H. Walker, "Contamination Control Using Production Test Data," *IEEE International Electronics Manufacturing Technology Symposium,* Austin, TX, pp. 70–76, 1995.

[16] J. B. Khare, W. Maly, S. Griep, and D. Schmitt-Landsiedel, "Yield-Oriented Computer-Aided Defect Analysis," *IEEE Transactions on Semiconductor Manufacturing,* vol. 8, no. 2, pp. 195–206, 1995.

[17] Y. J. Kwon and D. M. H. Walker, "Yield Learning via Functional Test Data," *IEEE International Test Conference,* Washington, DC, pp. 626–635, 1995.

[18] C. H. Stapper, "Defect Density Distribution for LSI Yield Calculations," *IEEE Transactions on Electron Devices,* vol. ED-20, no. 7, pp. 655–657, 1973.

[19] W. Maly and J. Deszczka, "Yield Estimation Model for VLSI Artwork Evaluation," *Electronics Letters,* vol. 19, no. 6, pp. 226–227, 1983.

[20] A. Ferris-Prabhu, "Modeling the Critical Area in Yield Forecasts," *IEEE Journal of Solid-State Circuits,* vol. SC-20, no. 4, pp. 874–878, 1985.

[21] A. V. Ferris-Prabhu, *Introduction to Semiconductor Device Yield Modeling.* Boston, Artech House, 1992.

[22] C. H. Stapper, "Modeling of Defects in Integrated Circuit Photolithographic Patterns," *IBM Journal of Research and Development,* vol. 28, no. 4, pp. 461–475, 1984.

[23] J. Khare, "Contamination-Defect-Fault Relationship–Modeling and Simulation," Ph.D. Thesis, Carnegie Mellon University, Department of Electrical and Computer Engineering, 1995.

[24] W. G. Oldham, S. N. Nandgankar, A. R. Neureuther, and M. O'Toole, "A General Simulator for VLSI Lithography and Etching Processes: Part I–Application to Projection Lithography," *IEEE Transactions on Electron Devices,* vol. ED-26, pp. 717–722, 1979.

[25] W. G. Oldham, A. R. Neureuther, C. Sung, J. L. Reynolds, and S. N. Nandgankar, "A General Simulator for VLSI Lithography and Etching Processes: Part II–Application to Deposition and Etching," *IEEE Transactions on Electron Devices,* vol. ED-27, pp. 1455–1459, 1980.

[26] J. Khare, C. S. Kellen, and W. Maly, Private Communication, 1994.

[27] P. Schvan, D. Y. Montuno, and R. Hadaway, "Yield Projection Based on Electrical Fault Distribution and Critical Structure Analysis," in *Defect and Fault Tolerance in VLSI Systems,* vol. 1, ed. by I. Koren, Amherst, MA, Plenum Press, 1989, pp. 117–127.

[28] I. A. Wagner and I. Koren, "An Interactive Yield Estimator as a VLSI CAD Tool," *IEEE International Workshop on Defect and Fault Tolerance in VLSI Systems,* Venice, Italy, pp. 167–174, 1993.

[29] D. M. H. Walker, "Yield Simulation for Integrated Circuits," Ph.D. Thesis, Carnegie Mellon University, Department of Computer Science, 1986.

[30] J. Ousterhout, "Corner Stitching: A Data-Structuring Technique for VLSI Layout Tools," *IEEE Transactions on Computer-Aided Design of Integrated Circuits and Systems,* vol. CAD-3, no. 1, pp. 87–100, 1984.

[31] D. Lang, Private Communication, 1979.

[32] I. Bubel, W. Maly, T. Waas, P. K. Nag, H. Hartmann, D. Schmitt-Landsiedel, and S. Griep, "AFFCCA: A Tool for Critical Area Analysis with Circular Defects and Lithography Deformed Layout," *IEEE International Workshop on Defect and Fault Tolerance in VLSI Systems,* Lafayette, LA, pp. 10–18, 1995.

[33] P. K. Nag and W. Maly, "Yield Estimation of VLSI Circuits," SRC Technical Conference, San Jose, CA, pp. 267–270, 1990.

[34] J. Pineda de Gyvez and C. Di, "IC Defect Sensitivity for Footprint-Type Spot Defects," *IEEE Transactions on Computer-Aided Design of Integrated Circuits and Systems,* vol. 11, no. 5, pp. 638–658, 1992.

[35] A. R. Dalal, P. D. Franzon, and M. J. Lorenzetti, "A Layout-Driven Yield Predictor and Fault Generator for VLSI," *IEEE Transactions on Semiconductor Manufacturing,* vol. 6, no. 1, pp. 77–82, 1993.

[36] H. Xue, C. Di, and J. A. G. Jess, "A Net-Oriented Method for Realistic Fault Analysis," *IEEE International Conference on Computer-Aided Design,* Santa Clara, CA, pp. 78–83, 1993.

[37] J. Pineda de Gyvez and J. A. G. Jess, "Systematic Extraction of Critical Areas from IC Layouts," *IEEE International Workshop on Defect and Fault Tolerance in VLSI Systems,* Tampa, FL, pp. 27–31, 1989.

[38] S. Gandemer, B. C. Tremintin, and J. J. Charlot, "Critical Area and Critical Levels Calculation in IC Yield Modeling," *IEEE Transactions on Electron Devices,* vol. 35, no. 2, pp. 158–166, 1988.

[39] M. Rivier, "Random Yield Simulation Applied to Physical Circuit Design," in *Yield Modelling and Defect Tolerance in VLSI,* ed. by W. R. Moore, W. Maly, and A. Strojwas, Bristol, UK, Adam Hilger Ltd., 1988, pp. 91–99.

[40] M. J. Lorenzetti, P. Magill, A. Dalal, and P. Franzon, "McYield: A CAD Tool for Functional Yield Projections for VLSI," *IEEE International Workshop on Defect and Fault Tolerance in VLSI Systems,* Grenoble, France, pp. 100–110, 1990.

[41] A. Jee and F. J. Ferguson, "Carafe: An Inductive Fault Analysis Tool for CMOS VLSI Circuits," *IEEE VLSI Test Symposium,* Atlantic City, NJ, pp. 92–98, 1993.

[42] J. J. H. T. de Sousa, F. M. Goncalves, and J. P. Teixeira, "Physical Design of Testable CMOS Digital Integrated Circuits," *IEEE Journal of Solid-State Circuits,* vol. 26, no. 7, pp. 1064–1072, 1991.

[43] J. Pineda de Gyvez, "IC Defect-Sensitivity–Theory and Computational Models for Yield Prediction," Ph.D. Thesis, Technische Universiteit Eindhoven, 1991.

[44] G. Han, J. P. de Gyvez, and E. Sanchez-Sinencio, "Optimal Manufacturable CNN Array Size for Time Multiplexing Schemes," *IEEE International Workshop on Cellular Neural Networks and Its Applications,* Seville, Spain, pp. 387–391, 1996.

[45] S. Perry, M. Mitchell, and D. Pilling, "Yield Analysis Modeling," *ACM/IEEE Design Automation Conference,* pp. 425–428, 1985.

[46] G. A. Allan and A. J. Walton, "Efficient Critical Area Algorithms and Their Applications to Yield Improvement and Test Strategies," *IEEE International Workshop on Defect and Fault Tolerance in VLSI Systems,* Montreal, Quebec, Canada, pp. 88–96, 1994.

[47] D. Feltham, J. Khare, and W. Maly, "A CAD Tool for Accurate Yield Estimation of Reconfigurable VLSI Circuits," CAD Center, Department of Electrical and Computer Engineering, Carnegie Mellon University CMUCAD-92-28, 1992.

[48] H. Walker and S. W. Director, "Yield Simulation for Integrated Circuits,"

IEEE International Conference on Computer-Aided Design, Santa Clara, CA, pp. 256–257, 1983.

[49] M. Chew and A. J. Strojwas, "Efficient Circuit Re-extraction for Yield Simulation Applications," *IEEE International Conference on Computer-Aided Design,* Santa Clara, CA, pp. 310–313, 1987.

[50] H. Walker and S. W. Director, "VLASIC: A Catastrophic Fault Yield Simulator for Integrated Circuits," *IEEE Transactions on Computer-Aided Design of Integrated Circuits and Systems,* vol. CAD-5, no. 4, pp. 541–556, 1986.

[51] D. D. Gaitonde, "Design and Applications of a Hierarchical Defect to fault Mapper," Ph.D. Thesis, Carnegie Mellon University, Department of Electrical and Computer Engineering, 1995.

[52] I. A. Wagner and I. Koren, "An Interactive VLSI CAD Tool for Yield Estimation," *IEEE Transactions on Semiconductor Manufacturing,* vol. 8, no. 2, pp. 130–138, 1995.

[53] C. H. Stapper, "Yield Model for Fault Clusters Within Integrated Circuits," *IBM Journal of Research and Development,* vol. 28, no. 5, pp. 636–639, 1984.

[54] I. Koren, Z. Koren, and C. H. Stapper, "Analysis of Defect Maps of Large Area VLSI ICs," *IEEE International Workshop on Defect and Fault Tolerance in VLSI Systems,* Dallas, TX, pp. 267–276, 1992.

[55] I. Chen and A. J. Strojwas, "RYE: A Realistic Yield Simulator for VLSIC Structural Failures," *IEEE International Test Conference,* pp. 31–42, 1987.

[56] D. Gaitonde and D. M. H. Walker, "Test Quality and Yield Analysis Using the DEFAM Defect to Fault Mapper," *IEEE International Conference on Computer-Aided Design,* Santa Clara, CA, pp. 202–205, 1993.

[57] M. Faust, "Oracle: A Design Rule Checking Program," Carnegie Mellon University, Department of Computer Science, VLSI Document V096, Aug. 1981.

[58] M. Blatt, "Yield Evaluation of a Soft-Configurable WSI Switch Network," *IEEE International Workshop on Defect and Fault Tolerance in VLSI Systems,* Tampa, FL, pp. 64–73, 1989.

[59] M. Blatt, "Effects of Switch Failure on Soft-Configurable WSI Yield," *IEEE International Conference on Wafer Scale Integration,* San Francisco, CA, pp. 152–159, 1990.

[60] A. Gupta and R. Hon, "HEXT: A Hierarchical Circuit Extractor," *Journal of VLSI and Computer Systems,* vol. 1, no. 1, pp. 23–39, 1983.

[61] J. Ousterhout, G. Hamachi, R. Mayo, W. Scott, and G. Taylor, "The Magic VLSI Layout System," *IEEE Design and Test of Computers,* vol. 2, no. 1, pp. 19–30, 1985.

[62] N. Hedenstierna and K. O. Jeppson, "The Halo Algorithm–An Algorithm for Hierarchical Design of Rule Checking of VLSI Circuits," *IEEE Transactions on Computer-Aided Design of Integrated Circuits and Systems,* vol. 12, no. 2, pp. 265–272, 1993.

[63] G. Taylor and J. Ousterhout, "Magic's Incremental Design-Rule Checker," *ACM/IEEE Design Automation Conference,* pp. 160–165, 1984.

References

[64] T. Whitney, "A Hierarchical Design-Rule Checking Algorithm," *Lambda*, vol. 2, no. 1, pp. 40–43, 1981.

[65] B. Hon, "The Hierarchical Analysis of VLSI Designs," Carnegie Mellon University, Department of Computer Science, VLSI Document V073, December 1980.

[66] M. Marek-Sadowska and W. Maly, "A Hierarchical Layout Description for Artwork Analysis of VLSI IC," *IEEE International Conference on Circuits and Computers*, New York, NY, pp. 419–422, 1982.

[67] L. Scheffer, "A Methodology for Improved Verification of VLSI Designs Without Loss of Area," *Second Caltech Conference on VLSI*, Pasadena, CA, pp. 299–309, 1981.

[68] G. Tarolli and W. Herman, "Hierarchical Circuit Extraction with Detailed Parasitic Capacitance," *ACM/IEEE Design Automation Conference*, pp. 337–345, 1983.

[69] M. Newell and D. Fitzpatrick, "Exploiting Structure in Integrated Circuit Design Analysis," *MIT Conference on Advanced Research in VLSI*, pp. 84–92, 1982.

[70] W. Maly and M. Marek-Sadowska, Private Communication, 1982.

[71] R. Hon, "The Hierarchical Analysis of VLSI Designs," PhD Thesis, Carnegie Mellon University, Department of Computer Science, 1983.

[72] M. Tucker and L. Scheffer, "A Constrained Design Methodology for VLSI," *VLSI Design*, vol. 3, no. 3, pp. 60–65, 1982.

[73] P. K. Nag and W. Maly, "Hierarchical Extraction of Critical Area for Shorts in Very Large ICs," *IEEE International Workshop on Defect and Fault Tolerance in VLSI Systems*, Lafayette, LA, pp. 19–27, 1995.

[74] P. K. Nag, "Yield Forecasting," Ph.D. Thesis, Carnegie Mellon University, Department of Electrical and Computer Engineering, 1996.

[75] G. A. Allan and A. J. Walton, "Hierarchical Critical Area Extraction with the EYE Tool," *IEEE International Workshop on Defect and Fault Tolerance in VLSI Systems*, Lafayette, LA, pp. 28–36, 1995.

[76] P. K. Nag and W. Maly, Private Communication, 1993.

[77] D. D. Gaitonde and D. M. H. Walker, "Hierarchical Mapping of Spot Defects to Catastrophic Faults—Design and Applications," *IEEE Transactions on Semiconductor Manufacturing*, vol. 8, no. 2, pp. 167–177, 1995.

[78] D. M. H. Walker, C. S. Kellen, D. M. Svoboda, and A. J. Strojwas, "The CDB/HCDB Semiconductor Wafer Representation Server," CAD Center, Carnegie Mellon University, Department of Electrical and Computer Engineering CMUCAD-91-48, July 1991.

[79] D. M. H. Walker, C. S. Kellen, D. M. Svoboda, and A. J. Strojwas, "The CDB/HCDB Semicondutor Wafer Representation Server," *IEEE Transactions on Computer-Aided Design of Circuits and Systems*, vol. 12, no. 2, pp. 283–295, 1993.

[80] R. S. Collica, X. J. Dietrich, R. Lambracht, and D. G. Lau, "A Yield Enhance-

ment Methodology for Custom VLSI Manufacturing," *Digital Technical Journal,* vol. 4, no. 2, pp. 83–99, 1992.

[81] F. J. Ferguson and J. P. Shen, "A CMOS Fault Extractor for Inductive Fault Analysis," *IEEE Transactions on Computer-Aided Design of Integrated Circuits and Systems,* vol. 7, no. 11, pp. 1181–1194, 1988.

[82] D. B. I. Feltham and W. Maly, "Behavioral Modeling of Physical Defects in VLSI Neural Networks," *IEEE International Workshop on Defect and Fault Tolerance in VLSI Systems,* Grenoble, France, pp. 228–251, 1990.

[83] S. Naik and W. Maly, "Computer-Aided Failure Analysis of VLSI Circuits using Iddq Testing," *IEEE VLSI Test Symposium,* Atlantic City, NJ, pp. 106–108, 1993.

[84] W. Maly and P. Nigh, "Built-In Current Testing–Feasibility Study," *IEEE International Conference on Computer-Aided Design,* Santa Clara, CA, pp. 340–343, 1988.

[85] S. Naik, F. Agricola, and W. Maly, "Failure Analysis of High-Density CMOS SRAMs Using Realistic Defect Modeling and Iddq Testing," *IEEE Design and Test of Computers,* vol. 10, no. 2, pp. 13–23, 1993.

[86] T. Storey and W. Maly, "CMOS Bridging Fault Detection," *IEEE International Test Conference,* pp. 842–851, 1990.

[87] D. D. Gaitonde, D. M. H. Walker, and W. Maly, "Accurate Yield Estimation of Circuits with Redundancy," *IEEE International Workshop on Defect and Fault Tolerance in VLSI Systems,* Lafayette, LA, pp. 155–163, 1995.

[88] G. Saucier, J. L. Patry, E. F. Kouka, T. Midwinter, P. Ivey, M. Huch, and M. Glesner, "Defect Tolerance in a Wafer Scale Array for Image Processing," in *Defect and Fault Tolerance in VLSI Systems,* vol. 1, ed. by I. Koren, Amherst, MA, Plenum Press, 1989, pp. 327–338.

[89] S. Lakkapragada and D. M. H. Walker, "Defect-Tolerant Processor Arrays," *IEEE International Conference on Wafer Scale Integration,* San Francisco, CA, pp. 228–237, 1995.

[90] D. B. I. Feltham and W. Maly, "Physically Realistic Fault Models for Analog CMOS Neural Networks," *IEEE Journal of Solid-State Circuits,* vol. 26, no. 9, pp. 1223–1229, 1991.

[91] R. Naik and D. M. H. Walker, "Large Integrated Crossbar Switch," *IEEE International Conference on Wafer Scale Integration,* San Francisco, CA, pp. 217–227, 1995.

[92] D. D. Gaitonde, W. Maly, and D. M. H. Walker, "Fault Probability Prediction for Array Based Designs," *IEEE International Workshop on Defect and Fault Tolerance in VLSI Systems,* Boston, MA, pp. 30–38, 1996.

[93] J. C. Harden, J. J. Wang, and B. W. Tebbs, "A High-Level WSI Yield Simulation System," *IEEE International Conference on Wafer Scale Integration,* San Francisco, CA, pp. 235–243, 1989.

[94] C. H. Stapper, "Spatial Fault Simulation and the Saturation Effect," *IEEE International Workshop on Defect and Fault Tolerance in VLSI Systems,* Dallas, TX, pp. 187–196, 1992.

References

[95] D. Gaitonde, J. Khare, D. M. H. Walker, and W. Maly, "Estimation of Reject Ratio in Testing of Combinatorial Circuits," *IEEE VLSI Test Symposium,* Atlantic City, NJ, pp. 319–325, 1993.

[96] J. J. T. Sousa, "Defect Level Estimation for Digital ICs," *IEEE International Workshop on Defect and Fault Tolerance in VLSI Systems,* Dallas, TX, pp. 32–41, 1992.

[97] P. Gangatirkar, R. Presson, and L. Rosner, "Test/Characterization Procedures for High Density Silicon RAMs," *IEEE International Solid-State Circuits Conference,* pp. 62–63, 1982.

[98] W. Maly, B. Trifilo, R. Hughes, and A. Miller, "Yield Diagnosis Through Interpretation of Tester Data," *IEEE International Test Conference,* pp. 10–20, 1987.

[99] J. Khare and W. Maly, "Inductive Contamination Analysis (ICA) with SRAM Application," *IEEE International Test Conference,* Washington, DC, pp. 552–560, 1995.

[100] W. Maly and S. Naik, "Process Monitoring Oriented Testing," *IEEE International Test Conference,* pp. 527–532, 1989.

[101] S. Griep, J. Khare, R. Lemme, U. Papenberg, D. Schmitt-Landsiedel, W. Maly, D.M.H. Walker, J. Winnerl, and T. Zettler, "Application of Defect Simulation as a Tool for More Efficient Failure Analysis," *Quality and Reliability Engineering International,* vol. 10, no. 4, pp. 297–302, 1994.

[102] J. Khare and W. Maly, "Rapid Failure Analysis Using Contamination-Defect-Fault (CDF) Simulation," *IEEE International Symposium on Semiconductor Manufacturing,* Austin, TX, pp. 136–141, 1995.

[103] J. Khare, S. Griep, H. D. Oberie, W. Maly, D. Schmitt-Landsiedel, U. Kollmer, and D.M.H. Walker, "Key Attributes of an SRAM Testing Strategy Required for Effective Process Monitoring," *IEEE Memory Technology, Design, and Testing Workshop,* San Jose, CA, pp.1–6, 1993.

[104] D. M. H. Walker, "Yield Analysis for Fault-Tolerant Arrays," CAD Center, Carnegie Mellon University, Department of Electrical and Computer Engineering, Technical Report CMUCAD-88-46, Nov. 1988.

[105] R. Razdan and A. J. Strojwas, "A Statistical Design Rule Developer," *IEEE Transactions on Computer-Aided Design of Integrated Circuits and Systems,* vol. CAD-5, no. 4, pp. 508–520, 1986.

[106] V. K. R. Chiluvuri and I. Koren, "New Routing and Compaction Strategies for Yield Enhancement," *IEEE International Workshop on Defect and Fault Tolerance in VLSI Systems,* Dallas, TX, pp. 325–334, 1992.

[107] V. K. R. Chiluvuri and I. Koren, "Topological Optimization of PLAs for Yield Enhancement," *IEEE International Workshop on Defect and Fault Tolerance in VLSI Systems,* Venice, Italy, pp. 175–182, 1993.

[108] S. Y. Kuo, "YOR: A Yield-Optimizing Routing Algorithm by Minimizing Critical Areas and Vias," *IEEE Transactions on Computer-Aided Design of Integrated Circuits and Systems,* vol. 12, no. 9, pp. 1303–1311, 1993.

[109] A. Pitaksanonkul, S. Thanawastien, C. Lursinsap, and J. A. Gandhi, "DTR: A Defect-Tolerant Routing Algorithm," *ACM/IEEE Design Automation Conference*, Las Vegas, NV, pp. 795–797, 1989.

[110] E. P. Huijbregts, H. Xue, and J. A. G. Jess, "Routing for Reliable Manufacturing," *IEEE Transactions on Semiconductor Manufacturing*, vol. 8, no. 2, pp. 188–194, 1995.

[111] K. Balachandran, "A Yield Enhancing Router," *IEEE International Symposium on Circuits and Systems*, pp. 1936–1939, 1991.

[112] H. Xue, E. P. Huijbregts, and J. A. G. Jess, "Routing for Manufacturability," *Design Automation Conference*, San Diego, CA, pp. 402–406, 1994.

[113] A. Tyagi, M. Bayoumi, and P. Manthravadi, "Yield Enhancement in the Routing Phase of Integrated Circuit Layout Synthesis," *IEEE International Conference on Wafer Scale Integration*, San Francisco, CA, pp. 52–60, 1994.

[114] N. Maldonado, G. Andrus, A. Tyagi, M. Madani, and M. Bayoumi, "A Post-Processing Algorithm for Short-Circuit Defect Sensitivity Reduction in VLSI Layouts," *IEEE International Conference on Wafer Scale Integration*, San Francisco, CA, pp. 288–297, 1995.

[115] V.K.R. Chiluvuri, I. Koren, and J. L. Burns, "The Effect of Wire Length Minimization on Yield," *IEEE International Workshop on Defect and Fault Tolerance in VLSI Systems*, Montreal, Quebec, Canada, pp. 97–105, 1994.

[116] Z. Chen and I. Koren, "Layer Assignment for Yield Enhancement," *IEEE International Workshop on Defect and Fault Tolerance in VLSI Systems*, Lafayette, LA, pp. 173–180, 1995.

[117] C. Bamji and E. Malavasi, "Enhanced Network Flow Algorithm for Yield Optimization," *Design Automation Conference*, Las Vegas, NV, pp. 746–751, 1996.

[118] D. Feltham, J. Khare, and W. Maly, "Design for Testability View of Placement and Routing," *European Design Automation Conference*, Hamburg, Germany, pp. 382–387, 1992.

[119] J. Khare, D. Feltham, and W. Maly, "Accurate Estimation of Defect-Related Yield Loss in Reconfigurable VLSI Circuits," *IEEE Journal of Solid-State Circuits*, vol. 28, no. 2, pp. 146–156, 1993.

[120] J. Khare, S. Mitra, P. K. Nag, W. Maly, and R. A. Rutenbar, "Testability Oriented Channel Routing," *IEEE International Conference on VLSI Design*, New Delhi, India, pp. 208–213, 1995.

6
Statistical Methods of Parametric Yield and Quality Enhancement

Maciej Styblinski

Manufacturing process variations and environmental effects (such as temperature) result in variations in the values of IC elements and parameters, and lead to the loss of manufacturing yield and inferior circuit performance. Manufacturing yield is defined as the percentage of the total number of products manufactured that fulfill both *functional* and *parametric* performance requirements.[1] *Functional* circuit performance represents the ability of the circuit to perform desired functions. Catastrophic (or "hard") circuit failures, such as shorts (or open-circuit faults) caused by particle wafer contamination, will completely eliminate some of the circuit functions, thus decreasing the part of the overall yield called the *functional* yield. Modeling of this part of yield was described in Chapter 4. *Parametric* circuit performance is a measure of circuit quality and is represented by measurable performance functions such as gain, delay, and bandwidth. The part of yield related to the parametric circuit performance is called *parametric* yield and is discussed in this section. The actual manufacturing yield is equal to the product of the functional and parametric yield [16], so it is smaller than either of the two. Usually, parametric yield is estimated and optimized during the IC electrical design, while functional yield is used to characterize the manufacturing process itself and to modify the IC layout in such a way that functional yield is improved.[2]

In this chapter, methods of parametric yield modeling and optimization are discussed, including yield generalization into a broader concept of the overall circuit

---

[1] For more detailed yield definitions involving different types of yield (e.g., design yield, wafer yield, probe yield, processing yield), see [16].
[2] Layout design (i.e., device spacing, location, size) also has influence on the mismatch of device parameter values (as will be discussed in Section 6.3), so it can affect the parametric yield as well.

quality. Statistical methods of parametric yield and quality enhancement take the IC element variations into account during the circuit design process. The major objective is to reduce the circuit performance sensitivity with respect to the element and parameter changes, that is, to reduce performance variability (which already leads to yield enhancement) and to further improve the parametric yield by design centering. Design for high yield and quality is part of the general area of Statistical Circuit Design.

6.1 PROBLEMS AND METHODOLOGIES OF STATISTICAL CIRCUIT DESIGN

A broad class of problems exists in this area: statistical *analysis* involves studying the effects of element and parameter variations on circuit performance. It applies statistical techniques, such as Monte Carlo (MC) simulation [42] and the Propagation of Variance (POV) method [49], to estimate performance variability. *Design Centering* attempts to find a center of the acceptability region [16] such that manufacturing yield is enhanced. *Direct methods* of yield optimization use parametric yield as the objective function and utilize various statistical (or mixed statistical/deterministic) algorithms to find the yield maximum in the space of designable circuit/process parameters. *Indirect* yield enhancement methods do not use parametric yield as the objective function. Instead, some other, easier-to-calculate objective function is defined, whose minimization (or maximization) leads to parametric yield *improvement* but not necessarily to the maximum possible yield. *Design Centering and Tolerance Assignment* (used mostly for discrete circuits) attempts to find the design center, with simultaneous optimal assignment of circuit element tolerances, minimizing some suitable cost function and providing 100% yield (worst-case design) [5, 10]. To solve this problem, mostly deterministic algorithms of nonlinear programming are used. Worst-case design is often too pessimistic and too conservative, leading to substantial *overdesign*. This fact motivates the use of *statistical techniques*, which provide a much more realistic estimation of the actual performance variations and lead to superior designs.

Stringent requirements of the contemporary VLSI design prompted an increased interest in the practical application of these techniques. The most significant philosophy introduced recently in this area is statistical Design for Quality (DFQ). It was stimulated by the practical appeal of the DFQ methodologies introduced by Taguchi [38, 75], oriented toward "on-target" design with performance variability minimization. This chapter discusses some of the techniques of parametric yield optimization and their Design for Quality generalizations.

6.2 CIRCUIT VARIABLES, PARAMETERS, AND PERFORMANCES

The majority of statistical circuit design problems can be formulated introducing three classes of parameters (or variables): designable parameters x, random parameters θ, and circuit variables e. The circuit variables can be more generally called *simulator variables,* since they represent the parameters and variables used directly as the input parameters to circuit, process, or system simulators.

6.2.1 Designable Parameters

Designable parameters, represented by the *n*-dimensional vector[3] $x = (x_1, \ldots, x_n) \in R^n$, are used by circuit designers as "decision" variables during circuit design and optimization. Typical examples include nominal values of passive elements, nominal MOS transistor mask dimensions, and process control parameters. Very often, x parameters are identical to a subset of the simulator variables e, but in some cases x are just *dummy optimization variables,* transformed to the actual values of e variables. For example, lengths L_i, $i = 1, \ldots, 4$ of four MOS transistors can be related to one dummy (e.g., x_7) optimization variable: $L_i = a_i x_7$, $i = 1, \ldots, 4$, where a_i is a coefficient, so the L_i values change proportionally (i.e., "track") during the optimization process.

6.2.2 Random Variables

The *t*-dimensional vector of random variables (or "noise" parameters in Taguchi's terminology [38]) is denoted as $\theta = (\theta_1, \ldots, \theta_t) \in R^t$. It represents statistical R, L, C element variations, disturbances or variations of manufacturing process parameters, variations of device model parameters, such as t_{ox} (oxide thickness), V_{TH} (threshold voltage), and environmental effects, such as temperature and supply voltages. Usually, θ represents *principal random variables,* selected to be *statistically independent* such that all other random parameters can be related to them through some *statistical models.* Statistical parameters of the θ vector, such as expectations and standard deviations, are often assumed independent of x and *fixed*, but dependent statistical models for θ can also be constructed. θ parameters can be defined in several ways: (a) they are often identical to some circuit (simulator) variables e, (b) they represent *process-level disturbances*, or (c) they are some dummy random variables. In the last two cases, the actual values of the e variables are obtained from suitable statistical models (see below).

The probability density function (p.d.f.) of θ parameters will be denoted as $f_\theta(\theta)$.

6.2.3 Circuit (Simulator) Variables

These variables represent parameters and variables used as input parameters to a circuit simulator, such as SPICE, or to a process simulator. They are represented as the *c*-dimensional vector $e = (e_1, \ldots, e_c) \in R^c$. Specific examples of e variables include R, L, C elements, gate widths W_j and lengths L_j of MOS transistors, device model parameters, and so on. If a process simulator is used, they are process related control, physical and random parameters, available to the user. The e vector contains

---

[3] Vectors are denoted by lower-case letters without subscripts or superscripts, so $e \in R^n$ denotes the vector $(e_1, \ldots, e_n)$ of real numbers in *n*-dimensional space ($\in$ means "belongs to").

only those variables that are *directly related* to the x and θ vectors.[4] This relationship is, in general, expressed as

$$e = e(x, \theta) \qquad (6\text{-}1)$$

The p.d.f. of θ is transformed into $f_e(e)$, the p.d.f. of e. This p.d.f. can be *singular*, that is, defined in a certain *subspace* of the e-space (see examples below). Moreover, it can be very complicated, with highly nonlinear statistical dependencies between different parameters, so it is very difficult to represent it directly as a p.d.f. of e. In the majority of cases, the analytic form of $f_e(e)$ is not known. For that reason, techniques of *statistical modeling* are used (see Section 6.3).

6.2.4 Circuit Performances

The vector of circuit performance indices (or simply performances) is defined as the m-dimensional vector $y = (y_1, \ldots, y_m) \in R^m$. Its elements can be gain, bandwidth, slew rate, signal delay, circuit response for a single frequency or time, and other characteristics of a circuit or system. Each of the performances y_j is a function of the vector of circuit elements e: $y_j = y_j(e) = y_j[e(x, \theta)]$. These transformations are most often not directly known in analytic form, and a circuit simulator (such as SPICE) must be used to find the values of y_j's corresponding to the given values of x and θ. The overall simulator time required to determine all the performances of y can be substantial for large circuits. This is the major limiting factor for the practical application of statistical circuit design techniques. To circumvent this problem, new *statistical macromodeling* and *behavioral modeling* techniques are being introduced [39, 40].

The y_i parameters can be obtained directly from a circuit simulator or by post-processing the simulator output (e.g., post-processing of a frequency response is needed to calculate the circuit bandwidth, which is one of the performance parameters). Dependencies between x, θ, e and y variables are shown schematically in Figure 6-1.

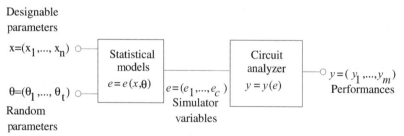

Figure 6-1 Representation of the dependencies between variables and parameters used in statistical circuit design. All parameters shown are vectors.

---

[4] This means that, for example, some SPICE parameters will always be *fixed*.

6.3 STATISTICAL MODELING OF CIRCUIT (SIMULATOR) VARIABLES

Statistical modeling is the process of finding a suitable transformation $e = e(x, \theta)$, such that given the distribution of θ, the distribution of e can be generated. The transformation $e = e(x, \theta)$ can be described by closed-form analytical formulas or by a computer algorithm.

6.3.1 Passive Discrete RLC Elements

For *discrete active RLC circuits* (e.g., such as a common emitter amplifier built with discrete components on a printed circuit board), the statistically perturbed RLC elements are represented by the model: $e_i = x_i + \theta_i$. In this formula, θ represents *absolute* element spreads whose expected (average) value $E\{\theta\} = 0$ and x_i is the *nominal* value of e_i, selected in what follows as the expected value of e_i. This implies that the variance of e_i, $\text{var}\{e_i\} = \text{var}\{\theta_i\}$, $E\{e_i\} = x_i$, and the distribution of e_i is the same as that of θ_i, with the expected value shifted by x_i. Alternatively, if θ_i represents *relative* element spreads, $e_i = x_i(1 + \theta_i)$, where $E\{\theta_i\} = 0$. Therefore, $E\{e_i\} = x_i$; $\text{var}\{e_i\} = x_i^2 \text{var}\{\theta_i\}$, that is, the standard deviations σ_{e_i} and σ_{θ_i} are related: $\sigma_{e_i} = x_i \sigma_{\theta_i}$, or $\frac{\sigma_{e_i}}{E\{e_i\}} = \frac{\sigma_{e_i}}{x_i} = \sigma_{\theta_i}$. This means that with fixed σ_{θ_i}, the *relative* standard deviation of e_i is constant, independent of the nominal value x_i. In practice, such models are often used for passive RLC elements with their standard deviations expressed as percentages of the element nominal values. Both forms of e_i indicate that each e_i is directly associated with its corresponding θ_i and x_i, and that there is one-to-one mapping between e_i and θ_i. These dependencies are important, since many of the early yield optimization algorithms were developed assuming that $e_i = x_i + \theta_i$. A typical p.d.f. for discrete elements is shown in Figure 6-2, before "binning" into different categories (the whole curve) and after binning into $\pm 1\%$, $\pm 5\%$ and $\pm 10\%$ resistors (the shaded and white areas: e.g., the $\pm 10\%$ resistors will have the distribution characterized by the external shaded areas, with a $\pm 5\%$ "hole" in the middle).

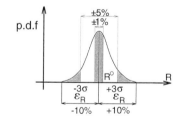

Figure 6-2 A typical probability function (p.d.f.) of a discrete resistor before and after "binning."

Usually, passive discrete elements are statistically independent[5] as shown in Figure 6-3. The cross section shown is often called a *level set, tolerance body, tolerance region,* or *norm body* [16] (since it is determined by various vector norms). Typical examples of that region include a hyper-box, a hypersphere, and a hyper-ellipsoid. For discrete circuits, it is defined as a set $L_\varepsilon(x, \alpha)$ of element values for which the element p.d.f. $f_\theta(\theta)$ is larger than a prescribed value α:

$$L_\varepsilon(x, \alpha) = \{e = x + \theta \in R^n \mid f_\theta(\theta) \geq \alpha\} \qquad (6\text{-}2)$$

---

[5] But if, for instance, $R_L(R_C)$ is a loss resistance of an inductor L (capacitor C), then L and $R_L(C$ and $R_C)$ *are* correlated.

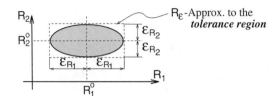

Figure 6-3 A level set (cross section) of a p.d.f. function for two discrete resistors after manufacturing.

The value of α is selected so that the probability of the element values falling into the tolerance body is equal to, for example, 95%, (i.e., the tolerance body represents 95% of the entire element population). Alternatively, the tolerance region can be approximated by a (hyper-) box R_δ shown in the figure, with the sides equal to $2\ \varepsilon_i$ ($2\ \varepsilon_{R_i}$ in the figure). Figure 6-3 also shows that the dependence $e_i = x_i + \theta_i$ is equivalent in this case to $R_i = R_i^0 + \Delta R_i$ (where $x_i = R_i^0$ is the nominal value and $\theta = \Delta R_i$).

6.3.2 Passive Integrated RLC Elements

All passive IC elements of the same type (e.g., diffused resistors, oxide capacitors, thin-film inductors in microwave ICs) are manufactured simultaneously in the same processing steps. Consequently, they are strongly correlated, as shown in Figure 6-4a for IC resistors. Statistical modeling in this case is performed in the same way as for active IC devices (including global variations and mismatches), described in Section 6.3.4.

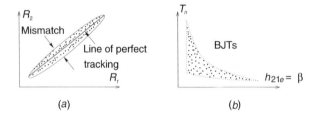

Figure 6-4 Statistical dependencies between circuit elements and device model parameters: (a) linear, between two integrated resistors R_1 and R_2; (b) nonlinear, between the base transit time T_n and the current gain $h_{21e} = \beta$ for bipolar junction transistors (same for discrete and integrated circuit).

6.3.3 Single Active Device Modeling for Discrete and Integrated Circuits

Statistical models for *single* active devices, such as bipolar or MOS transistors, are constructed in a similar way for the discrete and integrated implementations, since manufacturing of discrete and integrated devices is also quite similar. In this case, a subset of e parameters represents the device model parameters. Usually, they are strongly correlated for each single device, but for discrete circuits there are no correlations between the model parameters of *different* devices. This creates a problem for statistical design of discrete circuits, since each separate device must be represented by a separate statistical model, so the number of random variables that need to be handled is proportional to the number of devices used. For some types of ICs (e.g., a broad class of digital circuits), it is sufficiently accurate to assume that device mismatches are not important; that is, within the same chip there is 100% matching (correlation) between the model parameters of *different* devices of the same type (e.g., n-type MOS transistors). In such cases, the statistical modeling process is the simplest possible, since the same statistical model is used for all devices of the same type on the chip (e.g., for a digital CMOS implementation, only two common statistical models—for p-type and n-type devices—are used).

In contrast to this situation, for a broad class of analog ICs (e.g., op-amps, multipliers), a 100% correlation model is not acceptable, since mismatches between model parameters of different devices are critical for the circuit operation. Consequently, the mismatch modeling problem needs to be addressed, as discussed in Section 6.3.4.

Statistical modeling for a single discrete or integrated active device is performed as follows: each of the device model parameters e_d is related through a specific model $e_d = e_d(x, \theta)$ to the vector of θ parameters,[6] representing *principal random variables*, which are themselves often some device model parameters (such as substrate doping or oxide thickness of MOS transistors), or some *dummy* random variables. For example, in the Bipolar Junction Transistor (BJT) empirical statistical model introduced in [4], the base transient time T_n is modeled as follows:

$$e_d = e_d(\theta_1, \theta_2) \equiv T_n(\beta, X_{r5}) = \left(a + \frac{b}{\sqrt{3}}\right)(1 + cX_{r5}) \qquad (6\text{-}3)$$

That is, it is the function of the current gain $\theta_1 = \beta$ (the principal random variable, affecting the majority of the BJT model parameters) and $\theta_2 = X_{r5}$ (a dummy random variable, uniformly distributed in the interval $[-1, 1]$, independent of β and having no physical meaning); a, b, c are empirically selected constants (see Figure 6-4b).

A *systematic* method of constructing statistical device models was proposed in [13] and applied to statistical modeling of MOS transistors. Its main steps are as follows: First, the matrix of correlation coefficients is calculated from the measured (extracted) model parameter values for a specific device population,[7] and then, Principal Component Analysis (PCA) [45] is used. PCA involves coordinate rotation, which permits identification of a smaller number (eight in the MOS transistor case considered) of important, *uncorrelated* Principal Components. They accounted for 96% of the total variability of all model parameters. Factor Analysis (FA) [27, 45] involving eight (yet unknown) Common Factors and specific (local) variations was used next, to discover if a specific interpretation of the Common Factors could be found. Using the so-called varimax factor rotation (leading to strong correlation of just *one* Common Factor with *one* original parameter), the following Common Factors $F_1, \ldots, F_8$ were identified: t_{ox} (oxide thickness common to n- and p-type transistors) $N_{SUB,n}$, $N_{SUB,p}$ (n- and p-type substrate doping), ΔL_n, ΔL_p (length reduc-

---

[6] Observe that these models are parametrized by x: for example, the MOS transistor model parameters e_d will also depend on the device length L and width W.

[7] If needed, a nonlinear (usually power) transformation of some selected model parameters is performed, in order to obtain normal distributions for the transformed parameters; then the correlation matrix for the transformed parameters is constructed. In the specific case considered in [13], no transformations were required.

tion),[8] ΔW_n, ΔW_p (width reduction—for narrow transistors only), and XJ_p (p-type junction depth). All the other transistor model parameters were related to the $F_1, \ldots, F_8$ factors through quadratic (or linear in simplified models) regression formulas. The resulting models were able to represent—with a high level of accuracy—the strong nonlinear statistical dependencies existing between some model parameters.

6.3.4 Global and Local (Mismatch) Models for Integrated Circuits

For discrete circuits, only the passive elements are designable (in the majority of cases), and each designable parameter x_i has a random variable θ_i added to it; that is, x and θ are in *the same space* in which the p.d.f. $f_e(e) = f_\theta(x + \theta)$ is defined. This is not the case for *integrated circuits,* since most of the parameters involved are strongly correlated. Therefore, the distribution of θ parameters in the actual models used is often limited to a certain *subspace* of the entire e-space; that is, $f_e(e)$ is singular, and the formula $e_i = x_i + \theta_i$ does not hold.

EXAMPLE 6-1

Consider a subset of all e parameters representing the gate lengths $x_1 = L_1$, $x_2 = L_2$ (designable parameters) of two MOS transistors T_1, T_2 located in close proximity on a chip and an independent designable parameter $x_3 \equiv I_{\text{bias}}$, representing the transistor bias current. Assume also that $\theta \equiv \Delta L$ is the technological gate length reduction (a *common* or global random parameter), with $E\{\Delta L\} = 0$. ΔL changes the same way for T_1 and T_2 (since T_1 and T_2 are located very close to each other on the chip), which results in ideal matching of $\Delta L = \Delta L_1 = \Delta L_2$;[9] that is, $e_1 \equiv x_1 + \theta \equiv L_1 - \Delta L$, $e_2 \equiv x_2 + \theta \equiv L_2 - \Delta L$, $e_3 \equiv x_3 \equiv I_{\text{bias}}$. The only random variable in this model is ΔL, as shown in Figure 6-5. The p.d.f. $f_e(e)$ is in this case defined only in the *one-dimensional* subspace of the e-space (i.e., all the realizations of the vector e are located on the thick line in Figure 6-5). The major consequence of this difference is that, in general, many yield optimization algorithms developed for discrete circuits, using the explicit assumption that $e_i = e_i + \theta_i$, cannot be used for IC design.

In general, statistical IC modeling is concerned with representing both the *global* parameter variations and parameter *mismatches* that occur between different devices on the same chip. Though not critical for a broad class of digital circuits, element mismatches are of utmost importance for analog ICs. A mismatch between parameters, e_i, e_j is usually modeled introducing two additional independent *local mismatch variables* δ_i, δ_j, representing small local random deviations of the global variations from their ideal value θ, that is, $e_i \equiv x_i + \theta + \delta_i$. In this model, θ represents a global or *common factor,* affecting *both* parameters, and δ_i, δ_j represents *specific* (local) *variations*. The model just presented is a special (simple, one-factor) case of the Factor Analysis (FA) [27, 45] model of correlations between e_i and e_j. In practice, the variances of the global and local parts of the parameters e_i, e_j are determined indirectly by estimating (measuring) the variances var$\{e_i\}$, var$\{e_j\}$ of the two parameters e_i, e_j and the covariances cov$\{e_i, e_j\}$ (or the correlation coefficient ρ_{ij}) between them.

---

[8] For a CMOS process, $\Delta L = \Delta L_n = \Delta L_p$ is often assumed *common* to the n and p-type transistors.
[9] Such a model can be used for digital ICs [14], since it is of sufficient accuracy for digital applications.

6.3 ■ Statistical Modeling of Circuit (Simulator) Variables

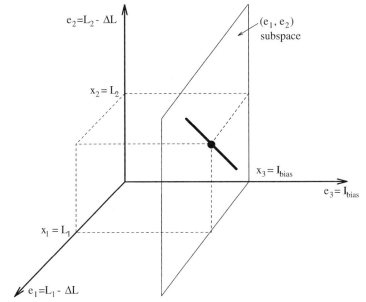

Figure 6-5 Singular distribution $f_e(e_1, e_2, e_3)$ in three-dimensional e-space, represented by the thick line in the (e_1, e_2)-subspace (plane). This is due to the assumed perfect matching of $\Delta L \equiv \Delta L_1 = \Delta L_2$ values.

EXAMPLE 6-2

Let us introduce mismatch between the two MOS transistors T_1, T_2 of Example 6-1, by adding two local mismatch variables: $\delta_1 \equiv \delta_{L_1}$, $\delta_2 \equiv \delta_{L_2}$, (with $E\{\delta_1\} = E\{\delta_2\} = E\{\Delta L\} = 0$), to the ideal (100% matching) model, to obtain:

$$e_1 = x_1 + \theta + \delta_i \equiv L_1 - \Delta L - \delta_{L_1}$$

$$e_2 = x_2 + \theta + \delta_i \equiv L_2 - \Delta L - \delta_{L_2}$$

Let the nominal (expected) transistor lengths be $\bar{e}_1 = \bar{e}_2 = E\{e_1\} = E\{e_2\} = x_1 = x_2 = L_1 = L_2 = L = 5\ \mu m$, the measured relative standard deviations: $\sigma_{e_1}/\bar{e}_1 = \sigma_{e_2}/\bar{e}_2 = 5\%$, and the correlation coefficient between e_1, e_2, $\rho_{12} = 0.95$. With this information available, our objective is to calculate the standard deviations of the three random variables, $\Delta L, \delta L_1, \delta L_2$, such that the model is correct. We calculate $\sigma_{e_1} = \sigma_{e_2} = 5 \times 0.05 = 0.25\ \mu m$, and $\text{cov}\{e_1, e_2\} = \rho_{12}\sigma_{e_1}\sigma_{e_2} = 0.95 \times 0.25 \times 0.25 = 0.05938\ \mu m^2$. But, by definition: $\text{cov}\{e_1, e_2\} = E\{(e_1 - \bar{e}_1)(e_2 - \bar{e}_2)\} = E\{(-\Delta L - \delta_{L_1})(-\Delta L - \delta_{L_2})\} = E\{(\Delta L)^2 + \Delta L(\delta_{L_1} + \delta_{L_2}) + \delta_{L_1}\delta_{L_2}\} = E\{(\Delta L)^2\} = \text{var}\{\Delta L\} = \sigma_{\Delta L}^2$, where the expectations of the mixed terms are zero, since $\Delta L_1, \delta_{L_1}, \delta_{L_2}$ are mutually independent. Therefore, the required standard deviation for the common (global) factor $\theta = \Delta L$ is $\sigma_{\Delta L} = \sqrt{\text{cov}\{e_1 e_2\}} = \sqrt{0.05938} = 0.2437\ \mu m$. With all random variables independent, we have $\sigma_{e_1}^2 = \sigma_{e_2}^2 = \sigma_{\Delta L}^2 + \sigma_{\delta_{L_1}}^2 = \sigma_{\Delta L}^2 + \sigma_{\delta_{L_2}}^2$, so $\sigma_{\delta_{L_1}}^2 = \sigma_{\delta_{L_2}}^2 = \sqrt{\sigma_{e_1}^2 - \sigma_{\Delta L}^2} = \sqrt{0.25^2 - 0.2437^2} = 0.05577\ \mu m$. Therefore, the required relative standard deviations (with respect to $\bar{e}_1, \bar{e}_2$) of the common (global) and local variables are $\sigma_{\Delta L}/\bar{e}_1 = 0.2437/5 = 4.874\%$ and $\sigma_{\delta_{L_1}}/\bar{e}_1 = \sigma_{\delta_{L_2}}/\bar{e}_2 = 0.05577/5 = 1.115\%$. Equivalently, we can *directly* specify (as it is also done in practice) the relative standard deviations of the common and mismatched parts of the model. Then, the total variations $\sigma_{e_i}/\bar{e}_i$, $\sigma_{e_j}/\bar{e}_j$ and the correlation coefficient ρ_{ij} can be calculated, reversing the procedure described in this example.

The simple "two-elements-at-a-time" mismatch models presented above are sufficient if there are clear dominant effects of matched transistor *pairs* (as is often the case for the contemporary analog IC functional blocks, such as operational amplifiers and multipliers). However, in some cases, correct statistical models can be obtained only if more complicated interactions involving *several* (rather than just two at a time) transistors and transistor parameters, are modeled. In those cases, a number of general methods are available for generating interdependent random variables, such as Principal Component Analysis, Factor Analysis, or a Cholesky decomposition-based method (leading to a linear model transforming uncorrelated random variables into a set of correlated random variables—see Section 6.5). These methods assume that the multidimensional parameter distribution is Gaussian (normal) and that the resulting models are linear. If the actual dependencies between different parameters are *nonlinear,* more complicated techniques have to be used, for example, techniques involving nonlinear regression models, as proposed in [13] and briefly discussed in Section 6.3.3.

The most recent results in physical mismatch modeling are the device area (or length) and distance dependencies of the mismatch level, first introduced in [37] and then modified in [30]. (For more references, see [30].) For example, the variance of the difference $e_1 - e_2$ between two MOS transistor model parameters e_1, e_2 is modeled as [30, 37]

$$\text{var}\{e_1 - e_2\} = \frac{a_p}{2W_1 L_1} + \frac{a_p}{2W_2 L_2} + s_p^2 d_{12}^2 \tag{6-4}$$

where W_1, L_1, W_2, L_2 are the widths and lengths of the two transistors, d_{12} is the distance between the transistor centers, and a_p, s_p are empirical coefficients adjusted individually for each model parameter. Using this model together with PCA and introducing other concepts, Michael and Ismail [30] proposed two quite complicated linear statistical models in the form $e_i = e(x_i, \theta, W_i, L_i, d)$. These models include the transistor separation distance information in two different forms. The models, constructed from on-chip measured data, were used for practical yield optimization. θ parameters were divided into two groups: the first group of *correlated* random variables, responsible for the *common* part of each parameter variation and *correlations* between model parameters of each *individual* transistor, and the second group of *local* (mismatch related) random variables, responsible for mismatches between *different* transistors. Additional dependencies, related to transistor spacing and device area-related coefficients, maintained proper mismatch relations.

It should be noted that the mismatch model (Equation 6-4) is only an approximation, valid for relatively short distances d_{12}. When d_{12} increases, the $\text{var}\{e_1 - e_2\}$ also increases in the model (theoretically to infinity), while in reality it is bounded by the value: $\text{var}\{e_1\} + \text{var}\{e_2\}$, since $\text{var}\{e_1 - e_2\} = \text{var}\{e_1\} + \text{var}\{e_2\} - 2\text{cov}\{e_1, e_2\}$, and $\text{cov}\{e_1, e_2\}$ tends to zero for large distances. Moreover, it can be shown that for more than two elements and when the distance term is dominant, a specific combination of a_p and s_p coefficients can lead to a nonpositive definite covariance matrix for the e vector. These issues need to be considered while using model 6-4 in practice.

6.4 ACCEPTABILITY REGIONS

The *acceptability region* A_y is defined as part of the space of performance parameters y (y-space), in which all inequality and equality constraints imposed on y are fulfilled. In the majority of cases, A_y is a hyper-box; that is, all the constraints are

6.4 ■ Acceptability Regions

of the form: $S_j^L \leq y_j \leq S_j^U; j = 1, \ldots, m$, where S_j^L, S_j^U are the (designer defined) lower and upper bounds imposed on y_j, also called *designer's specifications*. More complicated specifications, involving some relations between y_j parameters, or S_j^L, S_j^U bounds, can also be defined.

The *acceptability region A* in the e-space (in the case of y-space box constraints) is defined as[10]

$$A \triangleq \{e \in R^c \mid S_j^L \leq y_j(e) \leq S_j^U, \quad j = 1, \ldots, m\} \quad (6\text{-}5)$$

that is, such a set of e vectors in the c-dimensional space, $e \in R^c$, for which all inequalities $S_j^L \leq y_j(e) \leq S_j^U, j = 1, \ldots, m$ are fulfilled. This definition is illustrated in Figure 6-6 for the case where S^L and S^U are defined as specific functions ω (frequency): $S^L(\omega), S^U(\omega)$. For practical applications, ω is discretetized into a set of discrete frequencies $\{\omega_1, \omega_2, \ldots, \omega_m\}$, such that definition 6-5 holds. Figure 6-6 can be interpreted as the mapping of the acceptability region A_y from the y-space into the e-space. Acceptability regions A can be very complicated: they can be nonconvex and can contain internal infeasible regions (or "holes"), as shown in Figure 6-7 for a simple active RC filter [32].

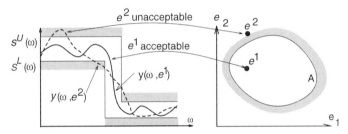

Figure 6-6 Illustration of the mapping of the $S^L(\omega), S^U(\omega)$ constraints imposed on $y(\omega, e)$ into the e-space of circuit parameters.

For discrete circuits, A is represented in the e-space, due to the simple relationships involved: $e_i = x_i + \theta_i$ [or $e_i = x_i(1 + \theta_i)$]. For integrated circuits, e is related to x and θ through the statistical model, x and θ are in different spaces (or subspaces), dimension of θ is lower than dimension of e, and the p.d.f. $f_e(e)$ is singular and usually unknown in analytic form. For these reasons, it is more convenient to represent A in the joint (x, θ)-space, as shown in Figure 6-8. For a fixed x, A can be defined in the θ-space and labeled as $A_\theta(x)$, since it is parametrized by the actual values of x, as shown in Figure 6-8. The shape and location of $A_\theta(x)$ change with x, as shown. For a fixed x, $A_\theta(x)$ is defined as such a region in the t-dimensional θ space, for which all the inequalities $S_j^L \leq y_j(e(x, \theta)) \leq S_j^U; j = 1, \ldots, m$ are fulfilled, that is,

$$A_\theta(x) \triangleq \{\theta \in R^t \mid S_j^L \leq y_j[e(x, \theta)] \leq S_j^U; \quad j = 1, \ldots, m\} \quad (6\text{-}6)$$

---

[10] A shorthand set notation for various regions in space is used. For instance: $A = \{x \in R^n \mid x \geq 0\}$, where $\{\ \}$ denotes a set, and $\mid$ reads "such that," is interpreted as: A is the set of real vectors x in n-dimensional space (i.e., a region in the x-space), such that for all components of x, the condition $x_i \geq 0, i = 1, \ldots, n$, holds.

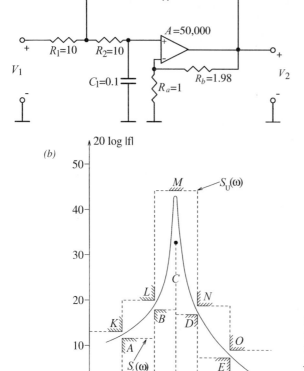

Figure 6-7 (a) A Sallen-Key active filter; (b) the lower and upper bounds imposed on the filter frequency response $|V_2(j\omega)/V_1(j\omega)| \equiv f(\omega)$; (c) two-dimensional cross sections of the acceptability region A. Capital letters in (b) and (c) indicate the correspondence of constraints to the boundaries of A.

In order to recognize if a given point $e(x, \theta)$ in the circuit parameter space belongs to A (or to $A_\theta(x)$), the *indicator function* $\phi(\cdot)$ is introduced

$$\phi[e(x, \theta)] = \begin{cases} 1 & \text{if } e(x, \theta) \in A \text{ (a successful, or ``pass'' point)} \\ 0 & \text{otherwise (a ``fail'' point)} \end{cases} \quad (6\text{-}7)$$

A *complementary* indicator function $\phi_F[e(x, \theta)] = \phi(e(x, \theta)] - 1$ is equal to 1 if $e(x, \theta) \notin A$ and equal to 0, if $e(x, \theta) \in A$. Both indicator functions will be used in what follows.

6.4.1 Methods of Acceptability Region Approximation

Except for some simple cases, the acceptability region A in the e-space (or the joint (x, θ)-space) is unknown, and it is impossible to define it *fully*. For yield optimization and other statistical design tasks, *implicit* or *explicit* knowledge of A and/or its boundary is required. If only the points belonging to A are stored, this can be considered a *point-based* "approximation" to A. Some of the point-based methods are Monte Carlo-based design centering, the Centers of Gravity method,

6.4 ■ Acceptability Regions

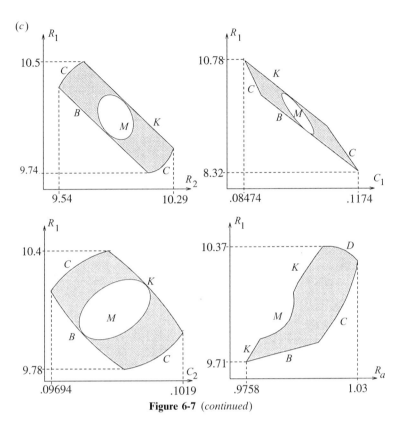

Figure 6-7 (*continued*)

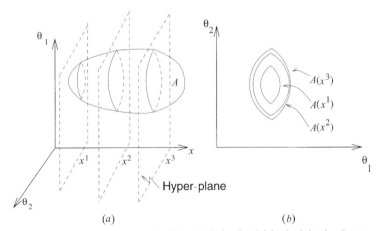

Figure 6-8 Acceptability region for integrated circuits: (*a*) in the joint (x, θ)-space, (*b*) in the θ-space, parametrized by different vectors: x^1, x^2, x^3. The hyper-planes shown represent t-dimensional subspaces of θ parameters.

point-based simplicial approximation and yield evaluation (see below), "Parametric Sampling-based" yield optimization [46], yield optimization with "reusable" points [49], and others.

In [34] the *acceptability segment*-based method of the A-region approximation was called a One-Dimensional Orthogonal Search (ODOS) technique leading to several yield optimization methods [32, 54]. Its basic principle is shown in Figure 6-9b, where line segments passing through the points e^i randomly sampled in the e-space and parallel to the coordinate axes are used for the approximation of A. ODOS is very efficient for *large linear circuits,* since the intersections with A can be *directly* found from analytical formulas. The two-dimensional cross sections of A shown in Figure 6-7 were obtained using this approach. The *surface* integral-based yield and yield gradient estimation and optimization method proposed in [19] also uses the segment approximation to the *boundary* of A. A variant of this method is segment approximation in one direction, as shown in Figure 6-9c. This method was then extended in [51] to the plane and hyper-plane approximation to A (Figure 6-9d). In another approach, called Radial Exploration of Space in [65], the segments approximating A are in *radial directions,* as shown in Figure 6-9e.

The techniques just described rely on the fact that "segment" yields (calculated in the subspace e) can be—for some special cases [34, 65]—calculated more efficiently than using a standard Monte Carlo method. This leads to higher efficiency and accuracy of yield estimation, in comparison to the point-based yield estimation.

The *simplicial approximation* proposed in [17] is based on approximating the boundary of A in the e-space, by a polyhedron, that is, by the union of those partitions of a set of c-dimensional hyper-planes which lie inside of the boundary of A or on it. The boundary of the A-region is assumed to be *convex* (see Figure 6-9f, g). The approximating polyhedron is a convex hull of points. Simplicial approximation is obtained by locating points on the boundary of A, by a systematic expansion of the polyhedron. The search for the next vertex is always performed in the direction passing through the center of the largest face of the polyhedron already existing and perpendicular to that face. In the Monte Carlo-oriented "point-based" version of the method [18] (see Figure 6-9a), subsequent simplicial approximations $\tilde{A}_i$ to A are not built using the points located on the boundary of A (which is computationally expensive) but the points e^i belonging to A and already generated during the MC simulation. After each new point is generated, it is checked if e^i belongs to $\tilde{A}^{i-1}$ (where $\tilde{A}^{i-1}$ is the latest simplicial approximation to A). If it does, the sampled point is considered successful *without* performing the circuit analysis. If e^i does not belong to $\tilde{A}^{i-1}$ and the next circuit simulation reveals that e^i belongs to A, the polyhedron is expanded to include the new point. Several versions of this general approach, leading to the substantial reduction of the computational effort of yield estimation, are described in the original paper.

The method of "cuts" proposed in [6] creates a "cutting-plane" approximation to A in the corners of the tolerance region R_ε assumed to be a hyper-cube, as shown in Figure 6-9h. The method was combined with *discretization* of the p.d.f. $f_e(e)$ and *multidimensional quadratic approximation* to circuit constraints, leading to simplified yield and yield-derivative formulas and yield optimization for arbitrary statistical distributions [1].

6.5 Parametric Yield

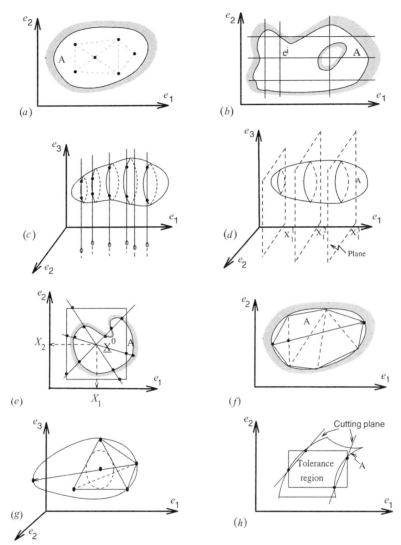

Figure 6-9 Various methods of acceptability region approximation in e-space: (a) "Point-based" simplicial approximation to the A-region. (b) Segment approximation to A in all directions along e_i axes. (c) Segment approximation to A in one direction. (d) (Hyper)-plane approximation to A. (e) Segment approximation to A in radial directions. (f) Simplicial approximation to A in two dimensions. (g) Simplicial approximation to A in three dimensions. (h) Cutting-plane approximation to A.

6.5 PARAMETRIC YIELD

Parametric yield is equal to the percentage of circuits that fulfill all parametric requirements; that is, it is equal to the probability that e belongs to the acceptability region A. Hence, it can be calculated as the integral of the p.d.f. of e, $f_e(e)$ over A, for a given vector of designable parameters x. Since $e = e(x, \theta)$ is a function of x, then $f_e(e) = f_e(e, x)$ (e.g., both $E\{e\}$ and var$\{e\}$ can be functions of x). Therefore[11]

$$Y(x) = P\{e \in A\} = \int_A f_e(e, x)\, de = \int_{R^c} \phi(e) f_e(e, x)\, de = E_e\{\phi(e)\} \qquad (6\text{-}8)$$

where $P\{\cdot\}$ denotes probability, $\phi(e)$ is the indicator function 6-7, and $E_e\{\cdot\}$ is the expectation with respect to (w.r.t.) the random variable e. The above formula is useful if $f_e(e, x)$ is a nonsingular p.d.f., which is usually the case for discrete circuits, for which $e_i = x_i + \theta_i$, [or $e_i = x_i(1 + \theta_i)$]. In a general case, however (e.g., for integrated circuits), the p.d.f. $f_e(e)$ is not known, since it has to be obtained from a complicated transformation $e = e(x, \theta)$, given the p.d.f. $f_\theta(\theta)$ of θ. Therefore, it is more convenient to integrate directly in the θ-space. Since parametric yield is also the probability that θ belongs to $A_\theta(x)$ (the acceptability region in the θ-space for any fixed x), parametric yield becomes

$$Y(x) = P\{\theta \in A_\theta(x)\} = \int_{A_\theta(x)} f_\theta(\theta)\, d\theta = \int_{R^t} \phi[e(x, \theta)] f_\theta(\theta)\, d\theta = E_\theta\{\phi[e(x, \theta)]\}$$
$$(6\text{-}9)$$

Equation 6-9 is general and is valid for both discrete and integrated circuits. An unbiased estimator of $E_\theta\{\phi(e(x, \theta))\} \equiv E_\theta\{\phi(\theta)\}$ (for fixed x) is the arithmetic mean, based on N points θ^i, sampled in θ-space with the p.d.f. $f_\theta(\theta)$, for which the function $\phi(\theta^i)$ is calculated (this involves circuit simulations). Thus, the yield estimator $\hat{Y}$ is expressed as

$$\hat{Y} = \frac{1}{N} \sum_{i=1}^{N} \phi(\theta^i) = \frac{N_S}{N} \qquad (6\text{-}10)$$

where N_S is the number of successful trials, that is, the number of circuits for which $\theta \in A_\theta(x)$ (all circuit constraints are fulfilled). Integral 6-9 is normally calculated using Monte Carlo (MC) simulations [42] and (10). The MC method is also used to determine statistical parameters of the p.d.f. $f_y(y)$ of $y = y(x, \theta)$. In order to sample the θ parameters with the p.d.f. $f_\theta(\theta)$, special numerical procedures, called *random number generators,* are used. The basic MC algorithm is as follows:

1. Set $i = 0$, $N_s = 0$ (i is the current index of a sampled point, and N_s is the total number of successful trials).
2. Substitute $i = i + 1$; generate the ith realization of θ: $\theta^i = (\theta_1^i, \ldots, \theta_t^i)$, with the p.d.f. $f_\theta(\theta)$.
3. Calculate the ith realization of $y^i = (y_1^i, \ldots, y_m^i) = y(x, \theta^i)$ with the aid of an appropriate circuit simulator and store the results.
4. Check if all circuit constraints are fulfilled, that is, if $S^L \leq y^i \leq S^U$; if yes, set $N_S = N_S + 1$.
5. If $i \neq N$, go to (2).

---

[11] Multiple integration performed in these formulas is over the acceptability region A, or over the entire c-dimensional space R^c of real numbers.

6. Otherwise, find the yield estimator $\hat{Y} = \frac{N_s}{N}$. If needed, also find some statistical characteristics of y-parameters (e.g., create histograms of y, find statistical moments of y, etc.).

To generate θ^i's with the p.d.f. $f_\theta(\theta)$, *uniformly* distributed random numbers are generated first and then transformed to $f_\theta(\theta)$. The most typical random number generator (r.n.g.), generating a sequence of pseudorandom, uniformly distributed integers θ_k in the interval [0, M) (M is an integer), is a multiplicative r.n.g., using the formula [42]:

$$\theta_{k+1} = c[\theta_k(\mathrm{mod}\, M)]$$

where c is an integer constant and $\theta_k(\mathrm{mod}\, M)$ denotes a remainder from dividing θ_k by M.[12] The initial value θ_0 of θ_k is called the "seed" of the r.n.g. and, together with c, should be chosen very carefully, to provide good quality of the random sequence generated. Several other r.n.g.'s are used in practice [42]. The r_k numbers in the [0, 1) interval are obtained from $r_k = \frac{\theta_k}{M}$.

Distributions other than uniform are obtained performing different operations on the uniformly distributed random numbers, such as using the inverse of the cumulative distribution function [42]. If $R \in (0, 1)$ is uniformly distributed, then $z = F^{-1}(R)$ has the p.d.f. $f_z(z) = \frac{dF}{dz}$, where $F(z)$ is the cumulative distribution function of z. Several other techniques, such as the function transformation method, composition method, rejection method, and acceptance-rejection method, can be used [42].

To generate the (column) vector of *correlated normal* variables $\theta = (\theta_1, \ldots, \theta_n)^T$, $E\{\theta\} = 0$, (where T denotes matrix transposition) with a given covariance matrix K^θ, the transformation $\theta = CZ$ is used, where $Z = (z_1, z_2, \ldots, z_n)^T$ is the vector of independent normalized Gaussian variables with $E\{z_i\} = 0$, $\mathrm{var}\{z_i\} = 1$, $i = 1, \ldots, n$, and C is the matrix obtained from the so-called Cholesky decomposition of the covariance matrix K^θ, such that $K^\theta = CC^T$. C is usually lower or upper triangular and can be easily constructed from a given matrix K^θ [42].

EXAMPLE 6-3

Assume that we have two IC resistors R_1, R_2 whose nominal values are $\overline{R}_1 = \overline{R}_2 = 1.3 k\Omega$ and measured standard deviations $\sigma_1 = 0.1 k\Omega$, $\sigma_2 = 0.2 k\Omega$, the correlation coefficient $\rho_{12} = 0.7$, and the joint p.d.f. is Gaussian. Create a Cholesky decomposition-based model of correlations to generate random numbers suitable for a Monte Carlo situation. The statistical model used is $e_i \equiv R_i = \overline{R}_i + \theta_i$, $E\{\theta\} = 0$, $i = 1, 2$.

The elements of the covariance matrix are calculated as follows: $\mathrm{var}\{R_1\} = 0.1^2 = 0.01$, $\mathrm{var}\{R_2\} = 0.2^2 = 0.04$, $\mathrm{cov}\{R_1, R_2\} = \rho_{12}\sigma_1\sigma_2 = 0.014$, so, the covariance matrix is

$$K^\theta = \begin{bmatrix} 0.01 & 0.014 \\ 0.014 & 0.04 \end{bmatrix} \qquad (6\text{-}11)$$

---

[12] In general, $a(\mathrm{mod}\, b) = a - b\, INT(\frac{a}{b})$, where $INT(\cdot)$ denotes an integer part of a real number.

The C matrix obtained from the Cholesky decomposition of K^θ is (it can be generated using any computer linear algebra package, e.g., the one available in MATLAB):

$$C = \begin{bmatrix} 0.1 & 0.0 \\ 0.14 & 0.1428 \end{bmatrix} \qquad (6\text{-}12)$$

(Check its correctness by calculating $K^\theta = CC^T$.) Therefore, the linear model to be used for MC sampling is $\theta = CZ$, or

$$\theta_1 = 0.1 z_1$$
$$\theta_2 = 0.14 z_1 + 0.1428 z_2 \qquad (6\text{-}13)$$

where z_i are independent normalized Gaussian random variables. As seen, this model is different from the mismatch model of Example 6-2 which used three random variables (one global and two local mismatches), rather than two only, as in this approach. The mismatch model has better physical representation, while the Cholesky decomposition-based model is more general, since it can be used for any covariance matrix. On the other hand, the mismatch model is very simple and is useful for representing pairwise correlations.[13] Theoretically, we can check the model correctness by calculating the (theoretical) covariance matrix that would result using MC sampling of z_1, z_2, and model (13): $K_{MC}^R = K_{MC}^\theta = E\{(\theta - \bar{\theta})(\theta - \bar{\theta})^T\} = E\{\theta\theta^T\} = E\{CZZ^TC^T\} = CE\{ZZ^T\}C^T = CC^T \equiv K^\theta$, where $E\{ZZ^T\} = \mathrm{diag}\{1, \ldots, 1\}$ (independent normalized random variables), so, indeed $K_{MC}^\theta \equiv K^\theta$. That is, the generated random variables have the covariance matrix equal to K^θ, as required. A direct check can also be performed using the actual model: $\mathrm{var}\{\theta_1\} = 0.1^2 \, \mathrm{var}\{z_1\} = 0.01$; $\mathrm{var}\{\theta_2\} = 0.14^2 \, \mathrm{var}\{z_1\} + 0.1428^2 \, \mathrm{var}\{z_2\} = 0.04$; $\mathrm{cov}\{\theta_1, \theta_2\} = E\{\theta_1, \theta_2\} = E\{0.1 z_1 (0.14 z_1 + 0.1428 z_2)\} = 0.014 \, \mathrm{var}\{z_1\} + 0.01428 \, \mathrm{cov}\{z_1, z_2\} = 0.014$ (since: $\mathrm{var}\{z_1\} = \mathrm{var}\{z_2\} = 1$; $\mathrm{cov}\{z_1, z_2\} = 0$). As shown, all the statistical parameters have correct values.

The yield estimator $\hat{Y}$ is a random variable, since performing different, independent MC simulations, we can expect different values of $\hat{Y} = N_S/N$. Variance or standard deviation of $\hat{Y}$ is normally used as a measure of $\hat{Y}$ variability. The process of finding $\phi(\theta^i)$ in subsequent simulations corresponds to Bernoulli trials [11], and the p.d.f. of N_s is binomial with $E\{N_S\} = NY$ and $\mathrm{var} N_S = NY(1 - Y)$. Therefore,[14]

$$E\{\hat{Y}\} \equiv E\left\{\frac{N_S}{N}\right\} = Y$$

and

$$\mathrm{var}\{\hat{Y}\} = \mathrm{var}\left\{\frac{N_S}{N}\right\} = \frac{1}{N^2} \mathrm{var}\{N_S\} = \frac{NY(1-Y)}{N^2} = \frac{Y(1-Y)}{N}$$

Therefore, the standard deviation of $\hat{Y}$, is equal to

$$\sigma_{\hat{Y}} = \sqrt{Y(1-Y)/N}$$

---

[13] The mismatch model can be generalized to several global variables and single local variables using Principal Component Analysis and Factor Analysis [45].

[14] It was found in [12] that in IC manufacturing the actual variance deviates from the one obtained for the true binominal distribution. For this reason, some corrections must be introduced to the variance formula to follow. See [12] for details.

That is, it is proportional to $\frac{1}{\sqrt{N}}$. Hence, to decrease the error of $\hat{Y}$ 10 times, the number of samples has to be increased 100 times. This is a major drawback of the MC method. However, the accuracy of the MC method (measured by $\sigma_{\hat{Y}}$) is *independent* of the dimensionality of the θ-space, which is usually a drawback of other methods of yield estimation. One method able to reduce the variance of $\hat{Y}$ is *importance sampling* [11, 42, 49]. Assume that instead of sampling θ with the p.d.f. $f_\theta(\theta)$, some other p.d.f. $g_\theta(\theta)$ is used. Then

$$Y = \int_{R^t} \phi(e(\theta)) \frac{f_\theta(\theta)}{g_\theta(\theta)} g_\theta(\theta)\, d\theta \equiv E\left\{\phi(e(\theta)) \frac{f_\theta(\theta)}{g_\theta(\theta)}\right\} \tag{6-14}$$

where $g_\theta(\theta) \neq 0$ if $\phi(\theta) = 1$. Yield Y can now be estimated as

$$\hat{Y} = \frac{1}{N} \sum_{i=1}^{N} \phi(e(\theta^i)) \frac{f_\theta(\theta^i)}{g_\theta(\theta^i)} \tag{6-15}$$

sampling N points θ^i with the p.d.f. $g_\theta(\theta)$. The variance of this estimator is

$$\text{var}\{\hat{Y}\} = E\{[\phi(\theta)f_\theta(\theta)/g_\theta(\theta) - Y]^2\}/N \tag{6-16}$$

If it is possible to choose $g_\theta(\theta)$ such that it mimics (or is similar to) $\phi(\theta)f_\theta(\theta)/Y$, the variability of $[\phi(\theta)f_\theta(\theta)/g_\theta(\theta) - Y]$ is reduced, and thus the variance of $\hat{Y}$. This can be accomplished if some *approximation* to $\phi(\theta)$, that is, to the acceptability region A is known. Some possibilities of using importance sampling techniques were studied (e.g., in [22]). A method called Parametric Sampling was used in [46], and other variants of important sampling were used in [3, 50] for yield optimization. There are several other methods of variance reduction, such as the method of control variates, correlated sampling, stratified sampling, antithetic variates, and others [11, 42, 49]. Some of them have been used for statistical circuit design [11, 49].

6.6 INDIRECT METHODS OF YIELD ENHANCEMENT

The objective of yield optimization is to find a vector of designable parameters $x = x_{opt}$, such that $Y(x_{opt})$ is maximized. This is illustrated in Figure 6-10 for the

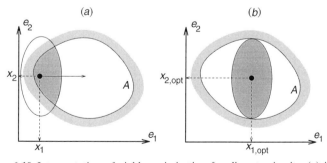

Figure 6-10 Interpretation of yield maximization for discrete circuits: (*a*) initial (low) yield, (*b*) optimized yield.

case of discrete circuits where $e_1 = x_1 + \theta_1$, $e_2 = x_2 + \theta_2$.[15] Case (*a*) corresponds to low initial yield, proportional to the area (hyper-volume, in general) weighted by the p.d.f. $f_e(e)$ and represented in the figure by the dark shaded part of the tolerance body shown. Case (*b*) corresponds to optimized yield, obtained by shifting the nominal point (x_1, x_2) to the vicinity of the *geometric* center of the acceptability region A. Because of this geometric property, the yield enhancement process is often referred to as *geometrical design centering*, or, simply, *design centering*.

Yield enhancement methods can be classified into *indirect* methods (to be discussed in this section) and *direct* methods (to be discussed in Section 6.7).

In the indirect yield enhancement methods, yield itself is not used as the objective function. Instead, some other objective function is selected, whose minimization (or maximization) leads to *improved,* but not necessarily maximum, possible yield. For this reason, the methods of this class are referred to as indirect yield *enhancement* (or improvement) methods rather than yield maximization methods.

Indirect methods are further divided into those that define the objective function to be optimized in the circuit *element* (or *parameter*) *space*, which can be *e*-space (for discrete circuits) or (x, θ)-space (for ICs), and those for which the objective function is defined in the *performance space y*. The optimization process involving the former group is often referred to as *geometric* design centering, since the objective is to locate the nominal point x in the "geometric center" of the acceptability region A, usually defined in the *e*-space. Since the definition of the "geometric design center" is not unique, several different approaches are possible. Two of them are briefly discussed next: the Simplicial Approximation-based (see Section 6.4.1) and the approach utilizing the worst-case distances from the boundaries of the acceptability region.

6.6.1 Simplicial Approximation-Based Design Centering

This is the best known geometric approach, introduced in [17]. It is based on *direct* approximation of the acceptability region in the *e*-space, as shown in Figure 6-9*f* and *g*. It was assumed in [17] that the p.d.f. $f_\theta(\theta)$ is a bounded quasi-concave function, and a p.d.f. level set, also called tolerance body or norm body in Section 6.3.1, is defined by Equation 6-2: $L_\varepsilon(x, \alpha) = \{e = x + \theta \in R^n \mid f_\theta(\theta) \geq \alpha\}$. The design centering problem is then defined as

$$\min_{x}\{\alpha\} \qquad (6\text{-}17)$$
$$\text{such that} \quad L_\varepsilon(x, \alpha) \in \tilde{A}$$

where $\tilde{A}$ is a Simplicial Approximation to the acceptability region, actually constructed during the optimization process. Design centering can be interpreted as an attempt to inscribe the largest level set (norm body) into the Simplicial Approximation approximation $\tilde{A}$ to the acceptability region A. Then the center of the largest norm body (e.g., a hyper-sphere) is taken as the optimal vector of parameters x. Several Simplicial Approximation yield optimization schemes have been proposed. The most typical is as follows.

---

[15] Alternatively, the model $e_i = x_i(1 + \theta_i)$ can be used, in which case the size of the tolerance body (see Section 6.3) will increase proportionally to x_i.

After a nominal point x belonging to the acceptability region is found, line searches are performed (usually along the coordinate axes first) from that point to obtain some points located on the boundary of A. Several circuit simulations may be required to find one boundary point. To form a polyhedron in an n-dimensional space, at least $(n + 1)$ boundary points need to be found. Once the first polyhedron approximation to A is obtained, the largest possible norm body is inscribed into it (using linear programming techniques), and its center is assumed as the first approximation to the center of A.

The next steps involve improvements to the current approximation $\tilde{A}$, by expanding the simplex as follows. The center of the largest polyhedron face[16] is found, as shown in Figure 6-9g, and a line search is performed from the center along the line passing through it in a direction orthogonal to the face considered, to obtain another vertex point on the boundary of A. The polyhedron is then inflated to include the new point generated. This process is repeated until no further improvement is obtained; that is, the design center and the distance from the center to each face of A do not significantly change between successive iterations.

Intuitively, this method should improve yield but, because of the approximation used, will not necessarily maximize it. The Simplicial Approximation will not be accurate if A is nonconvex, and it will fail if A is not simply connected, since some parts of the approximation $\tilde{A}$ will be outside the actual acceptability region. The computational cost of obtaining the Simplicial Approximation quickly increases for highly dimensional spaces e. Thus, the method is usually viewed as suitable for problems with a small number of designable parameters [49].

6.6.2 Worst-Case Distance-Driven Design Centering

The method proposed in [2] is also based on finding a geometrical center of A (defined differently than in the Simplicial Approximation method), but its computational cost increases only *linearly* with the number of designable parameters rather than exponentially, as was the case for Simplicial Approximation. The method can find the exact worst-case conditions, identifying the worst-case distances between the design center and the boundary of the acceptability region in the e-space. For design centering, the worst-case distances are used as objectives to be maximized so that the design center is moved as far away as possible from the boundary of A. The worst-case distances are also used as a key measure of performance, yield, and circuit robustness. The method uses an efficient, gradient-based deterministic optimization algorithm of the Sequential Quadratic Programming type for yield optimization and design centering. Yield estimation is performed by numerical integration (without performing any additional circuit analyses) within an approximation to A obtained using bounding hyper-planes (one per specification) passing through the points located on the boundary of A which are the most critical, that is, the *closest* to the design center. The hyper-planes are defined by the gradient

---

[16] Each face of the polyhedron is a simplex in the n-1 space; therefore, the center of the face is defined as the center of the largest n-1 dimensional hyper-sphere that can be inscribed into the face (simplex).

of specifications. For its operation, the method requires both the values of performances for each (x, θ) point and their derivatives. As a result, a large number of circuit analyses might be required, if derivatives are not available from the circuit simulator. As an example, for the switched-capacitor circuit presented in [2], with 12 statistical designable parameters and 73 specifications, the yield increase from 24 to 64% was obtained at the total expanse of 3120 circuit simulations, since all the derivatives were calculated by finite differences. The proposed method is mathematically sophisticated and universal, able to uniformly treat both discrete and integrated circuits within the same methodology.

6.6.3 Performance Space-Oriented Design Centering

The major difference between the circuit parameter e-space-oriented methods of design centering and the performance space-based methods is that generalized distances[17] between the lower S_j^L and upper S_j^L specifications and the actual response y_j are used to construct a suitable objective function to be optimized, rather than distances in the e-space. In this way, the costly iterative searches in the e-space (used to find the required intersections with the acceptability region A) are avoided. These methods are also much simpler, both computationally and algorithmically, but they do not lead, in general, to exact centering (in the geometrical sense) as the e-space-based methods, and no true yield maximum is normally obtained.

Individual scaled distances (also referred to as error functions) between the actual response y_j and the lower S_j^L and the upper S_j^U specifications are defined as $g_j^L = a_j^L(S_j^L - y_j)$ and $g_j^U = a_j^U(y_j - S_j^U)$ (where a_j^L, a_j^U are scaling coefficients). Thus, they are *negative* if e is located inside A and positive otherwise. In the simplest and least computationally involved case, the *nominal point* minimax centering (involving only the performance function evaluation at the nominal point x) is accomplished by solving the following minimax problem:

$$\min_{x \in D_x \subset R^n} \max_{j=1,...,m} \{g_j^L(x), g_j^U(x)\} \qquad (6\text{-}18)$$

where D_x is a constraint region in x-space, usually a hyper-box. Because of the properties of the optimal *minimax* solution, the optimal nominal point x_{opt} is pushed deeply into the A region to such a place where all the scaled distances for active constraints[18] between $y_j(x_{opt})$ and S_j^L and/or S_j^U are the same. This approach most often does not lead to the yield maximum, but it provides a good starting point for the application of more sophisticated optimization techniques. A class of such methods has been introduced in [7, 8, 9]. The basic feature of these methods is that *several* points are sampled in the e-space, according to the specific probability density function, rather than just one (nominal) point, as in the minimax approach discussed earlier. Because of that, the tolerance region is much better represented, and solutions much closer to the true yield optimum are obtained. The price to pay is the increased number of the required circuit simulations.

---

[17] In the sense of a specific vector norm or other measures of distance.
[18] That is, those that determine the boundary of the acceptability region.

6.6 ■ Indirect Methods of Yield Enhancement

Theoretically, the minimax approach [18] could be used to combine all the scaled distances into one joint objective function to be minimized. However, better objective functions were proposed in [7, 8, 9] and used successfully in practice. In [8], the following approach is proposed: a generalized l_p function is introduced, defined as

$$v(e^i) = \begin{cases} \left(\sum_{j \in J(e^i)} [g_j(e^i)]^p \right)^{\frac{1}{p}} & \text{if } J(e^i) \neq 0 \\ -\left(\sum_{j=1}^{m} [-g_j(e^i)]^{-\frac{1}{p}} \right) & \text{if } J(e^i) = 0 \end{cases} \quad (6\text{-}19)$$

where $J(e^i)$ is defined as the set of indices for all those error functions $g_j(e^i)$ (for a given point e^i sampled in the e-space), which are greater than or equal to zero:

$$J(e^i) = \{j \mid g_j(e^i) \geq 0\} \quad (6\text{-}20)$$

If any of the sampled points e^i is *outside* the A region, a standard pth norm appearing in (19) is used in order to push *all* the points into A. Once this happens, a better-centered design can be obtained by further pushing all the sampled points e^i away from the boundary of A. This is accomplished by using the second definition of $v(e^i)$ in Equation 6-19. The authors of the proposed approach observed that the l_1 norm (i.e., assuming $p = 1$ in 6-19) had better properties than other norms. This can be motivated as follows: let us use a *one-sided* version of 6-19 and a *one-sided* l_1 objective function defined as

$$U(x) = \sum_{i \in I} \alpha_i v(e^i) = \sum_{i \in I} \alpha_i v[e(x, \theta^i)] \quad (6\text{-}21)$$

where

$$I = \{i \mid v(e(x, \theta^i)) > 0\} \quad (6\text{-}22)$$

and $\alpha_i > 0$ are weighting coefficients. If α_i is selected as

$$\alpha_i = \frac{1}{|v(e(x^0, \theta^i))|}, \quad (6\text{-}23)$$

where x^0 represents vector x at the beginning of optimization, then the value of $U(x^0)$ is equal to the number of circuits located outside A; that is, yield is equal to $Y(x^0) = 1 - U(x^0)/N$, where N is the total number of points sampled with the p.d.f. $f_\theta(\theta)$. Therefore, yield maximization can be accomplished by minimizing the number of failures, that is, solving the problem

$$\min_{x \in D_x \subset R^n} \{U(x)\} \quad (6\text{-}24)$$

Coefficients α_i are fixed at their initial values at the beginning of optimization, expressed by Equation 6-23. That is, the original objective function is equal to the number of failures, and, when optimization progresses, it provides a continuous approximation to the number of failures. When all the sampled points are *inside* A, $U(x)$ is changed, such that the second definition of $v(e^i)$ in 6-19 is used to push the tolerance region deeper into A (since yield is already very close or equal to 100%).

To improve the behavior of $U(x)$ (e.g., to improve its "smoothness"), a continuous "yield probability function" was proposed in [8]: instead of directly using Equation 6-19 of the $v(\cdot)$ function, which can exhibit "jumps" of its values when points e^i move in and out of A, $v(\cdot)$ is mapped through a monotonic function, which eliminates the jumps. The level of smoothing can be controlled by an additional parameter. This procedure is approximately equivalent to replacing the step (indicator) function $\phi(e)$ (see Equation 6-7) appearing in the yield definition by a monotonic function of a finite slope.

Another improvement, proposed in [9], is to use a *Huber function* [23] as the replacement for the l_p norm. The Huber function is a combination of the l_2 norm for small errors, and the l_1 function (i.e., affine function) for large errors. The authors of [9] demonstrated that in the presence of both small and large errors, the new approach leads to a much more "robust" optimization process than when the standard l_1, l_2 or minimax norms are used.

6.7 STATISTICAL METHODS OF YIELD OPTIMIZATION

The major feature of *statistical* yield optimization methods, referred to as *statistical design centering,* is statistical sampling in either θ-space only or in both θ- and x-spaces. Sampling can also be combined with some geometrical approximation to the A region, such as the segment or radial-segment approximation. In what follows, some of these methods are discussed in more detail.

6.7.1 Problem Classification

The large variety of existing statistical yield optimization methods can be classified in many ways. The following classification and the practical use of a specific method are based on the overall available *information* characterizing the problem. The major classification criteria are as follows.

   A. The type of the transformation $e = e(x, \theta)$ from the θ-space to the circuit parameter space e, characterizing the type of the circuit (or problem) investigated. This is one of the most important criteria. The early, mostly heuristic yield optimization methods (developed for discrete circuits) were assuming the simple additive model $e_i = x_i + \theta_i$ [or $e_i = x_i(1 + \theta_i)$]. Because of those assumptions, these methods cannot be used, in general, for IC yield optimization. The major difference is that for ICs the number of θ parameters is most often smaller than the number of e parameters, and θ and x are in different spaces, or subspaces. As a result, the p.d.f. $f_e(e)$ is defined over some subspace of the e-space; that is, it is *singular*. This type of $e = e(x, \theta)$ transformation (for which the dimension of θ is less than the dimension of e) will be called θ-*singular*.

B. Information available about the performance indices $y_j(x, \theta)$: whether the values of y_j's are only available from the circuit simulator, or also the *derivatives* (*sensitivities*) of y_j with respect to x's and/or θ's can be calculated with little additional computational effort.

C. Availability of performance function approximation: whether some function $\bar{y}(x, \theta)$ approximating $y(x, \theta)$ can be efficiently constructed—either with respect to θ only (for a fixed x) or with respect to both x and θ. Calculations involving the approximating formulas should be orders of magnitude faster (as it usually the case) than the actual circuit simulation, and it should be much cheaper to create the approximation than to optimize the circuit directly. (This might not be the case for circuits with a large number of parameters.)

D. Knowledge and properties of the p.d.f. of θ: whether the analytical formulas for $f_\theta(\theta)$ are known and $f_\theta(\theta)$ is differentiable w.r.t. θ, or analytical formulas for $f_\theta(\theta)$ and its derivatives are not known. That is, only samples θ^i of θ are given (obtained from a numerical algorithm or from measurements).

Different combinations of the cases listed above require different optimization algorithms. The more general a given algorithm is, the larger number of cases it is able to cover; simultaneously, however, it is most likely to be less efficient than specialized algorithms covering only selected cases. An ideal ("optimal") algorithm would be the one that is *least restrictive* and could use the *minimum* information necessary. This corresponds to the most difficult case characterized by the general transformation $e = e(x, \theta)$ (possibly singular); the values of $y = y(x, \theta)$ available from a circuit simulator but without derivatives; no approximation to $y(x, \theta)$ available (large dimensions of x and θ); and unknown analytic formulas for $f_\theta(\theta)$ (only the samples of θ can be generated). Moreover, in this ideal case, we would expect the optimization algorithm to be reasonably efficient. The selected yield optimization algorithms discussed later in this chapter fulfill the listed criteria of the algorithm "optimality" to a quite different level of satisfaction. Owing to different assumptions made during the development of different algorithms and the statistical nature of the results, an entirely fair evaluation of the *actual* algorithm efficiency is quite difficult and is limited to some specific cases only. For this reason, no detailed algorithm comparison is attempted in what follows.

6.7.2 Large-Sample versus Small-Sample Methods

Yield optimization is concerned with maximizing the *regression function* $Y(x) = E_\theta\{\phi(x, \theta)\}$ with respect to x (see Equation 6-9). While solving problems of this type, a general function $w(\cdot)$ is used rather than $\phi(\cdot)$. In *large-sample* methods of optimizing $E_\theta\{w(x, \theta)\}$, the expectation (average) of w (and/or its gradient) w.r.t. θ is calculated for each $x^0, x^1, x^2, \ldots$ from a *large* number of θ^i samples. Therefore, the averages used are sufficiently accurate to take relatively large steps $x^{k+1} - x^k$ in each iteration. On the other hand, *small-sample* methods use just a few (very often just one) samples of $w(x, \theta^i)$ for any given point x and make relatively *small* steps in the x-space. However, to compensate for the loss of accuracy, they also utilize a special *averaging procedure,* which calculates the average of w or its gradient over a certain number of steps. Hence, in this case, the averaging in θ-space and progression in the x-space are combined, while in the large-sample methods they are separated. Both techniques have proven convergence to the solution x under certain (different) conditions. The vast majority of yield optimization methods belong to the large-sample category (but some of them can be modified to use a small number of samples per iteration). A class of small-sample yield optimization methods was proposed in [58] and is based on the well-known techniques of *stochastic approximation* [41, 42], to be discussed in Section 6.7.7.

6.7.3 Using Standard Deterministic Optimization Algorithms

Standard, nonderivative deterministic optimization algorithms, such as the simplex (polytope) method of Nelder and Mead, the Powell's search algorithm, and other methods discussed, for example, in [20], could theoretically be used for *direct* yield optimization. Yield (the objective function required by these algorithms) would have to be estimated from a large number of samples for each nominal point x^k of the sequence $\{x^k\}$ generated by the optimizer. This approach is actually quite appealing, since most of the conditions for the algorithm's "optimality" are fulfilled: it would work for any $e = e(x, \theta)$, and only the *values* of $y = y(x, \theta)$ and the *samples* of θ would be required. However, this "simple" approach is limited to some special cases only. Indeed, if no approximation to $y = y(x, \theta)$ is available, the method requires tens of thousands of circuit analyses, which normally is prohibitively expensive. Moreover, if the number of samples per iteration is reduced to increase efficiency, the optimizer is receiving highly noise-corrupted information, leading to poor algorithm convergence or divergence, since standard optimization algorithms work poorly with noisy data. (Specialized algorithms, able to work under *uncertainty*—such as Stochastic Approximation algorithms—have to be used.)

If some (efficient) approximating functions $\tilde{y} = \tilde{y}(x^k, \theta)$ are constructed *separately* for each x^k, a large number of MC analyses can be performed cheaply reducing the statistical error.[19] In practice, if the number of important θ parameters is large, such an approach is most often too expensive, owing to the high cost of obtaining the approximating formulas. As an illustration, consider the following example.

EXAMPLE 6-4

Let the number of θ parameters be equal to $t = 20$. (It can actually be much larger if *mismatches* are present in the IC case or many *discrete* transistors are used in a discrete circuit.) Then, the number of analyses required to construct a quadratic approximation at each iteration is equal to the number of coefficients of a quadratic polynomial, which is $N = (t + 1)(t + 2)/2 = 231$. A typical number of iterations required for a search-type algorithm to find a solution is about $M = 100-300$ (but can also be higher), depending on

---

[19] However, a *systematic* error, resulting from the error of approximating formulas, is introduced. If the approximating process is performed with sufficient error control over a specific range of variable values, the systematic error can be made reasonably small.

the function complexity and the number of optimized parameters x. Therefore, the total expected number of circuit analyses can be estimated as $M \times N \approx 23,100 - 69,300$. These numbers are much too large for any circuit of a reasonable size, and they can be much large if many mismatch random variables have to be taken into account.

The approximating functions $\tilde{y}_j(x, \theta)$ for each y_j can also be created in the *joint* (x, θ) space [62, 67, 72]. In [62], an efficient new approximating methodology was created, which is highly accurate for a relatively large range of the x_i values. However, also in this case the dimension of the joint space (x, θ) cannot be too large, since the cost of obtaining the approximating functions $\hat{y}_j(x, \theta)$ itself becomes prohibitively high. Because of these difficulties, several direct *dedicated* yield optimization methods have been developed, for which the use of function approximation is not required. Some of these methods are described in what follows.

6.7.4 Large-Sample Heuristic Methods for Discrete Circuits

These methods have been developed mostly for discrete circuits, for which $e_i = x_i + \theta_i$. Only the function values $y^i = y(x + \theta^i)$ are required, calculated for the samples θ^i obtained in an arbitrary way. Approximation functions in the e-space can be constructed to increase efficiency, but for discrete circuits the number of θ parameters can be large (proportional to the number of active devices, since no global parameters exist). Thus, the use of approximation is most often not practical. The most typical representative of this class of methods is the Centers of Gravity method [49]. The original method was based on a simple observation that if $\bar{x}_S$ is the center of gravity of "pass" points (as shown in Figure 6-11), defined as

$$\bar{x}_S = (x_A^1 + x_A^2 + \cdots + x_A^{N_A})/N_A$$

where N_A is the number of points x_A^i falling into the A region, then a step from x to $\bar{x}_S$ will improve yield. In [47], the center of gravity of the "fail" points $\bar{x}_F = x_F^1 + x_F^2 + \cdots + x_F^{N_F})/N_F$ was defined, and the direction of yield increase was taken as going from $\bar{x}_F$ through $\bar{x}_A$, as shown in Figure 6-12. Moving in this direction with the stepsize equal to $\mu(\bar{x}_S - \bar{x}_F)$, where $\mu \approx 0.2 - 2$ (often taken as $\mu = 1$), leads to a sequence of optimization steps, which is stopped if $\|\bar{x}_S - \bar{x}_F\|$ is less than a predefined small constant. This is based on the property (proved for a class of p.d.f.'s in [52]) that, under some conditions, at the yield maximum $\|\bar{x}_S - \bar{x}_F\| = 0$ (and $\bar{x}_S = \bar{x}_S = \hat{x}$, where $\hat{x}$ is the point of the yield maximum). It was also shown

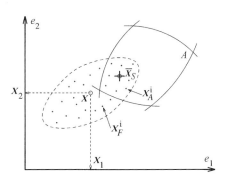

Figure 6-11 Interpretation of the original Centers of Gravity method.

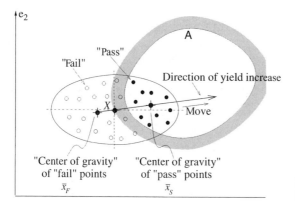

Figure 6-12 Interpretation of the modified Centers of Gravity method.

in [52] that for the normal p.d.f. $f_\theta(\theta)$ with zero correlations and all standard deviations $\sigma_{\theta_i} = \sigma_\theta$, $i = 1, \ldots, t$ equal, the Centers of Gravity direction coincides with the yield gradient direction. However, with correlations and σ_{θ_i}'s is not equal, the two directions can be quite different. Various schemes, aimed at reducing the total required number of analyses, were developed based on the concepts of "reusable" points [49].

In [25, 26] the original Centers of Gravity method was significantly improved, introducing a concept of Gaussian adaptation of the covariance matrix of the sampled point of θ^i, such that their distribution adopts (temporarily) to the shape of the acceptability region, leading to higher (possibly optimal) efficiency of the algorithm. The method was successfully used on large industrial design examples, involving as many as 130 designable parameters, not only for yield optimization but also for standard function minimization.

The radial exploration of space methods [65] (see Fig. 6-9e) and the One-Dimensional Orthogonal Searches (ODOS) technique [33, 34] (see Figure 6-9b, c) discussed in Section 6.4.1, have also been developed into yield optimization methods. In the radial exploration case, the *asymmetry vectors* were introduced, generating a direction of yield increase, and in the ODOS case a Gauss-Seidel optimization method was used. Both techniques were especially efficient for linear circuits, owing to the high efficiency of performing circuit analyses in radial and orthogonal directions.

6.7.5 Large-Sample, Derivative-Based Methods for Discrete Circuits

For discrete circuits, the relation $e_i = x_i + \theta_i$ holds for the part of the e vector related to the optimized passive RLC elements (active devices are normally not optimized), so, x_i and θ_i are in the same space. Moreover, the p.d.f. $f_\theta(\theta)$ is most often known, and $f_e(e, x)$ is the p.d.f. $f_\theta(\theta)$ transformed to the e-space (i.e., of the same shape as that of $f_\theta(\theta)$ but shifted by x). Then, from Equation 6-8, differentiating w.r.t. x_i, we obtain

$$\frac{\partial Y(x)}{\partial x_i} = \int_{R^n} \phi(e) \frac{\partial f_e(e, x)}{\partial x_i} \frac{f_e(e, x)}{f_e(e, x)} de = E_e \left\{ \phi(e) \frac{\partial \ln f_e(e, x)}{\partial x_i} \right\} \quad (6.25)$$

6.7 ■ Statistical Methods of Yield Optimization

where the equivalence $\frac{\partial f_e(e,x)/f_e(e,x)}{\partial x_i} \equiv \frac{\partial \ln f_e(e,x)}{\partial x_i}$ was used. Therefore, yield derivatives w.r.t. x_i can be calculated as the *average* of the expression in the braces of Equation 6-25, calculated from *the same* θ^i samples as those used for yield estimation, provided that the p.d.f. $f_e(e, x)$ is *differentiable* w.r.t. x (e.g., the normal or log-normal p.d.f's are differentiable, but the uniform p.d.f. is not). Notice that instead of sampling with the p.d.f. $f_e(e, x)$, some other (better) p.d.f. $g_e(e, x)$ can be used as in the *importance sampling* yield estimation (see Equation 6-14). Then

$$\frac{\partial Y(x)}{\partial x_i} = E_e \left\{ \phi(e) \frac{\partial f_e(e,x)}{\partial x_i} \left(\frac{f_e(e,x)}{g_e(e,x)} \right) \right\} \quad (6\text{-}26)$$

where sampling is performed with the p.d.f. $g_e(e, x) \neq 0$. This technique was used in [3, 46, 59] (to be discussed below). Consider the multivariate normal p.d.f., with the positive definite covariance matrix K

$$f_e(e) = \frac{1}{(2\pi)^{t/2}\sqrt{\det K}} \exp\left[-\frac{1}{2}(e-x)^T K^{-1}(e-x)\right] \quad (6\text{-}27)$$

where $e - x \equiv \theta$ (discrete circuits), $\det K$ is the determinant of K, and T denotes matrix transposition. Then, it can be shown that the yield gradient $\nabla_x Y(x)$ is expressed by

$$\nabla_x Y(x) = E\{\phi(e) K^{-1}(e-x)\} = Y(x) K^{-1}(\bar{x}_S - x) \quad (6\text{-}28)$$

where $\bar{x}_S$ is the center of gravity of "pass" points. If yield $Y(x)$ is a continuously differentiable function of x, then the necessary and sufficient condition for the yield maximum is $\nabla_x Y(\hat{x}) = 0$, which combined with Equation 6-28 means that the stationary point $\hat{x}$ for the yield function (the yield maximim if $Y(x)$ is also concave) is $\hat{x} = \bar{x}_S$, the center of gravity of the pass points. This result justifies (under the assumptions stated above) the Centers of Gravity method of yield optimization (since its objective is to make $\hat{x} = \bar{x}_S = \bar{x}_F$).

For $K = \text{diag}\{\sigma_{e_1}^2, \ldots, \sigma_{e_t}^2\}$, that is, with zero correlations, the yield gradient w.r.t. x is expressed as

$$\nabla_x Y(x) = E_e \left\{ \phi(e) \left[\frac{\theta_1}{\sigma_{\theta_1}^2}, \ldots, \frac{\theta_t}{\sigma_{\theta_t}^2} \right]^T \right\} \quad (6\text{-}29)$$

where $\sigma_{\theta_i} \equiv \sigma_{e_i}$ was used instead of σ_{e_i}. It can be readily shown that for all $\sigma_{\theta_1} = \ldots = \sigma_{\theta_t} = \sigma_\theta$ equal, the yield gradient direction coincides with the Center of Gravity direction [52] (Figure 6-13).

Observe that replacing x in all the yield gradient formulas presented above by the vector of arbitrary parameters describing the p.d.f. $f_e(e)$ (such as standard deviations), gradient formulas w.r.t. these parameters can be derived, provided that the p.d.f. $f_e(e)$ is differentiable. In the same way, relevant formulas for higher order derivatives can be found. Specifically, for the normal p.d.f., all higher order derivatives exist, so that all higher order yield derivatives can also be estimated from the same sampled points θ^i, as those used for yield estimation.

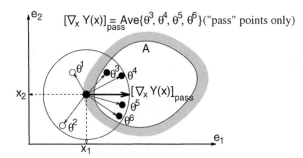

Figure 6-13 (Example 6-5) Interpretation of the yield gradient formula for normal p.d.f., with no correlations and $\sigma_{\theta_1} = \sigma_{\theta_2} = 1$.

The yield gradient can also be calculated from the "fail" points by simply using the $\phi_F(\cdot) = \phi(\cdot) - 1$ indicator function in all the expressions above. Then, the two resulting estimators can be combined into one joint average, as was done in the $\bar{x}_S, \bar{x}_F$-based Centers of Gravity method.

EXAMPLE 6-5

Consider the situation shown in Figure 6-13, where the normal distribution of the sampled points θ is assumed, no correlations, and $\sigma_{\theta_1} = \sigma_{\theta_2} = 1$. Under these conditions, from Equation 6-29, the yield gradient estimator for the "pass" points (black dots) is $[\nabla_x \hat{Y}(x)]_{\text{pass}} = (\theta^3 + \theta^4 + \theta^5 + \theta^6)/4$ and for the "fail" points (white dots): $[\nabla_x \hat{Y}(x)]_{\text{fail}} = (-\theta^1 - \theta^2)/2$. It is clearly seen that $x + [\nabla_x \hat{Y}(x)]_{\text{pass}} \equiv \bar{x}_S$ coincides with the center of gravity of the "pass" points and $x - [\nabla_x \hat{Y}(x)]_{\text{fail}} \equiv \bar{x}_F$ with the center of gravity of the "fail" points. The center of gravity yield improvement direction is obtained as: $\bar{x}_S - \bar{x}_F = x + [\nabla_x \hat{Y}(x)]_{\text{pass}} - (x - [\nabla_x \hat{Y}(x)]_{\text{fail}}) = (\theta^3 + \theta^4 + \theta^5 + \theta^6)/4 + (-\theta^1 - \theta^2)/2$, which shows that some kind of an overall average of all the sampled points is obtained. However, this average is not a regular average (i.e., $(-\theta^1 - \theta^2 + \theta^3 + \theta^4 + \theta^5 + \theta^6)/6$) but a *weighted average* of all the sampled points.

The specific way of combining the "pass" and "fail" gradient directions (consistent with the Centers of Gravity method) shown in Example 6-5 is not the only possible choice. Actually, there exists an *optimal* weighted combination of the two gradient estimators for any given problem, resulting in an overall gradient estimator of minimum variability. It is difficult, however, to determine it in practice both sufficiently and precisely. A general rule is that at the beginning of optimization, when x is far away from $\hat{x}$ (the optimal point), the yield gradient estimator based on the "pass" points should be more heavily weighted; the opposite is true at the end of optimization, when the "fail" points carry more precise gradient information (provided that the final yield is not too close to 1 but is relatively high).

In the majority of the large-sample derivative methods developed, it was assumed that $f_e(e)$ was Gaussian. An iterative step in those methods is typically made in the gradient direction

$$x^{k+1} = x^k + \alpha_k \nabla_x Y(x^k) \quad (6\text{-}30)$$

where α_k is most often selected empirically, since yield maximization along the gradient direction is too expensive, unless some approximating functions $\tilde{y} = \tilde{y}(x + \theta)$ are used. (Normally, this is not the case for these methods.) Since the numbers of points θ^i sampled for each x^k, is large, the main difference between various published algorithms is how to use the information already available most efficiently. The three methods to be briefly discussed (introduced almost at the same time [3, 46, 59], utilize some form of *importance sampling* for that purpose, discussed in Section 6.4.1.

6.7 ■ Statistical Methods of Yield Optimization

In [46] a Parametric Sampling technique was proposed in which the θ^i points were sampled with the p.d.f. $g_e(e, x)$ in a broader range than for the original p.d.f. (i.e., all the σ_{θ_i}'s were artificially increased). All points sampled were stored in a database, and the gradient-direction steps were made according to Equation 6-30. The importance sampling-based gradient formula (Equation 6-26) was used in subsequent iterations within the currently available database. Then, a new set of points was generated and the whole process was repeated.

The methods developed in [3] and [59] were also based on the importance sampling concept, but instead of using the gradient steps as in Equation 6-30, the yield gradient vector and the *Hessian*[20] matrix were calculated and updated using the currently available sampled points θ^i. Then, a more efficient Newton's direction was taken in [59], or a "yield prediction formula" was derived and used in [3]. In order to deal with the singularity or nonpositive definiteness of the Hessian matrix (which is quite possible owing to the randomness of data and the behavior of $Y(x)$ itself), suitable Hessian *corrections* were implemented, using different kinds of Hessian matrix decomposition (Cholesky-type in [59] and eigenvalue decomposition in [3]).

As was the case for the heuristic methods, the methods discussed here are relatively insensitive to the dimensionality of x- and θ-spaces.

For the *uniform* p.d.f., centered at $e = x$ (see Figure 6-14) and defined within a hyper-box $x_i - \varepsilon_i \leq e_i \leq x_i + \varepsilon_i$, $i = 1, \ldots, t$, where ε_i are element tolerances (see Figure 6-3), the yield gradient formula (Equation 6-25) cannot be used, since the uniform p.d.f. is nondifferentiable w.r.t. x_i. However [52], the derivatives can still be defined in the sense of *generalized* functions, though Dirac's delta functional $\delta(x)$ is defined as $\delta(x) = 0$, if $x \neq 0$ and $\delta(x) = \infty$, if $x = 0$, with the property that $\int_{-\infty}^{+\infty} \delta(x)\, dx = 1$. Observe that for the unit step function defined as $U(x) = 0$, if $x < 0$ and $U(x) = 1$, if $x \geq 0$, the generalized derivative of $U(x - a)$ is defined at $x = a$ as

$$dU(x - a)/dx = \delta(x - a)$$

Moreover, the basic property of the $\delta(x)$ functional is that

$$\int_{-\infty}^{+\infty} w(e)\delta(e - e_0)\, de = w(e_0)$$

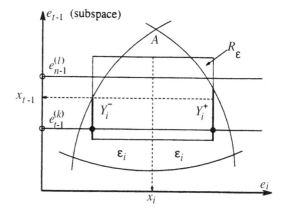

Figure 6-14 Yield and yield derivative estimation for uniform distribution. Y^+ and Y^- denote symbolically the "yields" calculated on the $t - 1$ dimensional faces of the tolerance hyper-cube R_ε. The $e_{t-1}^{(k)}$ points are sampled in the $t - 1$ dimensional subspace of e_{t-1} parameters, with the uniform p.d.f. $f_{e_{t-1}}(e_{t-1})$.

---

[20] Matrix of second derivatives.

It can be shown [33, 51, 52] that yield can be calculated by sampling in the $t - 1$ dimensional subspace of the e-space, represented by $e_{t-1} \equiv (e_1, \ldots, e_{i-1}, e_{i+1}, \ldots, e_t)$ and analytical integration in the one-dimensional subspace e_i, as shown in Figure 6-14. The p.d.f. $f_e(e)$ can be represented as

$$f_e(e) = f_{e_i}(e_i) \cdot f_{e_{t-1}}(e_{t-1})$$

where

$$f_{e_i}(e_i) = \frac{1}{2\varepsilon_i}[U(e_i - \{x_i - \varepsilon_i\}) - U(e_i - \{x_i + \varepsilon_i\})]$$

and $f_{e_{t-1}}(e_{t-1})$ is the joint p.d.f. for all the other variables e_k, $k \neq i$, Using this approach, we can further prove [52] that yield is expressed by

$$Y(x) = \int_{R^{t-1}} \left\{ \int_{-\infty}^{+\infty} \phi(e) f_{e_i}(e_i) \, de_i \right\} f_{e_{t-1}}(e_{t-1}) \, de_{t-1} = E_{e_{t-1}} \left\{ \int_{-\infty}^{+\infty} \phi(e) f_{e_i}(e_i) \, de_i \right\} \quad (6\text{-}31)$$

Differentiating this formula w.r.t. x_i and using all the properties of $\delta(x)$ listed above, we obtain

$$\frac{\partial Y(x)}{\partial x_i} = E_{e_{t-1}}\{\phi(e)[\delta(e_i - (x_i + \varepsilon_i)) - \delta(e_i - (x_i - \varepsilon_i))]\}\frac{1}{2\varepsilon_i} \quad (6\text{-}32)$$

$$= \frac{1}{2\varepsilon_i} E_{e_{t-1}}\{\phi(e_{t-1}, x_i + \varepsilon_i) - \phi(e_{t-1}, x_i - \varepsilon_i)\} = \frac{1}{2\varepsilon_i}(Y_i^+ - Y_i^-)$$

where Y_i^+, Y_i^- are "yields" calculated on the faces of the $t - 1$ tolerance hyper-box R_ε, corresponding to $x_i + \varepsilon_i$ and $x_i - \varepsilon_i$, respectively. Calculation of these "yields" is very expensive, so in [52][21] different algorithms improving efficiency were proposed. In [61], an approximate method using efficient three-level orthogonal array (OA) sampling on the faces of R_ε was proposed, in which (owing to specific properties of OAs) *the same* sampled points were utilized on different faces. (Actually, one-half or one-third of *all* sampled points were available for a single face.) This has led to substantial computational savings and faster convergence.

6.7.6 Large-Sample, Derivative-Based Method for Integrated Circuits

In this case, which is typical of IC yield optimization, yield gradient calculations cannot be performed in the e-space, as was the case for discrete circuits. In general, yield gradient can be calculated by differentiating the $\phi[e(x, \theta)] \equiv \phi(x, \theta)$ term in the θ-space-based yield formula (6-9), derived for the general case, where the general transformation $e = e(x, \theta)$ is used. Differentiation of $\phi(x, \theta)$, however, is not possible in the traditional sense, since $\phi(x, \theta)$ is a nondifferentiable unit step function determined over the acceptability region A, as was the case for the uniform p.d.f. discussed above. One possible solution was proposed in [21]. In what follows, a related, but more general, method proposed in [19] is discussed.

It was first shown in [19] that yield can be evaluated as a *surface-integral* rather than as the *volume-integral* (as is normally done using the MC method). To

---

[21] General formulas for yield gradient calculation for *truncated* p.d.f.'s were derived in [52].

understand this, observe that yield can be evaluated by sampling in the $t-1$ subspace of the t-dimensional θ-subspace, as shown in Figure 6-15, evaluating "local yields" along the lines parallel to the θ_i-axis, and averaging all the local yields.[22] Each "local yield" can be evaluated using the values of the cumulative (conditional) density function $F_{\theta,i}(\cdot)$ along each parallel, calculated at the points of its intersection with the boundary of the acceptability region A. For instance, for the case shown in Figure 6-15, the "local yield" along the parallel shown is equal to $F_{\theta,i}(\theta_i^{(2)}) - F_{\theta,i}(\theta_i^{(1)}) + F_{\theta,i}(\theta_i^{(4)}) - F_{\theta,i}(\theta_i^{(3)})$. This process is equivalent to evaluating the surface integral over the values of the cumulative density function calculated (with appropriate signs) on the boundary of the acceptability region.

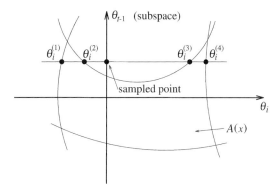

Figure 6-15 Interpretation of "local yield" calculation along a parallel.

The actual yield formula derived in [19] is

$$Y(x) = E_{\theta_{t-1}} \left\{ \sum_k s_k F_{\theta,i}(\theta_i^{(k)}) \right\} \qquad (6\text{-}33)$$

where the expectation is with respect to $\theta_{t-1} \equiv (\theta_1, \ldots, \theta_{i-1}, \theta_{i+1}, \ldots, \theta_t)$, since sampling is performed in the $t-1$ dimensional subspace of the t-dimensional θ-space as shown in Figure 6-15. The summation in Equation 6-33 is over all the points of intersection of all the parallel (as the one shown in Figure 6-15) with the boundary of $A_\theta(x)$, and s_k is a sign factor, assuming the value of 1 if the indicator function $\phi(x, \theta)$ changes from 1 to 0 at $\theta_i^{(k)}$, and the value of -1 if $\phi(x, \theta)$ changes from 0 to 1. As above, $F_{\theta,i}(\cdot)$ is the cumulative density function noise factor for the θ_i, corresponding to the *conditional* p.d.f. [51] $f_{\theta_i}(\theta_i|\theta_{t-1})$ (in general) and to the *marginal* p.d.f. f_{θ_i} of θ_i (as it was assumed in [19]), if the θ_i noise factors are independent. As a result, Equation 6-33 represents a surface integral of $s_k F_{\theta,i}(\theta_i^{(k)})$ calculated over the boundary of $A_\theta(x)$ for any given x.

Differentiating the above equation w.r.t. x_p, we obtain

$$\frac{\partial F_{\theta,i}(\theta_i^{(k)})}{\partial x_p} = \frac{\partial F_{\theta,i}(\theta_i)}{\partial \theta_i}\bigg|_{\theta_i = \theta_i^{(k)}} \cdot \frac{\partial \theta_i^{(k)}}{\partial x_p} = f_{\theta_i}(\theta_i^{(k)}) \cdot \frac{\partial \theta_i^{(k)}}{\partial x_p}$$

---

[22] Identical technique was used in the previous section [51, 52] to derive the yield formula 6-31 (see Figure 6-15): "local yields" are the one-dimensional integrals shown in the braces of Equation 6-31, calculated along the lines parallel to the e_i-axis shown in Figure 6-14.

where the p.d.f. $f_{\theta_i}(\theta_i^{(k)})$ (for single θ_i) has to be calculated at each intersecting point $\theta_i^{(k)}$. Notice that the $A_\theta(x)$ region is the function of x_p, so the intersecting points $\theta_i^{(k)}$ will move with the change of x_p. Recall that at the boundary of $A_\theta(x)$ the equation $y_a(x, \theta_{t-1}, \theta_i) = S_a$ (where $S_a \equiv S_a^L$ or $S_a \equiv S_a^U$) must be satisfied for *one specific*, ath, performance function.[23] Differentiating this equation, we obtain:

$$\frac{\partial y_a}{\partial x_p} + \frac{\partial y_a}{\partial \theta_i} \cdot \frac{\partial \theta_i^{(k)}}{dx_p} = 0$$

from which the required $\partial \theta_i^{(k)}/\partial x_p$ is found as

$$\frac{\partial \theta_i^{(k)}}{\partial x_p} = -\frac{\partial y_a/\partial x_p}{\partial y_a/\partial \theta_i}$$

The two derivatives on the right-hand side of the last equation are calculated for a specific x at the specific intersection point $\theta_i^{(k)}$, shown in Figure 6-15. After a few more steps, the following final yield gradient formula is obtained:

$$\nabla Y(x) = E_{\theta_{t-1}} \left\{ \sum_k f_{\theta_i}(\theta_i^{(k)}) \cdot \frac{\nabla_x y_a(x, \theta)}{|\partial y_a(x, \theta)/\partial \theta_i|} \bigg|_{x, \theta_{t-1}, \theta_i^{(k)}} \right\} \quad (34)$$

where summation is over all intersection points $\theta_i^{(k)}$ of all the parallels (as the one shown in Figure 6-15) with the boundary of $A_\theta(x)$. As mentioned above, y_a is the *specific* performance function $y_j(x, \theta)$, which actually determines the boundary of the acceptability region at the $\theta_i^{(k)}$ intersection point. The gradient $\nabla_x y_a(x, \theta)$ and the derivatives $\partial y_a(x, \theta)/\partial \theta_i$ have to be calculated for every fixed sampled point (x, θ_{t-1}), at each intersecting point $\theta_i^{(k)}$ shown in Figure 6-15. Observe that the derivative calculations in Equation 6-34 can (and often will have to) be performed in two steps: since $y_a(x, \theta) = y_a(e(x, \theta))$, so,

$$\frac{\partial y_a}{\partial x_p} = \sum_s \frac{\partial y_a}{\partial e_s} \frac{\partial e_s}{\partial x_p}$$

and

$$\frac{\partial y_a}{\partial \theta_i} = \sum_s \frac{\partial y_a}{\partial e_s} \frac{\partial e_s}{\partial \theta_i}$$

where the first derivatives appearing under the summation in both formulas are calculated using a circuit simulator and the second ones from a given statistical model $e = e(x, \theta)$.

In the practical implementation presented in [19], random points θ^r were sampled in the θ-space, in the same way as shown in Figure 6-9b but replacing e by θ. Then, searches for the intersections $\theta_i^{(k)}$ were performed in the directions of *all* the axes, the formula in the braces of Equation 6-34 was calculated for each intersection, and averaged out over all outcomes. Searches are performed along the parallel lines in *all* directions in order to increase the accuracy of the yield gradient estimator.

---

[23] Obviously, the a index will be different for different intersection point $\theta_i^{(k)}$, since different performances will be active at the boundary of $A_\theta(x)$.

Observe that this technique requires tens of thousands of circuit analyses to iteratively find the intersection points $\theta_i^{(k)}$, together with additional analyses (if needed) for the calculation of the gradient and the derivatives in 6-34. As an illustration, assume that the number of MC samples is $N = 1000$ (as it was usually assumed in the examples considered in [19]), the number of noise parameters θ, equal to $t = 10$, the average number of intersections with the boundary of the acceptability region per direction, $MI = 2$ (as is often the case for convex A regions), and the average number of circuit analyses required to find iteratively an intersection with the boundary of $A_\theta(x)$ equal, say, to $N\_AI = 4$. Under these assumptions, the method would require $N \times t \times MI \times N\_AI \cong 80{,}000$ circuit analyses per single x point, plus additional effort (not negligible!) for finding (and approximating) all the derivatives required in the gradient formula (6-34).

This problem has been circumvented in [19] by constructing approximating functions $\tilde{y} = \tilde{y}(x, \theta)$ w.r.t. θ for each x, together with approximating functions for all the derivatives.[24] Because of a high level of *statistical* accuracy obtained in evaluating yield and its gradients, an efficient, gradient-based deterministic optimization algorithm, utilizing the Sequential Quadratic Programming technique was used, requiring a small number of iterations (from 5 to 11 for the examples discussed in [19]). The gradient $\nabla_x y_a(x, \theta)$ and the derivatives in Equation 6-34 were either directly obtained from a circuit simulator or (if not available) using (improved) finite difference estimators. The method was quite efficient for a moderate size of the θ-space (10 to 12 parameters).

The resulting yield optimization method is independent of the form of $e = e(x, \theta)$. The derivatives of y_j w.r.t. to both x_k and θ_s are required, and the *analytical* form of $f_\theta(\theta)$ and its cumulative density function must be known. The method cannot work practically without constructing the approximating functions $\tilde{y} = \tilde{y}(x, \theta)$, for the reasons explained above. (Approximating functions in the joint space (x, θ) can also be used, if available).

6.7.7 Small-Sample, Stochastic Approximation-Based Methods for Discrete Circuits

Standard methods of nonlinear programming perform poorly solving problems in which statistical "noise" is added to the objective functions (i.e., yield) and to its derivatives.[25] In yield optimization, the statistical noise is the result of the unavoidable statistical variability of the yield and yield gradient estimators, especially high, if the number of samples used is small. One of the methods *dedicated* to the solution of such problems is the Stochastic Approximation (SA) approach [41] developed for solving the regression equations and adopted to the unconstrained and constrained optimization by several authors. These methods attempt

---

[24] Low-degree polynomials were used with very few terms, generated from a stepwise regression algorithm.

[25] This was the reason high levels of accuracy and consistency of the yield and yield gradient estimators were required (and were actually provided) using the method described in the previous section. Otherwise, the deterministic optimization algorithm used would diverge, as observed in [19].

to find a minimum (maximum) of a function corrupted by noise (i.e., a regression function). The SA methods were first applied to yield optimization and statistical design centering in [57, 58]. The theory of these methods is well established, so its application to yield optimization offers the theoretical background that is missing (e.g., in the heuristic methods of section 6.7.4). As compared to the large-sample methods, the SA algorithms to be discussed use *a few* (often *just one*) randomly sampled points per iteration, which is compensated for by a large number of iterations exhibiting a trend toward solution. The method tends to bring large initial improvements with a small number of circuit analyses, efficiently utilizing the high content of the deterministic information present at the beginning of optimization.

In 1951, in their pioneering work [41], Robins and Monro proposed a scheme for finding a root of a regression function, which they named the Stochastic Approximation procedure. The problem was to find a zero of a function $G(x)$, whose "noise-corrupted" values could be observed only, namely

$$G(x) = g(x) + \theta(x)$$

where $g(x)$ is a given (unknown) function of x, θ is a random variable, such that $E\{\theta\} = 0$, and $var\{\theta\} \leq L < \infty$. Therefore, a zero of the regression function $g(x) \equiv E\{G(x)\}$ was to be found. The SA algorithm proposed in [41] works as follows. Given a point $x^{(k)}$, set the next point as

$$x^{(k+1)} = x^{(k)} - a_k G(x^{(k)})$$

The sequence of $\{x^{(k)}\}$ points converges to the $\hat{x}$ under, for example, the following conditions:[26] $a_k > 0$, $\lim_{k \to \infty} a_k = 0$, $\sum_{k=1}^{\infty} a_k = \infty$, $\sum_{k=1}^{\infty} a_k^2 < \infty$. These conditions are satisfied, for instance, by the harmonic sequence $\{1, \frac{1}{2}, \frac{1}{3}, \frac{1}{4}, \ldots\}$. Assuming that $G(x) \equiv \xi^k$ is a "noise-corrupted" observation of the *gradient* of a regression function $f(x) = E_\theta\{w(x, \theta)\}$, we can use the algorithm to find a stationary point (e.g., a maximum) of $f(x)$, since this is equivalent to finding $\hat{x}$ such that $E\{G(\hat{x})\} = E\{\xi(\hat{x})\} = \nabla_x f(\hat{x}) = 0$. For yield optimization, $f(x) \equiv Y(x) = E\{\phi(x, \theta)\}$. The simplest scheme that can be used, if the yield gradient estimator is available, is

$$x^{k+1} = x^k + \tau_k \xi^k \tag{6-35}$$

where $\xi_k \equiv \widehat{\nabla_x Y}(x^k)$ is an estimate of the yield gradient, based on one (or more) points θ^i sampled with the p.d.f. $f_\theta(\theta)$ and $\tau_k > 0$ is the step-length coefficient selected such that the sequence: $\{\tau_k\} \to 0$, and $\sum_{k=0}^{\infty} \tau_k = \infty$, $\sum_{k=0}^{\infty} \tau_k^2 < \infty$ (e.g., the harmonic series $\{1, \frac{1}{2}, \frac{1}{3}, \ldots\}$ fulfills these conditions). For the convergence with probability one, the required condition for the conditional gradient expectation is $E\{\xi^k \mid x^1, x^2, \ldots, x^k\} = \nabla Y(x^k)$. The algorithm (6-35) is similar to the steepest ascent algorithms of nonlinear programming, so it is slowly convergent for ill-conditioned problems. A faster algorithm was introduced in [43] and used for yield optimization in [57, 58]. It is based on the following iterations:

$$x^{k+1} = x^k + \tau_k d^k \tag{6-36}$$

---

[26] The conditions stated are for the so-called convergence with probability one, which requires that the probability that $\{x^{(k)}\}$ converges to $\hat{x}$, $P\{x^{(k)} \xrightarrow[k \to \infty]{} \hat{x}\} = 1$. For other types of convergence, slightly different conditions apply.

6.7 ■ Statistical Methods of Yield Optimization

$$d^k = (1 - \rho_k)d^{k-1} + \rho_k \xi^k, \quad 0 < \rho_k \leq 1 \tag{6-37}$$

where ξ^k is a (one- or more-point) estimate of $\nabla_x Y(x^k)$ and $\{\tau_k\} \to 0$, $\{\rho_k\} \to 0$ are nonnegative coefficients. $d^{(k)}$ is a convex combination of the previous (old) direction d^{k-1} and the new gradient estimate ξ^k, so the algorithm is an analog of a more efficient, conjugate gradient method. Formula 6-37 provides *gradient averaging*. Observe that $d^k = d^{k-1} + (\xi^k - d^{k-1})\rho_k$, so that ρ_k controls the "memory" or "inertia" of the search direction d^k, as shown in Figure 6-16. If ρ_k is small, the "inertia" of the algorithm is large; that is, the algorithm tends to follow the previous gradient directions. For convergence with probability one, the same conditions must hold as those for Equation 6-35.

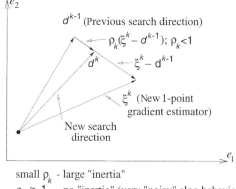

Figure 6-16 An illustration of the gradient averaging equation.

The coefficients τ_k and ρ_k are adaptively adjusted during the course of optimization, based on the heuristic statistical algorithms proposed in [43]. Qualitatively, the major ideas of these algorithms can be explained as follows. The step-length τ_k should be (on average) chosen so that the yield maximum in the current search direction d^k is reached. At the maximum, the current direction d^k and the *next* gradient estimate ξ^{k+1} (at the same point) are (on average) orthogonal; that is, the scalar product $\langle d^k, \xi^{k+1} \rangle$ should be (on average) equal to zero. If $\langle d^h, \xi^{k+1} \rangle \langle 0, \tau_k \rangle$ is too large, if $\langle d^k, \xi^{k+1} \rangle > 0$, τ_k is too small. The algorithm calculates the average value of the scalar products $\langle d^k, \xi^{k+1} \rangle$ within the "learning series" (usually from 5 to 30 iterations), during which τ_k remains constant. The student t-test is then performed to test the hypothesis that $E\{\langle d^k, \xi^{k+1} \rangle\}$ is equal to 0, versus the alternative hypothesis that it is not. Depending on the test results, τ_k is increased, decreased, or it remains constant.

A somewhat similar, but more sophisticated, strategy is used to control the coefficient ρ_k. For proper averaging, ρ_k has to *decrease* in subsequent iterations. This is based on the observation that in general any *arithmetic* mean $\bar{x}^k$ can be expressed after k trials as

$$\bar{x}^k = \frac{1}{k}\sum_{i=1}^{k} x_i = \frac{k-1}{k} \frac{1}{k-1} \sum_{i=1}^{k-1} x_i + \frac{1}{k} x_k = (1 - \rho_k)\bar{x}^{k-1} + \rho_k x_k$$

where $\rho_k = 1/k$. Notice also that the rate of decrease of ρ_k determines how heavily the new gradient information is weighted, as compared to the old averages. An adaptive rate of change of ρ_k is provided by the formula

$$\rho_k = \frac{\rho_{k-1}}{1 + \rho_{k-1} - R} \quad (6\text{-}38)$$

where $0 < \rho_k \leq 1$ and $0 < R < 1$. Parameter R (to be adaptively adjusted by the algorithm) stays constant within the learning series. It can be shown that $\rho_k \to R$ as $k \to \infty$. The closer R is to one, the slower the changes ρ_k are (there is no change of ρ_k, if $R = 1$). Large R values (and thus large values of ρ_k) weight heavily the most recent gradient estimates $\xi^{(k)}$ (low inertia). Usually, $\rho_0 = 1$ at the beginning of optimization. Since the new direction d^k is a convex combination of the old direction d^{k-1} and the current gradient estimate ξ^k, all possible directions d^k belong to the cone created by d^{k-1} and ξ^k, and a particular d^k depends on the actual value of ρ_k. It is intuitively reasonable to assume (a suitable mathematical proof is given in [43]) that the best direction d^k is the one leading to the region where the $Y(x)$ function reaches its maximum. In that region, there is a rapid change of gradient directions, and one of them is orthogonal to the old direction d^{k-1}. Thus, it is reasonable to look for such directions (i.e., for such coefficients R) for which (on average) $E\{\langle d^{k-1}, \xi^{k+1}\rangle\} = 0$ (averaging is performed within the learning series). If $E\{\langle d^{k-1}, \xi^{k+1}\rangle\} < 0$, R is too small, if $E\{\langle d^{k-1}, \xi^{k+1}\rangle\} > 0$, R is too large. Again, the student t-test is used to test the relevant hypothesis, and the value of R is suitably adjusted.

Several other enhancements were introduced to speed up the algorithm convergence, especially in its much refined version proposed in [43]. Yield optimization for discrete circuits (or, in general, for those cases where the condition $e_i = x_i + \theta_i$ is fulfilled) is performed using, in general, the (one- or two-point) yield gradient estimator (6-25); for the normal p.d.f., Equation 6-28 or 6-29 is used.

EXAMPLE 6-6

A simple two-dimensional example [57] will be considered to illustrate the properties of the SA algorithm. Consider a resistive voltage divider, composed of two resistors $e_1 = R_1$, $e_2 = R_2$. The design objective is to maintain the voltage division ratio between $0.45V/V$ and $0.55V/V$ for the random independent resistor variations, described by the model $e_i = x_i + \theta_i$, and the p.d.f. of θ_i is normal with $E\{\theta_1\} = E\{\theta_2\} = 0$, and $cov(\theta_1, \theta_2) = 0$. Moreover, the maximum resistor values are assumed to be equal to 1.2 and $1.3k\Omega$, for R_1 and R_2, respectively. As a result, the acceptability region A for the divider is defined in the two-dimensional space (e_1, e_2) as

$$A = \left\{ e \in R^2 \,\middle|\, 0.45 \leq \frac{e_2}{e_1 + e_2} \leq 0.55; \ 0 \leq e_1 \leq 1.2; \ 0 \leq e_2 \leq 1.3 \right\} \quad (6\text{-}39)$$

The A region is shown in Figure 6-17, together with a typical algorithm trajectory. The initial yield was low (7.4%). Very few (25 to 30) iterations (involving one circuit analysis per iteration) were required to bring the nominal point close to the final solution. This is because at the beginning of optimization, a high content of *deterministic* gradient information is available even from a few sampled points, so that the algorithm progress is rapid. Close

6.7 ■ Statistical Methods of Yield Optimization

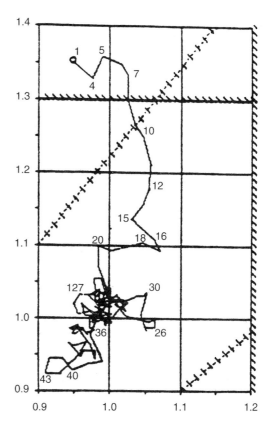

Figure 6-17 A typical trajectory of the Stochastic Approximation algorithm (Example 6-6).

to the solution, however, the yield gradient estimator is very noisy, and the algorithm has to filter the directional information out of noise, which takes the majority of the remaining iterations. After the total of 168 circuit analyses, the yield increases to 85.6%. Other starting points resulted in a similar algorithm behavior. It can be noticed that the yield optimum is "flat" in the diagonal direction, reflected in the observed random nominal point movement in this direction (iterations 30 through 43).

EXAMPLE 6-7

Parametric yield for the Sallen-Key active filter of Figure 6-7a (often used in practice as a test circuit) with the specifications on its frequency response shown in Figure 6-7b was optimized. (Recall that the shape of the acceptability region for this case was very complicated, containing internal nonfeasible regions or "holes.") All R, C variables were assumed designable. The relative standard deviations were assumed equal to 1% for each element. The p.d.f. was normal, the correlation coefficients between the like elements (R's or C's) were assumed equal to 0.7, and zero correlations were assumed between different elements. The initial yield was 6.61%. Based on the SA algorithm as in the previous example, 50 iterations (with one analysis per iteration) brought yield to 46.2%; the next 50 iterations to 60%; and the remaining 132 iterations increased yield to only 60.4% (again the initial convergence was very fast). The algorithm compared favorably with the Hessian matrix-based, large-sample method discussed in Section 6.7.5 [59], which required about 400 circuit analyses to obtain the same yield level and whose initial convergence was also much slower. It has to be stressed, however, that the results obtained are *statistical*, and it is difficult to draw strong general conclusions. One observation (confirmed also by other authors) is that the SA-type algorithms provide, in general, fast initial convergence into the *neighborhood* of the optimal solution, as shown in the examples just investigated.

6.7.8 Small-Sample, Stochastic Approximation Methods for Integrated Circuits

The methods of yield gradient estimation developed for discrete circuits cannot be used for ICs because of the form of the singular $e = e(x, \theta)$ transformation, as discussed in Section 6.3.4. In Section 6.7.6, a complicated algorithm was described for gradient estimation and yield optimization in such situations. This section describes a simple method proposed in [55, 56], based on *random perturbations* in x-space. It is especially useful in those cases where the conditions for the application of the yield gradient formula (34) are hard to meet, namely: the cost of constructing the approximating functions $\tilde{y} = \tilde{y}(x, \theta)$ for fixed x is high, and calculating the gradient of y w.r.t. x is also expensive (which is the case, e.g., if the number of important x and θ parameters is large), and analytical form of $f_\theta(\theta)$ is not available (as it is required in Equation 6-34).

The method applications go beyond yield optimization: a general problem (to be discussed below) is to find a minimum of a general regression function $f(x) = E_\theta\{w(x, \theta)\}$, using the Stochastic Approximation method, in the case where the gradient estimator of $f(x)$ is *not directly available*. Several methods have been proposed to estimate the gradient *indirectly*, all based on adding some *extra perturbations* to x parameters. Depending on their nature, size, and the way the perturbations are *changed* during the optimization process, different interpretations result, and different problems can be solved. In the simplest case, some extra deterministic perturbations (usually double-sided) are added individually to each x_k (one-at-a-time), while random sampling is performed in the θ-space, and the estimator of the derivative of $f(x)$ w.r.t. x_k is found from the different formula

$$\hat{\xi}_k = \widehat{\frac{\partial f(x)}{\partial x_k}} = \frac{1}{N} \sum_{i=1}^{N} \{[w(x + a_k e_k, \theta_1^i) - w(x - a_k e_k, \theta_2^i)]/(2a_k)\} \qquad (6\text{-}40)$$

where $a_k > 0$ is the size of the perturbation step and e_k is the unit vector along the x_k coordinate axis. Usually, $\theta_1^i \equiv \theta_2^i$ to reduce the variance of the estimator. Normally, $N = 1$. Other approaches use *random direction* derivative estimation, by sampling points randomly on a sphere of radius a [42], randomly at the vertices of a hypercube in the x-space [48], or at the points generated by orthogonal arrays [61], commonly used in the design of experiments. Yet another approach, dealing with *nondifferentiable* p.d.f.'s, was proposed in [66].

In what follows, a random perturbation approach resulting in *convolution function smoothing* is described for a more general case, where a *global* rather than a local minimum of $f(x)$ is to be found (i.e., $f(x)$ is a multi-extremal function). In order to better understand the working of the method, assume that the multi-extremal regression function $f(x)$ is characterized by a strong dominant trend toward a global minimum with a number of "shadow" local minima superimposed on it. Such a function can be considered a superposition of a function having just one minimum (i.e., a uni-extremal function) with some other multi-extremal functions of much smaller range of variations than the uni-extremal function. As a result, relatively small deterministic "noise" is added to the uni-extremal function. Moreover, $f(x) = E_\theta\{w(x, \theta)\}$ is the regression function, so the original function $w(x, \theta)$ is itself corrupted by the statistical noise due to θ.

The objective of convolution smoothing can be visualized as "filtering out" the *deterministic* noise (while the random noise is eliminated by the SA algorithm itself) and performing minimization of the "smoothed" uni-extremal function (or, sequentially, a family of such functions), in order to reach the global minimum. Since the minimum of a specific smoothed uni-extremal function does not, in general, coincide with the global function minimum, a sequence of minimization runs is required, with the amount of smoothing eventually reduced to zero in the neighborhood of the global minimum. The smoothing process is performed by averaging $f(x)$ over some region of the n-dimensional parameter space x using a proper smoothing (kernel) function $\hat{h}_\eta(\cdot)$, defined below.

Let the n-dimensional vector $\eta \in R^n$ denote the vector of random perturbations added to x to create the convolution function [42]:

$$\tilde{f}(x,\beta) = \int_{R^n} \hat{h}_\eta(\eta,\beta) f(x-\eta)\, d\eta \qquad (6\text{-}41)$$
$$= \int_{R^n} \hat{h}_\eta(x-\eta,\beta) f(\eta)\, d\eta = E_\eta\{f(x-\eta)\}$$

where $\tilde{f}(x, \beta)$ is the *smoothed approximation* to the original multi-extremal function $f(x)$, and the kernel function $\hat{h}_\eta(\eta, \beta)$ is the p.d.f. used to sample η. Note that $\tilde{f}(x, \beta)$ can be interpreted as an *averaged* version of $f(x)$ weighted by $\hat{h}_\eta(\eta, \beta)$. Parameter β controls the dispersion of η, that is, the degree of $f(x)$ smoothing (e.g., β can control the standard deviations of $\eta_1 \ldots \eta_n$). $E_\eta\{f(x - \eta)\}$ is the expectation with respect to the random variable η. Therefore, an unbiased estimator $\hat{f}(x, \beta)$ of $\tilde{f}(x, \beta)$ is the average: $\tilde{f}(x, \beta) = \frac{1}{N}\sum_{i=1}^N f(x - \eta^i)$, where η is sampled with the p.d.f. $\hat{h}_\eta(\eta, \beta)$. The kernel function $\hat{h}_\eta(\eta, \beta)$ should have the following properties [42]:

(a) $\hat{h}_\eta(\eta, \beta) = \frac{1}{\beta^n} h_\eta(\eta/\beta)$ is piecewise differentiable with respect to η
(b) $\lim_{\beta \to 0} \hat{h}_\eta(\eta, \beta) = \delta(\eta)$ (Dirac's delta functional)
(c) $\hat{h}_\eta(\eta, \beta)$ is a p.d.f.

Under these conditions: $\lim_{\beta \to 0} \tilde{f}(x, \beta) = \int_{R^n} \delta(\eta) f(x - \eta)\, d\eta = f(x - 0) = f(x)$ (based on the properties of Dirac's δ functions discussed in Section 6.7.5). These conditions are fulfilled by several p.d.f.s, for example, the Gaussian and uniform.

For the function $f(x) = x^4 - 16x^2 + 5x$, which has two distinct minima, the smoothed functionals—obtained using Equation 6-41—are plotted in Figure 6-18 for different values of $\beta \to 0$, for: (a) Gaussian and (b) uniform kernels. As seen, smoothing eliminates the local minima of $\tilde{f}(x, \beta)$ if β is sufficiently large. If $\beta \to 0$, then $\tilde{f}(x, \beta) \to f(x)$.

The objective now is to solve the following optimization problem: minimize the smoothed functional $\tilde{f}(x, \beta)$ with $\beta \to 0$ as $x \to \hat{x}$, where $\hat{x}$ is the global minimum of the original function $f(x)$. The modified optimization problem can be written as

$$\min_{x \in R^n} \tilde{f}(x, \beta)$$

with $\beta \to 0$ as $x \to \hat{x}$. Differentiating Equation 6-41 and using variable substitution, we obtain the gradient formula

$$\nabla_x \tilde{f}(x, \beta) = \int_{R^n} \nabla_\eta \hat{h}_\eta(\nu, \beta) f(x - \eta) \, d\eta = \frac{1}{\beta} \int_{R^n} \nabla_\eta h_\eta(\eta) f(x - \beta\eta) \, d\eta \qquad (6\text{-}42)$$

where $\hat{h}_\eta(\cdot)$ is as defined above and $h_\eta(\cdot)$ is a normalized version of $\hat{h}_\eta(\cdot)$, as described by (a)–(c) above. For a normalized multinormal p.d.f. with zero correlations selected for h_η, the gradient of $\tilde{f}(x, \beta)$ is

$$\nabla_x \tilde{f}(x, \beta) = \frac{-1}{\beta} \int_{R^n} \eta f(x - \beta\eta) h_\eta(\eta) \, d\eta$$

$$= \frac{-1}{\beta} E_\eta \{\eta f(x - \beta\eta)\} = \frac{-1}{\beta} E_{\eta,\theta} \{\eta w(x - \beta\eta, \theta)\} \qquad (6\text{-}43)$$

where sampling is performed in the x-space with the p.d.f. $h_\eta(\eta)$, and in the θ-space with the p.d.f. $f_\theta(\theta)$; $E_{\eta,\theta}$ denotes expectation w.r.t. *both* η and θ, and it was taken into account that $f(x)$ is a noise-corrupted version of some function of interest

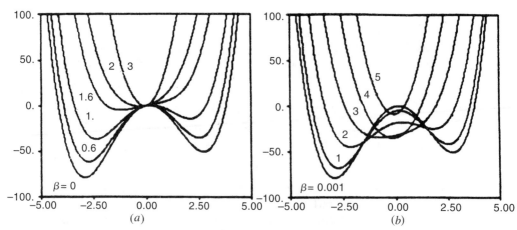

Figure 6-18 Smoothed functional $\tilde{f}(x, \beta)$ for different β's using: (a) Gaussian kernel, (b) uniform kernel.

$w(x, \theta)$, that is, $f(x) = E_\theta\{w(x, \theta)\}$. The unbiased *single-sided* gradient estimator is, therefore

$$\hat{\nabla}_x \tilde{f}(x, \beta) = \frac{-1}{\beta} \frac{1}{N} \sum_{i=1}^{N} \eta^i w(x - \beta\eta^i, \theta^i) \tag{6-44}$$

In practice, a *double-sided* estimator [42] of smaller variance is often used

$$\hat{\nabla}_x \tilde{f}(x, \beta) = \frac{1}{2\beta} \frac{1}{N} \sum_{i=1}^{N} \eta^i [w(x + \beta\eta^i, \theta_1^i) - w(x - \beta\eta^i, \theta_2^i)] \tag{6-45}$$

Normally, $N = 1$ for best overall efficiency. Statistical properties of these two estimators (such as their variability) were studied in [60]. To reduce variability, *the same $\theta_1^i = \theta_2^i$ are used in Equation 6-45 for positive and negative $\beta\eta^i$ perturbations.* For yield optimization, $w(\cdot)$ is simply replaced by the indicator function $\phi(\cdot)$. For multi-extremal problems, β values should originally be relatively large and then systematically reduced to some small number rather than to zero. For single-extremal problems (this might be the case for yield optimization), it is often sufficient to perform just a *single* optimization with a relatively small value of β, as was done in the examples that follow.

6.7.9 Case Study: Process Optimization for Manufacturing Yield Enhancement

The objective of the work presented in [56] was to investigate how to modify the MOS control process parameters together with a *simultaneous* adjustment of transistor widths and lengths to maximize parametric yield.[27] To make it possible, process/device simulator(s) (such as FABRICS, SUPREM, and PISCES) have to be used together with a circuit simulator. In what follows, an example is shown in which FABRICS [16, 29, 31] is used as a process/device simulator. IC technological process parameters are statistical, but variations of some of the parameters (e.g., times of operations, implant doses) might have small relative variations, and some parameters are common to several transistors on the chip. Because of that, the transformation $e = e(x, \theta)$ (where θ are now process related random variables) is such that standard methods of yield optimization developed for discrete circuits cannot be used. Let $Z_1 = z_1 + \xi_1$ be the process control parameters (doses, times, temperatures), where ξ_1 are random parameter variations and z_1 are deterministic *designable* parameters. Let $Z_2 = z_2 + \xi_2$ be the designable layout dimensions, where z_2 are designable and ξ_2 are random variations (common to several transistors on a chip). $P = p + \psi$ are process physical parameters (random, nondesignable) such as diffusivities and impurity concentrations (as above, p is the nominal value and ψ is random). All random perturbations are collected into the vector of random parameters θ, also called the vector of *process disturbances*: $\theta = (\xi_1, \xi_2, \psi)$. The vector of designable parameters $x = (z_1, z_2)$ is composed of the process z_1 and lay-

---

[27] This might be important, for example, for the process refinement and IC cell *redesign*.

out z_2 designable parameters. Therefore, x and θ are in *different* spaces (subspaces).

There are also other difficulties: the analytical form of the p.d.f. of θ is most often not known, since θ parameters are hierarchically generated from a numerical procedure [16], the derivatives of the performances y w.r.t. x and θ are not known from FABRICS and can only be estimated by finite differences, and the θ- and x-spaces are large (see below). So, creating approximation or finding derivatives using finite differences would be expensive. Because of these difficulties, the smoothed-functional approach discussed in this section will be used, as shown in the following example.

EXAMPLE 6-8

The objective is to maximize parametric yield for the NMOS NAND gate shown in Figure 6-19 [56], by automated adjustment of process and layout parameters. Specifications are: $V_0 = V_{out}(t = 0) \leq 0.7V$, $V_{out}(t_1 = 50ns) > 6.14V$, circuit area $\leq 2500 \mu m^2$. There are 45 designable parameters: all 39 technological process parameters, 6 transistor dimensions, and about 40 noise parameters, so it is a large problem, suitable for use of the Stochastic Approximation-based random perturbation method described above.

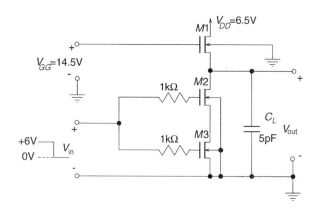

Figure 6-19 NMOS NAND gate of Example 6-8.

The initial yield was $Y = 20\%$. The method of random perturbations was used with 2% relative standard deviations of the perturbations added to each designable parameter. After the first optimization run, involving the total of 110 FABRICS/SPICE analyses, yield increased to 100% and the nominal area decreased to 2138 μm^2. Then, the specs were tightened to $V_{out}(t_2 = 28ns) > 6.14V$ with the same constraints on V_0 and area, causing yield to drop to 10.1%. After 60 FABRICS/SPICE analyses, using the perturbation method, yield increased to: $Y = 92\%$, and area = 2188 μm^2. These much improved results produced the nominal circuit responses shown in Figure 6-20. Several technological process parameters were changed during optimization in the range between 0.1% and 17%: times of oxidation, annealing, drive-in, partial pressure of oxygen, and others, while the transistor dimensions changed in the range between 0.8% and 6.3%. The cost of obtaining these results was quite reasonable: the total of 170 FABRICS/SPICE analyses, in spite of the large number of optimized and noise parameters. Other related examples are discussed in [56].

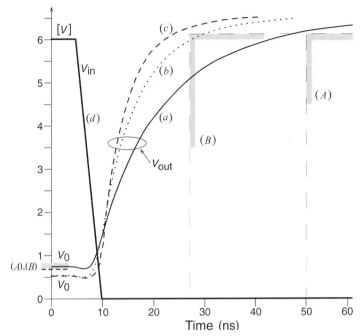

Figure 6-20 Nominal transient responses for Example 6-8: (a) Initial. (b) After first optimization. (c) After tightening the specs and second optimization.

6.8 DESIGN FOR QUALITY

Parametric yield is not the only objective that should be considered during statistical circuit design. Often more important is the need to minimize variabilities of circuit performances, caused by various manufacturing and environmental disturbances. Variability minimization has been an important issue in circuit design for many years [24] (see also [24] for earlier references). It also leads (indirectly) to parametric yield improvement. Circuits characterized by low-performance variability are regarded as high-quality products. Most recently, variability minimization has been reintroduced into practical industrial design through the work of Taguchi [38]. Utilizing some of the Design of Experiments [38, 75] techniques, Taguchi has successfully popularized the idea of Off-line Quality Control (rather than the traditional In-Line Quality Control). His overall strategy, whose major objective is to design products characterized by low-performance variability and performances y_j tuned to their "target" values S_j^T, has been extensively applied in practice. Yield optimization, variability minimization, Taguchi techniques, and other generalized methodologies, are often referred to as the area of Design for Quality (DFQ).

In the next section, a generalized approach proposed in [53] is discussed, in which a broad range of various problems can be solved, including yield optimization and Taguchi's variability minimization as special cases. A statistical approach, utilizing Stochastic Approximation optimization and the general gradient formulas developed in Sections 6.7.7 and 6.7.8, is discussed. Then, in Section 6.8.2 a brief introduction to the *deterministic* DFQ techniques, based on Propagation of Variances, is presented.

6.8.1 Generalized Formulation of Yield, Variability, and Taguchi Circuit Optimization Problems

It is convenient to introduce an M-dimensional vector $g(e) \in R^M$ of *scaled* constraints, which is composed of the vectors $g^L(e)$, and $g^U(e)$, defined as

$$g_k^L(e) = \frac{y_k(e) - S_k^T}{S_k^L - S_k^T}, \quad k = 1, \ldots, M_L \tag{6-46}$$

$$g_r^U(e) = \frac{y_r(e) - S_r^T}{S_r^U - S_r^T}, \quad r = 1, \ldots, M_U \tag{6-47}$$

where $e = e(x, \theta)$ and, $M = M_L + M_U \leq 2m$. (Note that in general $M < 2m$, since some of the lower or upper specs might not be defined.) These constraints are linear functions of $y_j(\cdot)$; they have important properties: if $y_j = S_j^T$, $g_j^L = g_j^U = 0$; if $y_j = S_j^L$, $g_j^L = 1$; if $y_j = S_j^U$, $g_j^U = 1$. For $S_j^L < y_j < S_j^U$ and $y_j \neq S_j^T$, either g_j^L or g_j^U is greater than zero but never both.

For any "good" design, we would like to make *each* normalized constraint introduced above equal to zero, which might not be possible because of various existing design tradeoffs. Therefore, this is a typical example of a *multi-objective optimization problem*. The proposed scaling helps to consider various tradeoff situations. The Taguchi "on-target" design with variability minimization is formulated as follows [for a *single*-performance function $y(e)$]:

$$\min_{x \in R^n} \{M_{TAG}(x) = E_\theta\{[y(e(x,\theta)) - S^T]^2\} = \text{var}\{y(x)\} + [\bar{y}(x) - S^T]^2\} \tag{6-48}$$

where $\bar{y}(x)$ is the mean (average) value of y (w.r.t. θ), and M_{TAG} (the Taguchi measure) is called the expected quality loss function. (In actual implementations, Taguchi uses related but different performance statistics). As seen, the Taguchi measure M_{TAG} is composed of the *variability* term $\text{var}\{y(x)\}$ and the *bias* term $[\bar{y}(x) - S^T]^2$. Generalization of Equation 6-48 to *several* objectives can be done in many different ways. One possible approach is to introduce $u(e) = \max_{s=1,\ldots,M}\{g_s(e)\}$ and define the Generalized (Expected) Loss Function (GLF) as

$$M_{GLF}(x) = E_\theta\{[u(e(x, \theta))]^2\} \tag{6-49}$$

This generalization is meaningful, since $u(e) = \max_{s=1,\ldots,M}\{g_s(e)\}$ is a scalar, and for the proposed scaling $u(\cdot) > 0$, except for the case where all the performances assume their target values S_j^T, so it is equal to zero. Since 6-46 and 6-47 are less than or equal to 1 any time $S_j^L \leq y_j(e) \leq S_j^U$, then $e = e(x, \theta)$ belongs to A (*the acceptability region* in the e-space), if $u(e) < 1$. Let us use the complementary indicator function, defined as: $\phi_F[u(e)] = 1$, if $u(e) \geq 1$ [i.e., $e \notin A$, or $\theta \notin A_\theta(x)$] and equal to zero otherwise. Maximization of the parametric yield $Y(x)$ is equivalent to the minimization of the probability of failures $F(x) = 1 - Y(x)$, which can be formulated as the following minimization process, w.r.t. x

$$\min_{x \in R^n} \{F(x) = P\{\theta \notin A_\theta(x)\} = \int_{R^t} \phi_F(u(x, \theta)) f_\theta(\theta) \, d\theta = E_\theta[\phi_F(u(x, \theta))]\} \quad (6\text{-}50)$$

Often, some additional (e.g., box) constraints are imposed on x. The scalar "step" function $\phi_F(\cdot)$ (whose argument is also a scalar function $u(e) = \max_{1,\ldots,M}\{g_s(e)\}$) can now be generalized into a scalar weight function $w(\cdot)$, in the same spirit as in the Zadeh fuzzy set theory [73, 74]. For further generalization, the original p.d.f. $f_\theta(\theta)$ used in Equation 6-50 is parametrized, multiplying θ by the *smoothing parameter* β to control the dispersion of e, which leads to the following optimization problem, utilizing a *Generalized Measure* $M_w(x, \beta)$

$$\min_{x \in R^n} \{M_w(x, \beta) = E_\theta\{w(u(e(x, \beta\theta)))\}\} \quad (6\text{-}51)$$

where $u(e(x, \beta\theta)) = \max_{s=1,\ldots,M}\{g_s(e(x, \beta\theta))\}$. If $0 \leq w(\cdot) \leq 1$, $M_w(\cdot)$ corresponds to the *probability measure* of a *fuzzy event* introduced by Zadeh [74]. The choice of $w(\cdot)$ and β leads to different optimization problems and different algorithms for yield/variability optimization. The standard yield optimization problem results if $w(\alpha) = \phi_F(\alpha)$ and $\beta = 1$ (Equation 6-50) (Figure 6-21a). Yield can also be approximately optimized *indirectly*, using a "smoothed" (e.g., sigmoidal) membership function $w(\alpha)$ (Figure 6-21e). For variability minimization, we have a whole family of possible approaches: (1) The generalized Taguchi approach with $w(\alpha) = \alpha^2$ (see Equation 6-49) and $\beta = 1$ (see Figure 6-21c). (2) If $w(\alpha) = \alpha$ and $\beta = 1$ is kept *constant*, we obtain a *statistical* minimax problem, since the expected value of the max function is used; this formulation will also lead to performance variability reduction. (3) If $w(\alpha)$ is piecewise constant (Figure 6-21d), the approach is equivalent to the *Income Index Maximization* with separate *quality classes*, introduced in

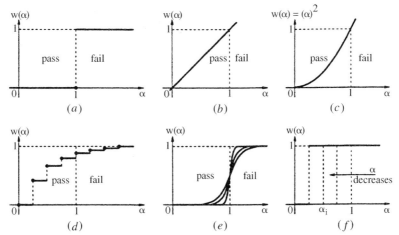

Figure 6-21 Various cases of the weight function $w(\alpha)$.

[35] and successfully used there for the increase of the percentage of circuits belonging to the best classes, that is, those characterized by small values of $u(e)$. This approach also reduces the variability of circuit performance functions. For all the cases, optimization is performed using the proposed Stochastic Approximation approach. The smoothed functional gradient formulas (6-44) or (6-45) have to be used, in general, since the "discrete-circuit" type-gradient formulas are valid only for a limited number of situations.

EXAMPLE 6-9

The objective of this example [39] was to reduce performance variability and to tune to target values the performances of the MOSFET-C filter shown in Figure 6-22. The MOSFETS are used in groups of four transistors, implementing equivalent resistors with improved linearity. Design specifications are: f_0—center frequency, with the S^L, S^T, S^U values specified as {45, 50, 55}kHz; H_0—the voltage gain at f_0, with the specs {15, 20, 25}dB; and Q—the pole quality factor, with the specs {8, 10, 12}. The Taguchi-like "on-target" design and tuning was performed by minimizing the generalized measure 6-51 with $w(x) = \alpha^2$, using the Stochastic Approximation approach with convolution smoothing, since the designable and random parameters are in different (sub) spaces. The circuit has the total number of 90 transistors, so its *direct* optimization using transistor-level simulation would be too costly. Therefore, in [39] the operational amplifiers (Op-Amps) shown in Figure 6-22 were modeled by a *statistical macromodel*, representing the most relevant Op-Amp characteristics: dc gain

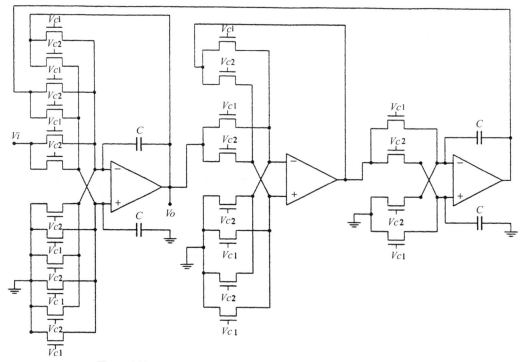

Figure 6-22 The MOSFET-C bandpass filter optimized in Example 6-9.

A_0, output resistance R_0, and the -20dB/dec frequency roll-off. Transistor-model parameters were characterized by a statistical model developed in [13], based on six *common factors* (see Section 6.3): t_{ox} (oxide thickness), ΔL_n, ΔL_p (length reduction), $N_{SUB,n}$, $N_{SUB,p}$ (substrate doping), and x_{jp} (junction depth). All other transistor model parameters were calculated from the six common factors using second-order regression formulas developed in [13]. The major difficulty was to create *statistical* models of the Op-Amp macromodel parameters A_0, R_o, and f_{3dB} as the functions of the six common factors listed above. A special extraction procedure (similar, in principle, to the Factor Analysis model [27]) was developed, and the relevant models were created. The ideal matching between the transistor model parameter of the 22 transistors of each Op-Amp was assumed for simplicity.[28] (Correlational dependencies *between* individual parameters for *individual* transistors were maintained using the statistical model of [13], described earlier.) Moreover, ideal parameter matching between the three macromodels (for the 3 Op-Amps) was also assumed. Mismatches between the threshold voltages and K_p (gain) coefficients of all the transistors in Figure 6-22 were taken into account, introducing an additional 48 noise parameters. So, including the 6 global noise parameters discussed above, the overall number of θ parameters was 54. Moreover, it was found that even if a large number of noise parameters had small individual effect, their *total* contribution was significant. Thus, no meaningful reduction of the dimension of the θ-space was justified. Because of that, it was difficult to use approximating models in the θ-space, and even more difficult in the joint (x, θ) space of $7 + 54 = 61$ parameters. Therefore, no approximation was used.

The Monte Carlo studies showed that the proposed statistical macromodel provided quite reasonable statistical accuracy for f_0 (less than 1% errors for both the mean and the standard deviation) and less than 9.5% errors for both H_0 and Q. The macromodel-based analysis was about 30 times faster than using the full device-level SPICE analysis. A total of 25 designable parameters were selected, including transistor channel length of the "resistor-simulating transistors," 4 capacitors, and one of the gate voltages. Because of the tracking of the designable parameters for transistor quadruples, the actual number of optimized parameters x was reduced to 7: $x = (V_{c2}, C_1, C_2, L_2, L_3, L_4, L_5)$, where V_{c2} denotes the gate voltage shown in Figure 6-22, and C_i, L_j represent the capacitors and channel lengths, respectively. The original yield was 43.5%, relative standard deviations for H_0, f_0, and Q were 3.98%, 7.24%, and 22.06%, respectively, and all nominal values were close to their target values. After Stochastic Approximation optimization with convolution smoothing, involving 5% perturbations for each of the x parameters, and the total of about 240 circuit analyses, yield increased to 85.0%, and H_0, f_0, and Q relative standard deviations were reduced to 3.87%, 6.35%, and 13.91%, respectively. Therefore, the largest variability reduction (for Q) was about 37%, with simultaneous significant yield increase.

This example demonstrates a typical approach that has to be taken in the case of large analog circuits: the circuit has to be (hierarchically) macromodeled first,[29] and suitable *statistical* macromodels have to be created, including mismatch modeling.

6.8.2 Propagation of Variance Method

An important class of approximate DFQ methods, aimed at estimating performance variability and its minimization, utilizes linear approximation to the performance functions $y = y(x, \theta)$ with respect to (w.r.t.) θ for fixed x. This approach is

---

[28] A more sophisticated model, taking the mismatches into account, was later proposed in [40].
[29] Behavioral models can also be used.

usually cheaper than the Monte Carlo-based methods and offers a reasonable approximation to the covariance matrix K^y of the y parameters. The knowledge of K^y allows calculation of approximate values of various variability measures, such as the Taguchi Measure (6-48), C_p/C_{pk} indices [44] (often used to control the IC manufacturing process), and other measures [64, 75]. Linear approximation can be especially attractive when a circuit simulator is available which can directly (and cheaply) calculate circuit sensitivities (derivatives), used to construct the approximation. Otherwise, derivatives are estimated by finite difference approximation, perturbing θ along the coordinate axes about its nominal point and using the circuit simulator to calculate the relevant differences in performances.

Linear approximation to the performance functions $y = y(x, \theta)$ for fixed x can be obtained from a truncated Taylor series expansion of $y = y(\theta)$. Assuming that the changes $\Delta\theta = \theta - \overline{\theta}$ ($\theta \in R^t$) are small

$$y = y(\overline{\theta} + \Delta\theta) \cong y^0 + J(x)\Delta\theta \qquad (6\text{-}52)$$

where $y^0 \stackrel{\Delta}{=} y(\overline{\theta})$ and $\overline{\theta} \equiv E\{\theta\}$. $J(x)$ is the Jacobian matrix w.r.t. θ for a specific, fixed x, defined as

$$J(x) = \left.\frac{\partial y(x, \theta)}{\partial \theta}\right|_{\substack{x \text{ fixed} \\ \theta = \overline{\theta}}} = \frac{\partial y(\theta)}{\partial \theta} \qquad (6\text{-}53)$$

that is, the jith element of $J(x)$ is expressed as $[J(x)]_{ji} \stackrel{\Delta}{=} \partial yy_j(\theta)/\partial \theta_i$. The expected value of y can be approximated from (52) as follows:

$$\overline{y} = E\{y\} \cong E\{y^0\} + J(x)E\{\Delta\theta\} = y^0 \qquad (6\text{-}54)$$

since $E\{\Delta\theta\} = E\{\theta\} - E\{\overline{\theta}\} = 0$. An approximate expression for the covariance matrix K^y of y can be found by applying the covariance formula

$$K^y = E\{(y - E\{y\})(y - E\{y\})^T\} \cong E\{J(x)\Delta\theta\Delta\theta^T J(x)^T\} \qquad (6\text{-}55)$$
$$= J(x)E\{\Delta\theta\Delta\theta^T\}J(x)^T$$

where, as seen from Equations 6-52 and 6-54, $y - E\{y\} \approx J(x)\Delta\theta$, and T denote matrix transposition. From the definition of the covariance matrix, it is seen that $E\{\Delta\theta\Delta\theta^T\}$ is the covariance matrix K^θ for θ. Thus,

$$K^y \cong J(x)K^\theta J(x)^T \qquad (6\text{-}56)$$

Equation 6-56 is often referred to as the Propagation of Variance-covariance (POV) formula and is frequently used in practice. It can be readily shown from 6-56 that the variance of y_j, $var\{y_j\}$ can be approximated as

$$var\{y_j\} \cong \sum_{k=1}^{t}\sum_{i=1}^{t} \frac{\partial y_j}{\partial \theta_k}\frac{\partial y_j}{\partial \theta_i} cov\{\theta_k, \theta_i\} \qquad (6\text{-}57)$$

6.8 ■ Design for Quality

In practice, it is often more convenient to use the normalized derivatives, defined as

$$S_{\theta_i}^{y_j} \triangleq \frac{\partial \ln y_j}{\partial \ln \theta_i} \equiv \frac{\partial y_j(\overline{\theta})}{\partial \theta_i} \frac{\overline{\theta}_i}{\overline{y}_j}; \quad \overline{\theta}_i \neq 0, \overline{y}_j \neq 0 \tag{6-58}$$

which is called the (relative) *sensitivity* of y_j w.r.t. θ_i, so the *relative* variance of y_i is expressed as

$$\frac{\text{var}\{y_j\}}{\overline{y}_j^2} \cong \sum_{k=1}^{t} \sum_{i=1}^{t} S_{\theta_k}^{y_j} S_{\theta_i}^{y_j} \frac{\sigma_{\theta_k}}{\overline{\theta}_k} \frac{\sigma_{\theta_i}}{\overline{\theta}_i} \rho_{ki} \tag{6-59}$$

where σ_{θ_k} is the standard deviation of θ_k and $\rho_{ki} = cov\{\theta_k, \theta_i\}/(\sigma_{\theta_k}\sigma_{\theta_i})$ is the correlation coefficient between θ_k and θ_i. The POV formula is used to estimate the standard deviations of all the circuit performances and to construct various performance variability/tuning measures, frequently used in practice. Their form depends on a particular application.

As an example, let us consider *capability indices* C_p/C_{pk}, commonly used in semiconductor industry [44]. These measures account for both performance variability relative to the lower and upper specifications S^L and S^U (for a single performance y) and performance tuning to the inside of the specification range. For a single performance y, the *potential capability index* C_p [44] is defined as

$$C_p = \frac{S^U - S^L}{6\sigma_y} \tag{6-60}$$

In an idealized case, where the distribution of y is Gaussian, C_p relates the width of the specification limits $\pm 3\sigma_y$ to the width of the y performance function variations. That is, $C_p = 1$ indicates that the width of the specification limits is the same as the $\pm 3\sigma_y$ width of the performance function variations. In Figure 6-23, two cases of $C_p = 1$ and $C_p = 2$ are shown, with the mean $\overline{y}$ allowed to vary within the $\pm 1.5\sigma_y$ interval about the midpoint between S^L and S^U.

It has been established [44] that $C_P = 2$ is often a very desirable goal of process quality control. It ensures that at least 99.99966% of the products (or parts) will be

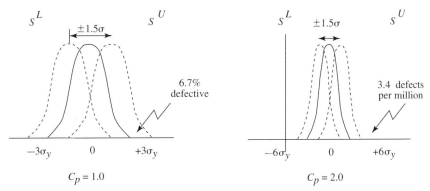

Figure 6-23 Interpretation of C_p capability indices.

nondefective, if the mean is centered within $\pm 1.5\sigma_y$ from the midpoint of the specifications (fewer than 3.4 defects per million).

In order to include the centering effects, the *mean-based capability indices* are used [44], namely,

$$C_{pu} = \frac{S^U - \bar{y}}{3\sigma_y} \tag{6-61}$$

$$C_{pl} = \frac{\bar{y} - S^L}{3\sigma_y} \tag{6-62}$$

combined together into the C_{pk} *performance capability index*

$$C_{pk} = \min\{C_{pu}, C_{pl}\} \tag{6-63}$$

where $\bar{y}$ is the actual mean of the y performance (process) parameters. C_{pk} relates the distance of the actual mean $\bar{y}$ from the *nearest* (most critical) specification limit to the $\pm 3\sigma_y$ interval. The interpretation of C_{pk} is shown in Figure 6-24. Even if $C_p = 2$, $C_{pk} = 1.25$ due to the lack of centering. If $\bar{y} = \frac{S^L + S^U}{2}$, then $C_p = C_{pk}$. In [63] capability indices were combined into a composite performance/variability measure, defined for the jth performance as

$$\Phi_j^L(x) = \frac{S_j^U - S_j^L}{6\sigma_{y_j}} + \lambda \frac{\frac{S_j^U + S_j^L}{2} - \bar{y}_j}{3\sigma_{y_j}} \tag{6-64}$$

$$\Phi_j^U(x) = \frac{S_j^U - S_j^L}{6\sigma_{y_j}} + \lambda \frac{\bar{y} - \frac{S_j^U + S_j^L}{2}}{3\sigma_{y_j}} \tag{6-65}$$

$$\Phi_j(x) = \min\{\Phi_j^U(x), \Phi_j^L(x)\} \tag{6-66}$$

where $0 \leq \lambda \leq 1$. By selecting a suitable value of λ, either variability minimization (λ close to zero) is stressed or tuning *and* variability minimization (λ close to one) are simultaneously performed. For $\lambda = 0$, the measure is equal to C_p, and for $\lambda = 1$, equal to C_{pk}. Often, as suggested by Taguchi [38, 75], it is advantageous to start with the small values of λ (mostly variability minimization) and then increase λ to one, such that tuning to the specifications is also accomplished. This process is quite

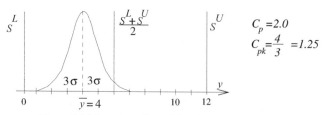

Figure 6-24 Interpretation of the C_{pk} capability index.

often more successful than direct minimization of C_{pk} (which should theoretically accomplish the same goal).

The total measure to be optimized is obtained by combining the individual measures for all m performances using the *min* norm[30]

$$\Phi(x) = \max_{x} \min_{j=1,\ldots,m} \{\Phi_j(x)\} \qquad (6\text{-}67)$$

The first term in the measure definitions, 6-64, 6-65, is $C_{p,j}$; in order to reduce it, the standard deviation σ_{y_j} must be made smaller. The quantity $\frac{S_j^U + S_j^L}{2}$, the midpoint of the specification limits, is often the same as the desirable performance target value S_j^T. The second (bias) term of the measure is a weighted (by $3\sigma_{y_j}$) departure of $\bar{y}$ from the midpoint (target value) of the specification limits. Parameter λ can be thought of as a weight coefficient to the bias term to be selected by the user (or automatically by a suitable algorithm). The entire POV methodology was implemented in GOSSIP[31] [36], a specialized DFQ optimization system.

All the derivatives required for the POV formulas (used to calculate σ_{y_j}'s) are most often approximated using the finite difference approach. This process can also be expensive, if the dimension of θ is large.

In order to improve efficiency, the advanced function approximation methods developed in [62] were used in [63] to approximate the performance functions $y = y(x, \theta)$ in the joint space (x, θ). An overall design methodology utilizing the measures 6-64 through 6-67 was proposed in [63]. Quite good results were obtained, leading to significant performance variability reduction and much improved overall circuit performance. The approximation used reduced the time of circuit simulations by a factor of about 3 to 5, in comparison to the circuit simulation time required by the original approach using the direct finite-difference calculation of the required sensitivities.

In [64] a new approach to efficient POV-based circuit characterization and optimization was proposed, using linear algebra concepts, including the Singular Value Decomposition (SVD) of the system Jacobian matrix. It was shown that several measures of the overall system performance variability can be related to the sum of squares of the singular values of the Jacobian matrix and thus to its Frobenius norm [15]. Since some of the singular values are often very small, a reduced-space characterization of the system variability was obtained, which is important for better circuit understanding by circuit designers. Moreover, an efficient, SVD-based method of updating variability measure was proposed (a measure related to the inverses of the C_p indices was actually used), often eliminating about 60 to 80% of the required Jacobian matrix calculations during the optimization process. This has led to significant computational savings, with the optimal results quite similar to those obtained re-calculating the full Jacobian matrix at each iteration.

---

[30] Other norms (such as the Euclidean norm) can also be used here.
[31] Generic Optimization System for Statistical Improvement of Performance.

6.9 CONCLUSION

Selected techniques of parametric yield optimization, circuit quality enhancement, and related topics of statistical circuit design (such as statistical modeling) have been presented in this chapter. Since this area has been under research and development for at least the last three decades, it is not surprising that other methods not described here have been proposed. Indeed, a vast literature on the subject exists. Other techniques of yield estimation and optimization can be found, for instance, in [28] (process optimization) and in [21, 68, 70, 72, 71], including other approaches to yield generalization/Taguchi design/Design for Quality developed in [19, 35, 69]. DFQ methodologies are used in many areas: for IC design, IC manufacturing process optimization, and other areas of manufacturing [38]. The major objective of DFQ is not to maximize the parametric yield as the only design criterion, but to minimize the performance variability around the designer-specified target performance values. This has been a subject of research for several years using a sensitivity-based approach, and (most recently) using the Taguchi methodology based on Design of Experiments techniques rather than on the use of sensitivities. Some of the automated methods implementing Taguchi-like design were also described.

In general, statistical circuit design remains an active research area, but several mature techniques already exist and have been practically applied to sophisticated industrial IC design. This is of great importance to the overall manufacturing cost reduction, circuit quality improvement, and reduction of the overall IC design cycle.

REFERENCES

[1] H. L. Abdel-Malek and J. W. Bandler, "Yield Optimization for Arbitrary Statistical Distributions: Part I—Theory," *IEEE Transactions on Circuits and Systems,* CAS-27, no. 4, pp. 245–253, Apr. 1980.

[2] K. J. Antreich, H. E. Graeb, and C. U. Weiser, "Circuit Analysis and Optimization Driven by Worst-case Distances," *IEEE Transactions on Computer-Aided Design,* vol. 13, no. 1, pp. 57–71, Jan. 1994.

[3] K. J. Antreich and R. K. Koblitz, "Design Centering by Yield Prediction," *IEEE Transactions on Circuits and Systems,* CAS-29, pp. 88–95, Feb. 1982.

[4] P. Balaban and J. J. Golembeski, "Statistical Analysis for Practical Circuit Design." *IEEE Transactions on Circuits and Systems,* CAS-22, no. 2, pp. 100–108, Feb. 1975.

[5] J. W. Bandler, "Optimization of Design Tolerance Using Nonlinear Programming," *Journal of Optimization Theory and Applications,* vol. 14, pp. 99, 1974. Also in: Proceedings of Princeton Conference on Information Sciences and Systems. Princeton, NJ, p. 655, Feb. 1972.

[6] J. W. Bandler and H. L. Abdel-Malek, "Optimal Centering, Tolerancing and Yield Determination Via Updated Approximations and Cuts," *IEEE Transactions on Circuits and Systems,* CAS-25, pp. 853–871, 1978.

[7] J. W. Bandler and S. H. Chen, "Circuit Optimization: The State of The Art,"

IEEE Transactions on Microwave Theory Techniques, vol. 36, no. 2., pp. 424–443, Feb. 1988.

[8] J. W. Bandler, S. H. Chen, and R. M. Biernacki, "A New Formulation for Yield Optimization." In *IEEE MTT-S International Microwave Symposium Dig.,* vol. 36, no. 2, pp. 1465–1468, Albuquerque, NM, June 1992.

[9] J. W. Bandler, S. H. Chen, R. M. Biernacki, K. Madsen, L. Gao, and H. Yu. Robustizing Circuit Optimization Using Huber Functions," *IEEE International Microwave Symposium Dig.,* Atlanta, GA, 1993.

[10] J. W. Bandler, P. C. Liu, and H. Tromp, "A Nonlinear Programming Approach to Optimal Design Centering, Tolerancing and Tuning," *IEEE Transactions. CAS,* CAS-23, p. 155, Mar. 1976.

[11] P. W. Becker and F. Jensen, *Design of Systems and Circuits for Maximum Reliability or Maximum Production Yield,* New York: McGraw-Hill, 1977.

[12] B. B. Cantell, "Estimation of Overdispersion in Binomial Models: An Application to Wafer Yield Data," Master's thesis, Worcester Polytechnic Institute, Dec. 17, 1993.

[13] J. Chen and M. A. Styblinski, "A Systematic Approach of Statistical Modeling and Its Application to CMOS Circuits," In *Proceedings of IEEE International Symposium on Circuits and Systems '93,* pp. 1805–1808, Chicago, May 1993.

[14] P. Cox, P. Yang, S. S. Mahant-Shetti, and P. Chatterjee, "Statistical Modeling for Efficient Parametric Yield Estimation of MOS VLSI Circuits," *IEEE Transactions on Electron Devices,* ED-32: 471–478, Feb. 1985.

[15] J. E. Dennis and R. B. Schnabel, *Numerical Method for Unconstrained Optimization and Nonlinear Equations,* Englewood Cliffs, NJ: Prentice-Hall, 1983.

[16] S. W. Director, W. Maly, and A. J. Strojwas, *VLSI Design for Manufacturing: Yield Enhancement,* Boston: Kluwer Academic Publishers, 1990.

[17] S. W. Director and G. D. Hachtel, "The Simplicial Approximation Approach to Design Centering," *IEEE Transactions on Circuits and Systems,* CAS-24, no. 7, pp. 363–372, July 1977.

[18] S. W. Director and G. D. Hachtel, "A Point Basis for Statistical Design," In *Proc. of the IEEE International Symposium on Circuits and Systems,* ISCAS-78. New York, 1978.

[19] P. Feldman and S. W. Director, "Integrated Circuit Quality Optimization Using Surface Integrals," *IEEE Transactions on CAD*-12, no. 12, pp. 1868–1879, Dec. 1993.

[20] P. H. Gill, W. Murray, and M. H. Wright, *Practical Optimization,* New York: Academic Press, 1981.

[21] D. E. Hocevar, P. F. Cox, and P. Yang, "Parametric Yield Optimization for MOS Circuit Blocks," *IEEE Transactions on Computer-Aided Design,* 7, no. 6, pp. 645–658, June 1988.

[22] D. E. Hocevar, M. R. Lightner, and T. N. Trick, "A Study of Variance Reduction Techniques for Estimating Circuit Yields," *IEEE Transactions on Computer-Aided Design,* CAD-2, no. 3, pp. 180–192, July 1983.

[23] P. Huber, ed., *Robust Statistics,* New York: John Wiley, 1981.

[24] A. Ilumoka, N. Maratos, and R. Spence, "Variability Reduction: Statistically Based Algorithms for Reduction of Performance Variability of Electrical Circuits," *IEEE Proceedings,* 129, Pt. G(4), pp. 169–180, Aug. 1982.

[25] G. Kjellstrom and L. Taxen, "Stochastic Optimization in System Design," *IEEE Transactions on Circuits and Systems,* CAS-28, pp. 702–715, July 1981.

[26] G. Kjellstrom, L. Taxen, and P. O. Lindberg, "Discrete Optimization of Digital Filters Using Gaussian Adaptation and Quadratic Function Minimization," *IEEE Transactions on Circuits and Systems,* CAS-34, no. 10, pp. 1238–1242, Oct. 1987.

[27] D. N. Lawley and A. E. Maxwell, *Factor Analysis as a Statistical Method,* New York: Elsevier Publishing Co., 1971.

[28] K. K. Low and S. W. Director, "An Efficient Methodology for Building Macromodels of IC Fabrication Processes," *IEEE Transactions on Computer-Aided Design,* 8, no. 12, pp. 1299–1313, Dec. 1989.

[29] W. Maly and A. J. Strojwas, "Statistical Simulation of IC Manufacturing Process," *IEEE Transactions on CAD,* CAD-1, July 1982.

[30] C. Michael and M. Ismail, *Statistical Modeling for Computer-Aided Design of MOS VLSI Circuits,* Boston: Kluwer Academic Publishers, 1993.

[31] S. R. Nassif, A. J. Strojwas, and S. W. Director, "FABRICS II," *IEEE Transactions on CAD,* CAD-3, pp. 40–46, Jan. 1984.

[32] J. Ogrodzki, L. Opalski, and M. A. Styblinski, "Acceptability Regions for a Class of Linear Networks," *Proc. of the IEEE International Symposium on Circuits and Systems,* Houston, TX, May 1980.

[33] J. Ogrodzki and M. A. Styblinski, "Optimal Tolerancing, Centering, and Yield Optimization by One-Dimensional Orthogonal Search (ODOS) Technique," In *Proc. European Conference Circuits Theory and Design (ECCTD),* vol. 2, pp. 480–485, Warsaw, Poland, Sept. 1980.

[34] L. Opalski, M. A. Styblinski, and J. Ogrodzki, "An Orthogonal Search Approximation to Acceptability Regions and Its Application to Tolerance Problems," *Process Conference SPACECAD,* Bologna, Italy, Sept. 1979.

[35] L. J. Opalski and M. A. Styblinski, "Generalization of Yield Optimization Problem: Maximum Income Approach," *IEEE Transactions on Computer Aided Design of ICAS,* CAD-5, no. 2, pp. 346–360, Apr. 1986.

[36] L. J. Opalski and M. A. Styblinski, "GOSSIP—a Generic System for Statistical Circuit Design," *European Design Automation Conference,* Hamburg, Germany, Sept. 7–10 1992. EuroDAC'92.

[37] M.J.M. Pelgrom, A. C. J. Duinmaijer, and A.P.G. Welbers, "Matching Properties of MOS Transistors," *IEEE Journal of Solid State Circuits,* 24, pp. 1433–1439, Oct. 1989.

[38] M. S. Phadke, *Quality Engineering Using Robust Design,* Englewood Cliffs, NJ: Prentice-Hall, 1989.

[39] Ming Qu and M. A. Styblinski, "Hierarchical Approach to Statistical Perfor-

mance Improvement of CMOS Analog Circuits," *SRC TECHCON '93,* Atlanta, GA, Sept. 1993.

[40] Ming Qu and M. A. Styblinski, "Statistical Characterization and Modeling of Analog Functional Blocks," *IEEE International Symposium on Circuits and Systems,* London, GB, May–June 1994.

[41] H. Robins and S. Munro, "A Stochastic Approximation Method," *Annal. Math. Stat.,* 22, pp. 400–407, 1951.

[42] R. Y. Rubinstein, *Simulation and the Monte Carlo Method,* New York: John Wiley, 1981.

[43] A. Ruszczynski and W. Syski, "Stochastic Approximation Algorithm with Gradient Averaging for Constrained Problems," *IEEE Transactions on Automatic Control,* AC-28, pp. 1097–1105, Dec. 1983.

[44] A. P. Sage, Y. Fasser, and D. Brettner, *Process Improvement in Electronic Industry,* New York: John Wiley, 1992.

[45] G.A.E. Seber, *Multivariate Observations,* New York: John Wiley, 1984.

[46] K. Singhal and J. F. Pinel, "Statistical Design Centering and Tolerancing Using Parametric Sampling, *IEEE Transactions on Circuits and Systems.,* CAS-28, pp. 692–702, July 1981.

[47] R. S. Soin and R. Spence, "Statistical Exploration Approach to Design Centering," *Proceedings of the Institute of Electronic Engineers,* 127, part G(6), pp. 260–262, 1980.

[48] J. C. Spall, "Multivariate Stochastic Approximation Using a Simultaneous Perturbation Gradient Approximation," *IEEE Transactions on Automatic Control,* vol. 37, no. 3, 1992.

[49] R. Spence and R. S. Soin, *Tolerance Design of Electronic Circuits,* Electronic Systems Engineering Series, Great Britain: Addison-Wesley Publishers Ltd., 1988.

[50] W. Strasz and M. A. Styblinski, "A Second Derivative Monte Carlo Optimization of the Production Yield," *Proceedings of the European Conference on Circuit Theory and Design,* vol. 2, pp. 121–131, Warsaw, Poland, Sept. 1980.

[51] M. A. Styblinski, "Estimation of Yield and Its Derivatives by Monte Carlo Sampling and Numerical Integration in Orthogonal Subspaces," *Proceedings of the European Conference on Circuits Theory and Design (ECCTD),* vol. 2, pp. 474–479, Warsaw, Poland, Sept. 1980.

[52] M. A. Styblinski, "Problems of Yield Gradient Estimation for Truncated Probability Density Functions," *IEEE Transactions on Computer Aided Design of ICAS,* CAD-5, no. 1, pp. 30–38, Jan. 1986. (Special Issue on Statistical Design of VLSI Circuits).

[53] M. A. Styblinski, "Generalized Formulation of Yield, Variability, Minimax and Taguchi Circuit Optimization Problems," *Microelectron Reliability,* vol. 34, no. 1, pp. 31–37, 1994.

[54] M. A. Styblinski, J. Ogrodzki, L. Opalski, and W. Strasz, "New Methods of Yield Estimation and Optimization and Their Application to Practical Prob-

lems (invited paper)," *Proceedings of the International Symposium on Circuits and Systems,* pp. 131–134, Chicago, 1981.

[55] M. A. Styblinski and L. J. Opalski, "A Random Perturbation Method for IC Yield Optimization with Deterministic Process Parameters, *Proceedings of IEEE International Symposium on Circuits and Systems,* pp. 977–980, Montreal, Canada, May 7–10 1984.

[56] M. A. Styblinski and L. J. Opalski, "Algorithms and Software Tools for IC Yield Optimization Based on Fundamental Fabrication Parameters," *IEEE Transactions on Computer Aided Design of ICAS* (*special issue on Statistical Design of VLSI Circuits*), CAD-5, no. 1, pp. 79–89, Jan. 1986.

[57] M. A. Styblinski and A. Ruszczynski, "Stochastic Approximation Approach to Production Yield Optimization," *Proceedings of the 25th Midwest Symposium on Circuits and Systems,* Houghton, Mich., Aug. 30–31, 1982.

[58] M. A. Styblinski and A. Ruszczynski, "Stochastic Approximation Approach to Statistical Circuit Design," *Electronics Letters,* vol. 19, no. 8, pp. 300–302, Apr. 14, 1983.

[59] M. A. Styblinski and W. Strasz, "A Second Derivative Monte Carlo Optimization of the Production Yield," *Proceedings of the European Conference on Circuit Theory and Design '80,* vol. 2, pp. 121–131, Warsaw, Poland, Sept. 1980.

[60] M. A. Styblinski and T.-S. Tang, "Experiments in Nonconvex Optimization: Stochastic Approximation with Function Smoothing and Simulated Annealing," *Neural Networks Journal,* vol. 3, no. 1, p. 467, 1990.

[61] M. A. Styblinski and J. C. Zhang, "Orthogonal Array Approach to Gradient Based Yield Optimization," *Proceedings of the International Symposium on Circuits and Systems,* pp. 424–427, New Orleans, LA, May 1990.

[62] M. A. Styblinski and S. A. Aftab, "Combination of Interpolation and Self Organizing Approximation Techniques—A New Approach to Circuit Performance Modeling," *IEEE Transactions on Computer-Aided Design,* vol. 12, no. 11, pp. 1775–1785, Nov. 1993.

[63] M. A. Styblinski and S. A. Aftab, "IC Variability Minimization Using a New C_p and C_{pk} Based Variability/Performance Measure," *IEEE International Symposium on Circuits and Systems,* London, England, May 5–June 2 1994. ISCAS'94.

[64] M. A. Styblinski, J. Vandewalle, and M. Senupta, "Statistical Characterization and Optimization of Integrated Circuits Based on Singular Value Decomposition," *The 3rd IEEE International Conference on Electronics, Circuits and Systems,* Rhodes, Greece, Oct. 13–16, 1996. ICECS'96.

[65] K. S. Tahim and R. Spence, "A Radial Exploration Algorithm for the Statistical Analysis of Linear Circuits," *IEEE Transactions on CAS,* CAS-27, no. 5, pp. 421–425, May 1980.

[66] T.-S. Tang and M. A. Styblinski, "Yield Optimization for Non-differentiable Density Functions Using Convolution Techniques," *IEEE Transactions on CAD of IC and Systems,* vol. 7, no. 10, pp. 1053–1067, 1988.

References

[67] W. J. Welch, T.-K. Yu, S. M. Kang, and J. Sacks, "Computer Experiments for Quality Control by Parameter Design," *Journal of Quality Technology,* vol. 22, no. 1, pp. 15–22, Jan. 1990.

[68] P. Yang, D. E. Hocevar, P. F. Cox, C. Machala, and P. K. Chatterjee, "An Integrated and Efficient Approach for MOS VLSI Statistical Circuit Design," *IEEE Transactions CAD of VLSI Circuits and Systems,* CAD-5, pp. 5–14, Jan. 1986.

[69] D. L. Young, J. Teplik, H. D. Weed, N. T. Tracht, and A. R. Alvarez, "Application of Statistical Design and Response Surface Methods to Computer-aided VLSI Device II: Desirability Functions and Taguchi Methods," *IEEE Transactions on Computer-Aided Design,* vol. 10, no. 1, pp. 103–115, Jan. 1991.

[70] T.-K. Yu, S. M. Kang, I. N. Hajj, and T. N. Trick, "Statistical Performance Modeling and Parametric Yield Estimation of MOS VLSI," *IEEE Transactions on CAD of VLSI Circuits and Systems,* CAD-6, no. 6, pp. 1013–1022, Nov. 1987.

[71] T. K. Yu, S. M. Kang, I. N. Hajj, and T. N. Trick," iEDISON: An Interactive Statistical Design Tool for MOS VLSI Circuits," *IEEE International Conference on Computer-Aided Design,* vol. ICCAD-88, pp. 20–23, Santa Clara, CA, Nov. 7–10 1988. IEEE Computer Society Press.

[72] T. K. Yu, S. M. Kang, J. Sacks, and W. J. Welch, "Parametric Yield Optimization of MOS Integrated Circuits by Statistical Modeling of Circuit Performances," Technical Report 27, Department of Statistics, University of Illinois, Champaign, IL, July 1989.

[73] L. A. Zadeh, "Fuzzy Sets," *Information Control,* vol. 8, pp. 338–353, 1965.

[74] L. A. Zadeh, "Probability Measures of Fuzzy Events," *Journal of Mathematical Analysis Applications,* 23:421–427, 1968.

[75] J. C. Zhang and M. A. Styblinski, *Yield and Variability Optimization of Integrated Circuits,* Boston: Kluwer Academic Publishers, 1985.

7

Architectural Fault Tolerance*

S. K. Tewksbury

7.1 INTRODUCTION

This chapter reviews a variety of techniques through which a microelectronic system can be made tolerant of faults appearing in portions of the circuitry [1–5]. Some background on the general issues surrounding such fault tolerance is covered below. Fault tolerance is a complex topic, covering a wide range of issues and areas of applicability. The general preference in VLSI components is to provide fault-free components by developing technologies and imposing limits on the IC area consistent with a good yield of fault-free components. Previous chapters have covered the issues of defects and fault models which apply whether or not fault tolerance is used. As we will see, despite the emphasis on providing fault-free IC components, there are significant applications in which fault tolerance is already routinely provided or in which fault tolerance will become increasingly important. At the same time, in many applications and conditions fault tolerance is not appropriate. By properly understanding the conditions under which the overhead imposed by fault tolerance is acceptable, we can better identify practical applications of fault tolerance.

Before considering some of these general conditions, what do we mean by the terminology *fault tolerance*? In general, we assume that we are provided with a correctly designed IC (i.e., free of design faults), fabricated using a manufacturing facility that routinely produces good ICs. As noted in earlier chapters, the fabrication

---

* Portions of this chapter are based on [5].

facility does not produce precisely perfect device and interconnect structures. Tolerances on fabrication processes and unavoidable contamination of materials and the fabrication environment lead to a nonvanishing probability that a small area of an IC will contain a defect, with an increasing probability of a defect within the IC as the circuit area increases. Routine process control procedures seek to provide a high "yield" of fault-free ICs, as long as the IC area is properly bounded. The expectation is that, upon completion of the IC fabrication, some fraction of the ICs will display faulty behavior that can be identified during testing. By simply discarding the faulty ICs and providing to the market only fault-free components, there would seem to be little need for fault tolerance.

In several cases, however, advantages (generally seen at the overall system level) result from the addition of fault tolerance capabilities within the IC. For example, it may be desirable to use an IC area substantially higher than the bound above which acceptable yields are obtained. The discarding of a highly complex IC, containing pehaps millions of transistors and vast highways of interconnections, due to a single microscopic defect seems to be a severe penalty when confronted with a serious architectural need to move beyond the normal yield-limited area bounds to achieve a higher performance system. In particular, the architectural penalty of avoiding even a single microscopic defect can be extensive when viewed from the perspective of partitioning the functionality of the overall system among the several ICs needed to realize that system. In such cases, it is natural to consider fault tolerance, though the actual decision to add fault tolerance will require assessment of a vast number of cost functions.

Even assuming that microelectronic systems are constructed using only components that are known to be functioning properly when first installed in a system, there remains the serious issue of a component failing during service. In particular, use of an initially good component does not eliminate the potential need for a fault tolerant design since faults can appear while the system is in service. A decision to add fault tolerance despite initial use of fault-free components depends on the importance of uninterrupted service when an in-service failure occurs in a device or interconnect.

7.1.1 Use of Known-Faulty Components When Manufacturing an Electronic System

The title of this section has the appearance of a seriously faulty manufacturing approach. If a component during initial testing following fabrication is found to be faulty, are there any cases in which it might still be advantageous to use that component (assuming it can be made to function properly using fault tolerance)? In general, the answer is clearly that the component should not be placed in the system while it is not functional. However, is there a role for repair of a faulty component and then placement of the repaired component in a system? This question sounds more reasonable. Two examples are used to illustrate the approach.

- If a minor and repairable defect appears on a printed circuit board (PCB), it may very well be better to repair the defect and use the PCB (which has already incurred a significant production cost) rather than simply discarding the defective PCB. The primary differences between this PCB example and an IC are (1) the inability to do simple physical repair of an IC and (2) the lower incurred cost lost by discarding the IC. In particular, use of known-faulty components in an electronic system is not unusual (as seen in the PCB example), but the practical justifications will be quite different for an IC compared to a PCB. An example of repair of faulty ICs for use in systems is seen in the routine use of fault tolerance in memory ICs, where spare rows or columns of memory cells are provided to replace one or more fault memory cells of the memory IC. In this case, the fault tolerance imposes little area or speed penalty on the memory IC, the repair is easily completed, and the increased area of functional memory ICs (using repair) has significant economic advantages in the market.

- If a minor region of a hard disk for a computer has faults, then it is possible to declare their associated blocks as defective and bypass those blocks when mapping data files onto available blocks of the disk. In this case, the system can exploit the functionality of most of the component and accept a certain level of defects, without significantly impacting the performance of the component. A similar possibility might arise in a semiconductor memory system using memory management schemes. In particular, if a particular isolated cell is defective (and doesn't cause faults outside its block of memory), then the memory mapping operation (combined with a good/bad bit in the block information file) could simply avoid use of that defective block in the memory system.

The first example concerns a repaired component, whereas the second concerns a component whose defect can be easily avoided through higher level system management (using features that are already used, for example, in the mapping of data blocks into a virtual memory structure). These examples suggest that it is indeed reasonable and practical to use components with faults to advantage in real systems, under the appropriate system-level conditions.

7.1.2 Large-Area Integrated Circuits

The design of multiple IC microelectronic systems generally requires that the overall functionality of the system be divided into (or partitioned among) separable functions that can be implemented on difference ICs such that inter-IC interconnection limits do not seriously degrade the performance of the overall system. However, there are many input/output barriers to increased performance (e.g., the well-known memory/processor I/O bandwidth limit) illustrating that such partitioning of functionality among different ICs is not always capable of avoiding significant performance degradations. There may be cases in which an increase in the area of an IC provides a substantial performance benefit, primarily as a result of limits imposed by the number and performance of the IC-to-IC interconnections. The penalty for such an increase of the IC area beyond the normal yield-limited area bound is severe, since the yield decreases exponentially (rather than linearly) with increasing area beyond the normal area bound. The objective of higher performance by using a larger than normal IC area is therefore confounded by the rapid decrease in yield (and the corresponding exponential increase in the cost) per good, large-area ICs. Fault tolerance plays a critical role in this case, effectively raising the yield when one has moved to areas beyond the normal yield-limited bound to satisfy performance objectives of the overall multiple-IC system. A recent example of this approach is seen in the Digital Signal Processing (DSP) reported by Brewer et al. [6]. In this case, a powerful DSP core (within the normal area bound imposed by yield) is augmented by a large static memory, dominating the overall IC area and potentially leading to a very low yield of the overall IC. However, by adopting the memory repair approaches, that large area devoted to memory can be made fault tolerant, while the DSP core remains non-fault tolerant. If the yield of the static memory section can be made substantially greater, through fault tolerance, than the yield of the DSP core, then the yield of the full IC is comparable to the yield of the DSP by itself.

This example illustrates a "large-area IC." By extending the area to include a substantial portion of an overall silicon wafer, one extends from the large-area, fault tolerant IC to the arena of Wafer Scale Integration (WSI), with full wafer-area, fault tolerant "ICs." Much of the work on fault tolerant integrated circuits derives from several years of research and exploratory development of schemes to allow full wafer-area circuits to be used despite the virtual certainty that each such circuit will contain faults. Many of the references in this chapter derive from this area of research, which evolved largely during the 1980s.

7.1.3 In-Service Failures

Even assuming that a component is functional when it is first installed into a system, that component may fail at some point in service. Standard procedures (i.e., the many accelerated testing proceduers such as burn-in) applied to ICs seek to identify and eliminate those ICs that are likely to fail early during service. Such accelerated testing can only act as a screen to identify some subset of the components that may fail since degradations which are accelerated but do not complete the transition to a fault have not been eliminated. (Indeed, they have been "aged" to fail earlier than if the accelerated testing procedures had not been applied.)

Given a correctly operating microelectronic system that fails in service, one confronts a different set of criteria impacting the advantages of fault tolerance. For example, discarding a defective component before it is used in a system is a lower cost action than repairing a defective system by replacing the defective component. In this sense, the cost comparisons for systems with and without fault tolerance (for in-service faults) relate to the balance between the component cost increase to provide fault tolerance relative to the repair cost of the overall system when failure occurs (and the probability that a system will fail during its normal in-service lifetime). In addition, several applications may require continued operation despite failure of an element until such time that a scheduled test and repair action can be taken. Large-scale fault tolerant computers have been developed to handle such applications. Other applications are "mission critical" systems (space, military, automotive, etc.) which must continue operating until the mission has been completed (e.g., spacecraft has returned, automobile has had time to move to a safe location, etc.).

There are also systems that are expected to operate in harsh environments and that cannot be expected to operate with high reliability in those environments without some form of fault tolerance. Harsh environments include systems operating in high-radiation environments (requiring radiation-hardened components and generally some fault tolerance), operating in high-temperature environments (e.g., automotive electronics), and so on. In such cases, the failure probabilities of individual components increase substantially beyond the levels seen in less harsh environments, but the system is expected to provide reliable operation despite the harsher environment.

The "harsh environment" may extend beyond the normal view of physically harsh environments. For example, the bit error rates (BER) in mobile communication systems are generally much higher than can be tolerated in standard data networks (ethernet, etc.), requiring refinements in services provided over such channels. In such cases, the "fault tolerance" is built into the specific application. For example, consider the case of a mobile computer running programs that interact with a distant supercomputer but suffering a high BER on its mobile communications. By requiring that the mobile computer's local program execute on a land-based fixed computer, with mobile communications displaying execution results on the monitor of the mobile computer, the visual presentation of information on the mobile monitor will mask transmission bit errors along the mobile link.

7.1.4 Difficulty of Repair

As IC technologies have evolved, traditional approaches to testing ICs have become less effective, leading to important strategies for "built-in testing" of complex ICs (e.g., JTAG, etc.). In a similar spirit, as microelectronic system packaging technologies evolve to include approaches such as Multichip Modules (MCMs) and stacked, three-dimensional ICs [7] and MCMs, it becomes increasingly difficult to "pull out" a defective component and "plug in" a functional replacement. The result is that the "throwaway" component is becoming a larger portion of the microelectronic system (i.e., containing a larger number of ICs and other components) and a more costly component to replace. Such trends suggest that, for reasons similar to those that led to built-in testing, traditional repair techniques will evolve to "built-in repair" techniques. It is useful to regard fault tolerance as the mechanism for "built-in repair" and to view it from a perspective similar to testing.

7.1.5 Selective Fault Tolerance

Figure 7-1 illustrates an overall system function, consisting of circuit functions which can be tested (and the fault isolated and corrected) and other circuit functions (including interconnections between functional cells) which cannot be tested and generally cannot tolerate faults. The mixture of different types of circuit functions

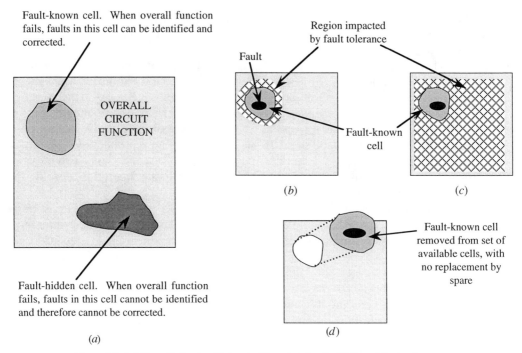

Figure 7-1 (*a*) Overall circuit with internal functional cells. (*b*) Local fault tolerance approach. (*c*) Global fault tolerance approach. (*d*) Operational fault tolerance approach.

within the overall circuit impacts the various fault tolerance approaches discussed later.

The "fault-known cell" in Figure 7-1*a* is an irreducible circuit function that is regarded as a unit when a defect causing a fault appears within the function. There is a close connection between the irreducible, testable cell (i.e., the smallest circuit unit that can be determined to be faulty or fault-free during testing) and the fault-known cell. Most fault tolerance approaches require some knowledge of the location of the fault (to within a portion of the overall circuit function), and the fault-known cells provide this location information.

The "fault-hidden cell" in Figure 7-1*a* is a cell that cannot be determined to be either faulty or fault-free, causing conditions under which a failure of the overall circuit cannot be isolated to a particular circuit cell (making application of fault tolerant procedures impossible). The presence of fault-hidden cells in the overall circuit merely reflects the practical difficulty of providing cost-effective fault tolerance and testability throughout all circuit elements of the overall circuit.

From the perspective of such an overall circuit function, there are three distinct approaches to fault tolerance. (1) Localized fault tolerance (Figure 7-1*b*) corrects the operation of a fault-known cell without changing or affecting the circuitry beyond that fault-known cell. (2) Global fault tolerance (Figure 7-1*c*) requires alterations in the behavior of the overall circuit at locations well removed from the position of the actual faulty cell. Finally, (3) there may be useful classes of fault-insensitive circuits (Figure 7-1*d*), which can operate correctly (i.e., have operational fault tolerance) by simple deletion of the faulty cell from the overall circuitry (e.g., a parallel computer with one fewer computer nodes). Operational fault tolerance is not discussed in detail in this chapter. However, important examples are readily seen, as in the following examples.

- A variety of regular networks have been discussed for parallel computing. In several cases, there exist multiple paths in a nonredundant network across which nodes can be connected. Although such alternative paths may not be optimal, use of the alternative paths in the event of a link failure can lead to a gracefully degraded performance (e.g., [8–10]). In addition, fault tolerance for crosspoint switches [11] and other data network switching structures have been studied.
- Field Programmable Gate Arrays (FPGAs) provide a user-specified interconnection of cells of the FPGA. In the event of a fault within one of the cells (or in an interconnection link of the FPGA), there are cases in which the programming step can bypass the faulty elements and achieve the desired function. Electronically reprogrammable FPGAs can be reconfigured after an already programmed FPGA has been found faulty. One-time programmable FPGAs present a more serious set of conditions, addressed for example in [12]. Use of fault tolerance to achieve significantly larger FPGA area has also been studied (e.g. [13]).
- Computer systems provide an interesting option to full hardware fault tolerance. For example, Saxena et al. [14] describe a memory management system in which hardware provides detection of errors and software provides correction of the errors.
- There have been some suggestions that neural networks can provide an intrinsic fault tolerance for faults in the circuitry. Although this does not appear to be a general characteristic of neural networks, some interesting work exploring such intrinsic fault tolerance has been done (e.g. [15, 16]).

7.2 LOCAL FAULT TOLERANCE

Local fault tolerance expands the circuitry of a circuit cell to include features such that the cell itself will perform correctly despite the presence of fault(s) within that cell. No major actions outside this locally fault tolerant cell are needed to create an overall function which is fault tolerant. In several techniques, the addition of fault tolerance to the cell also leads to a built-in self-testing (BIST) feature, as illustrated later in some of the examples. Perhaps the most common examples of local fault tolerance are triple modular redundancy and error correcting codes.

7.2.1 Modular Redundancy

Triple Modular Redundancy (TMR) [2] is a member of a set of fault tolerance techniques in which the circuit cell is replicated, those replicated cells being bound into a single cell performing the same function as the original cell.

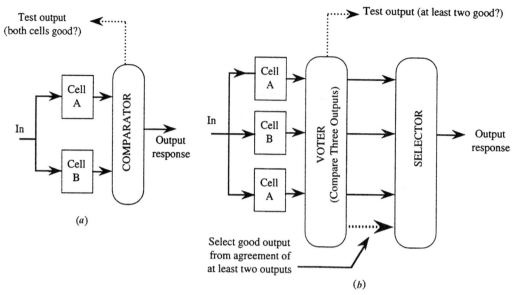

Figure 7-2 Modular redundancy. (*a*) Use of a single spare for fault detection only. (*b*) Use of two spares (TMR) for both fault detection and correction.

Figure 7-2 illustrates the case of a single spare element (double redundancy) as well as the case of two spare elements (triple redundancy, as used in TMR). The presence of a single spare combined with circuitry to select one of the two copies of the basic circuit cell seen in Figure 7-2a provides fault tolerance if testing can identify when the currently selected cell is faulty. Although a simple comparison of the outputs from the replicated cells can determine whether a fault exists (i.e., provides fault detection), it is not possible to determine which of the two cells is fault.

TMR, illustrated in Figure 7-2b, provides two spare copies of the cell, along with a *voter* circuit that can compare the simultaneous outputs of all three cells. If all cells are fault-free, then all produce identical outputs (since all receive the same inputs and, if a sequential circuit, are reset at some time to identical known states). If only one cell fails, then the voter can determine (1) which two cells are producing the same output and designate that output as the correct output, and (2) which cell is producing a different output and designate that cell as a faulty cell. If, after the failure of one cell, a second cell fails, then the voter can determine that the fault tolerance in the cell has itself failed. In this sense, the TMR approach includes a built-in testing function to detect a fault as well as capabilities of isolating a detected fault to one of the three duplicate cells. It is the combination of fault detection, fault isolation, and fault tolerance of the system of replicated cells in a very general form that makes TMR a popular and straightforward approach.

This fault tolerance provided by the replicated cells is compromised by the possibility of faults occurring within the voter circuit. A variety of techniques have been developed to provide "fault-resistant" voters. In addition, faulty interconnections on the common input lines to each replicated cell can lead to uncorrectable

7.2 ■ Local Fault Tolerance

failure of the overall function (e.g., a hard "stuck-at fault" on a common input line). Since all copies of the cell receive the same (faulty) input signal, they produce identical outputs and the fault is not detectable. This coverage of only a subset of possible faults is typical of fault tolerance methods. Rather than tolerating every possible fault, the objective is to achieve designs in which the probabiity of occurrence of a nontolerated fault is much smaller than the probability of tolerated faults.

The primary drawback of TMR is the high overhead in circuit area and speed. The penalty in the area (or amount of circuitry if a multiple IC function is involved) is at least a factor of 3 (due to the triplication of the basic cell), with an additional area penalty imposed by the additional routing of input wires to the three copies of the cell and by the voter's circuitry and interconnections. There is also a speed overhead. Assuming that the individual cells produce correct outputs in time τ_{cell}, the voter must evaluate those outputs; this evaluation will impose an additional delay τ_{vote} before the correct output reaches the output of the voter. Design of fast voters for in-line comparison and correction, where "fast" is relative to the speed of the replicated cells, is therefore an important design objective. Such in-line comparison and voting provides an important capability, namely, detection and correction of transient faults (faults that appear only briefly, including both data-dependent faults and transient faults that are not correlated with the data).

If this tolerance to transient faults is not necessary, then the steps of voting and selection can be separated, with testing by the voter performed during occasional test periods and the selection of the correct output remaining in line with the data flow. Such separate testing, illustrated in Figure 7-3, sets the state of the selector, and that state is held until the next test period, at which time a new state for the selector may be established. In this configuration, the delay τ_{vote} through the voter

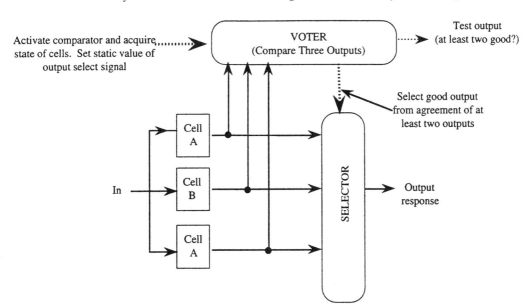

Figure 7-3 Off-line (perhaps sampled) voting to detect static failures and then set an output selector to accept the correct response.

is largely eliminated during normal operation. The delay penalty due to the TMS circuitry is then merely that imposed by the selection switch.

A somewhat different approach to redundancy, decreasing the number of replicated cells, can sometimes be achieved by providing a spare that is shared by more than one cell. This is seen, for example, in the case of fault tolerant tree structures [17], as illustrated in Figure 7-4. In Figure 7-4a, each note of the tree is provided with a local spare, providing 1-out-of-2 redundancy. In Figure 7-4b, two nodes of the tree (the lower solid circuits) share a single spare, creating a 2-out-of-3 redundancy. Finally, Figure 7-4c illustrates a single spare serving three nodes, providing 3-out-of-4 redundancy. The switch-level layout [17] for the example in Figure 7-4c is illustrated in Figure 7-5. Such n-out-of-m redundancy incurs a decreasing area overhead as the ratio $m/n \rightarrow 1$. Fault tolerant trees are discussed, for example, in [17–22].

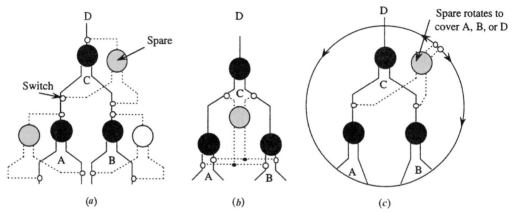

Figure 7-4 Binary tree sections with spare cells (e.g., for tree array of processor elements). (a) 1-out-of-2 redundancy. (b) 2-out-of-3 redundancy. (c) 3-out-of-4 redundancy. Open circles are switching points—not all are shown in (c). Lightly shaded circles are spare cells. Darkly shaded circles are normal cell of tree.

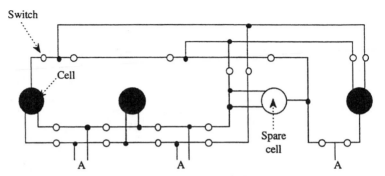

Figure 7-5 Representative switch organization [17] for 3-out-of-4 redundancy scheme in Figure 7-4.

7.2 ■ Local Fault Tolerance

Two other interesting applications are seen in fault tolerant busses and fault tolerant bit-sliced architectures. These are included here as locally redundant cells, even though their use requires changes in connections at several points removed from the actual defective component.

Fault Tolerant Busses: Fault tolerant busses (a parallel array of interconnection lines) can be made fault tolerant by adding one or more spare bus lines, combined with reconfiguration switches at the access points of the parallel data bus. If a bus interconnection is found defective, the connection normally made to that line and normal connections to lines above the defective line (in the case of a spare line at the top of the bus) are switched to the next higher bus line. Examples of this approach are discussed in [23, 24].

Fault Tolerant Bit-Sliced Architectures: A "bit-sliced" architecture is organized as a set of replicated functions operating on groups of data bits within a larger set of data bits. Figure 7-6a illustrates the general architecture, with a spare row of bit-slice elements added at the top to provide redundancy. In the case of an architecture using several bit slices, the area overhead due to the spare row is modest. Figure 7-6b illustrates a case in which one of the bit slices is defective. In much the same manner as the rerouting of data on a bus with a defective interconnection discussed above, the defective bit slice is removed (and the spare inserted) by redirecting all connections at and above the defective slice to the next higher slice. This is one of the approaches used in the study of a fault tolerant microprocessor reported in [25]. This study also evaluated fault tolerant control units using PLAs. Reconfigurable PLAs are also considered in [26].

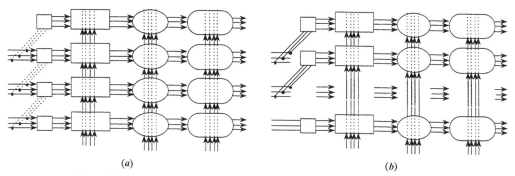

(a)   (b)

Figure 7-6 Bit-sliced function with spare slice. (*a*) Full function (including sequence of connected bit-sliced functions). (*b*) Faulty row bypassed to use spare.

7.2.2 Error Correcting Codes

Figure 7-7 illustrates the general approach of error correcting codes [27–35] for memory systems and for communications systems. In these applications, the K-bit "data word," with 2^K valid values, is expanded to a $(K + L)$-bit "code word" with 2^{K+L} distinct code words. The mapping of the data words to the code words

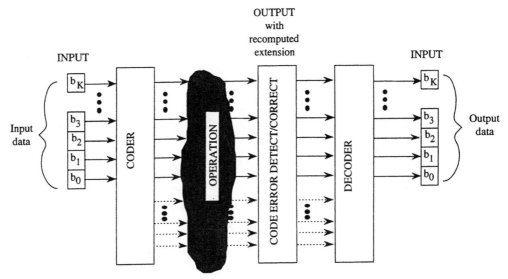

Figure 7-7 Coding extensions to data inputs for error detection and correction.

is such that each valid data word is mapped to a single valid code word (with the result that there are far more code words than needed to represent the valid data words). If an error appears between the origination of the code word and the receipt of the code word, then the error detecting codes are designed such that (for some bounded number of errors in the code word received) the received code word is not a member of the set of code words representing valid data words. In the case of error correcting codes, the received code word also contains sufficient information to reconstruct the correct data word from the faulty code word. The approach here can be expressed in a variety of mathematical forms allowing detailed design and optimization of the codes used. However, the basic approach for error correcting codes is clear. We illustrate the approach of error correcting codes by a few representative examples.

7.2.2.1 Block Parity Error Correcting Codes. Block parity error correcting codes are an extension of the simple parity concept of error correction. Consider a 32-bit data word. Rather than representing the data word as a linear array of digits, instead represent the 32-bit word as four rows, each row containing 8 bits, as illustrated in Figure 7-8a. Next, calculate the parity check bit for each row, resulting in the addition of one bit (the parity bit) to each of the four rows (and generating a total of 9 bits per row). In addition, calculate the parity check bit for each column of 4 bits and form a fifth row containing those parity bits. We now have an array of 9×5 bits representing the original 32-bit data word.

With the added parity (check) bits, Figure 7-8b illustrates detection and correction of a single bit error in the 32-bit word. In particular, on receipt of the coded data, the parity check bits are recomputed and compared with the check bits in the coded data. A single faulty bit $b_{j,k}$ will lead to a faulty check bit for its row (i.e., row j) and a faulty check bit for its column (i.e., column k). The check bits then

7.2 ■ Local Fault Tolerance

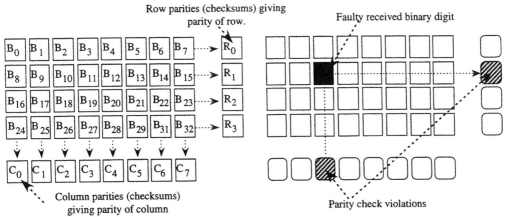

Figure 7-8 (*a*) Organization of binary digits as a two-dimensional array of digits, providing row and column parity bits for error detection and isolation. (*b*) Faulty digit and its location by intersection of parity violations in row and column checksums.

define the specific digit in the array which is faulty. Correction of a faulty binary digit merely requires that it be complemented, a simple correction operation. The approach could allow correction of some multiple faults, though this is generally not useful since not all multiple faults are covered. For example, if two faulty digits are in different columns but the same row, the column check sums will detect the occurrence of two faults (multiple fault detection), but the row check sum will not indicate a fault (and prevents isolation of the faults). The technique shown in Figure 7-8*b* therefore provides multiple-fault detection and single-fault correction.

7.2.2.2 Block Error Correcting Codes. The preceding example is a case of a "block error correcting code." Here, we consider a more general view of block error correcting codes. In particular, a k-bit data word represents 2^k distinct "data" words. We can regard the k bits as a "block" of binary digits. In general, such a block could represent a concatenated set of several computer words (as in the case of packets assembled for data communications). Consider then a recoding that transforms each distinct k-bit "data" block into a distinct n-bit "code" block (with $n > k$). The added $n - k$ bits are redundancy bits (also called *check bits*). This recoding is an (n, k) coding of the original k-bit data block (a one-to-one mapping of each data word to a code word). This mapping is "sparse" in the sense that 2^{n-k} code words are not assigned (i.e., "not allowed") in this mapping. By appropriate choice of the mapping (coding), one seeks codes for which an error in a digit of the sparsely populated code words leads to a nonassigned code word. At the far end, the presence of an error is detected by the received code word not being an assigned code word. To determine what the correct code word should be, one uses a measure of the "distance" of the received code from the allowed codes. Typically, the code correction corresponds to selecting the allowed code word nearest the received code.

The *Hamming distance* between two code words is the number of bit positions in which the two code words have different values of their binary digits. We can seek to construct a redundant (n, k) code such that there is some minimum distance d_{min} between any two allowed members of the code. Suppose that we have a redundancy code with $d_{min} = 5$. All the valid codes will fall on the expected values. If any single bit in the redundant code is erroneous, that faulty code will be at a distance 1 from a valid code. If two bits are erroneous, then the faulty code will be at a distance 2 from a valid code. Therefore, even if up to two bits are in error, we can identify the "nearest" valid code (i.e., with minimum Hamming distance between the received faulty code and the valid code).

The *Hamming codes* [28, 34, 35] are redundant codes with the (n, k) structure above and $n = 2^m - 1$, $k = 2^m - 1 - m$. Here, m is the number of check bits added to the original data word. For $m = 5$, $k = 2^5 - 1 - 5 = 26$ and $n = 31$. For $m = 4$, $k = 2^4 - 1 - 4 = 11$ and $n = 15$. This illustrates the coding overhead. Figure 7-9 illustrates the details of a Hamming code with $d_{min} = 3$ and an 11-bit data word. The parity bits C_j are calculated according to the set of digits Y_k included in the first four rows. The rows listing the possible values of (C_0, C_1, C_2, C_3) show the simple binary number sequence of values, that binary value C pointing directly to the subscript of the faulty digit Y_C. The check bits are organized among the digits of the original data word X when forming the extended data word Y such that faulty check bits, as well as faulty data bits, are covered by the scheme. As was the case for the TMR approach, the error detection/correction codes provide not only the correct output (tolerating faulty data bits) but also in-line testing and fault isolation (i.e., the value of C identifies the faulty digit).

The overhead for such redundant codes includes the additional interconnection area required for the redundant data bits (i.e., wider data path), the additional circuit area for generation (coding) of the redundant code from the original data word, additional circuit area for evaluation and correction of the received code word, and additional circuitry (decoder) to extract the data word from the code

Data word $X_1 X_2 X_3 X_4 X_5 X_6 X_7 X_8 X_9 X_{10} X_{11}$
Code word $Y_1 Y_2 Y_3 Y_4 Y_5 Y_6 Y_7 Y_8 Y_9 Y_{10} Y_{11} Y_{12} Y_{13} Y_{14} Y_{15}$
$= C_0 C_1 X_1 C_2 X_2 X_3 X_4 C_3 X_5 X_6 X_7 X_8 X_9 X_{10} X_{11}$

Parity(C_0)	-	Y_1	-	Y_3	-	Y_5	-	Y_7	-	Y_9	-	Y_{11}	-	Y_{13}	-	Y_{15}
Parity(C_1)	-	-	Y_2	Y_3	-	-	Y_6	Y_7	-	-	Y_{10}	Y_{11}	-	-	Y_{14}	Y_{15}
Parity(C_2)	-	-	-	-	Y_4	Y_5	Y_6	Y_7	-	-	-	-	Y_{12}	Y_{13}	Y_{14}	Y_{15}
Parity(C_3)	-	-	-	-	-	-	-	-	Y_8	Y_9	Y_{10}	Y_{11}	Y_{12}	Y_{13}	Y_{14}	Y_{15}
Value(C_0)	0	1	0	1	0	1	0	1	0	1	0	1	0	1	0	1
Value(C_1)	0	0	1	1	0	0	1	1	0	0	1	1	0	0	1	1
Value(C_2)	0	0	0	0	1	1	1	1	0	0	0	0	1	1	1	1
Value(C_3)	0	0	0	0	0	0	0	0	1	1	1	1	1	1	1	1
Value C	0	1	2	3	4	5	6	7	8	9	10	11	12	13	14	15
Faulty Y_i	-	Y_1	Y_2	Y_3	Y_4	Y_5	Y_6	Y_7	Y_8	Y_9	Y_{10}	Y_{11}	Y_{12}	Y_{13}	Y_{14}	Y_{15}
or	-	C_0	C_1	X_1	C_2	X_2	X_3	X_4	C_3	X_5	X_6	X_7	X_8	X_9	X_{10}	X_{11}

Figure 7-9 Hamming code ($d = 3$ and 11-bit data word).

word. Evaluation of the received code to detect a data error, allowing correction of the error, can significantly degrade the throughput rate of data. However, this additional latency can be reduced if one does not place the error detection operation in the path of the data. For the case of fixed faults (i.e., faults that remain continually active), the error detection operation can be performed in parallel with the data path (as illustrated in Figure 7-10) to locate the faulty line and a more efficient approach is used to correct the faulty digit (e.g., monitor the specific code bits pointing to whether that particular faulty digit is incorrect for the given received data word). It remains necessary to evaluate the parity bits of the code to determine whether a received word is correct. For example, if the line is stuck at 1, then a received value of 1 may be either correct (far end sending a 1) or incorrect (far end sending a 0).

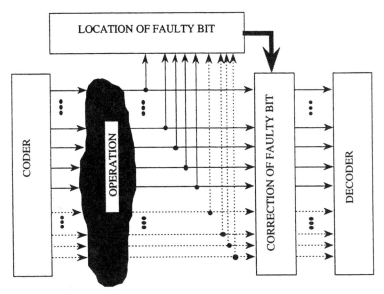

Figure 7-10 Organization of evaluation circuitry to detect static faults on lines without requiring in-line evaluations. The evaluation circuitry specifies a specific digit which is faulty, with that faulty digit complemented when a fault is identified through the check bits.

The single-bit error correction codes are quite well suited for parallel data links in microelectronic systems, including both data busses and storage of data in memory. However, the approach is less well suited for serial data communications environments, in which the set of data digits are transmitted serially over a single connection. Faults in that connection (even if transient) can lead to multiple bits of a data word being corrupted, limiting the effectiveness of the single-bit error correction for serial communication in the presence of "burst" noise.

Many of the commonly referenced error detection and correction codes have been developed for data transmission systems in which the length of the block (data packet) is large relative to data words used within computational units. Reed-Solomon and BCH (Bose-Chadhuri-Hocquenghem) codes [28, 34, 35] are representative examples. Within microelectronic systems and components, the emphasis is on minimizing the complexity of the coders and decoders for the redundant codes for the data words and the complexity of implementing the correction of any received, faulty code words.

7.2.3 Algorithm-Based Fault Tolerance

Algorithm-based fault tolerance [36–42] codes multiword data for error detection and correction in parallel array arithmetic computations. Matrix operations [36], a common example, are used below to illustrate the general approach.

Figure 7-11 illustrates a general $N \times M$ matrix A whose elements $a_{i,j}$ are data words. Constructed from this matrix A is a checksum-augmented matrix containing an extra row, an extra column, and an extra corner matrix element. The additional row elements $a_{N+1,j}$ of the row checksum are formed by summing all the elements in column j of the original matrix A, that is, $a_{N+1,j} \equiv \sum_{i=1}^{N} a_{i,j}$. The additional column elements $a_{j,M+1}$ of the column checksum are formed by summing all the words in row j of the original matrix A, i.e., $a_{j,M+1} \equiv \sum_{i=1}^{M} a_{j,i}$. The corner element $a_{N+1,M+1}$ is defined as the checksum of the row (equivalently column) checksum elements, that is, $\sum_{k=1}^{M} a_{N+1,k}$ (equivalently $\sum_{k=1}^{N} a_{k,M+1}$).

In Figure 7-12, the input matrices A and B are converted to checksum-augmented matrices and passed through some matrix operation $\mathcal{O}$ producing an output matrix C with an extra row and extra column, that is, elements $\hat{c}_{N+1,j}$ in the extra

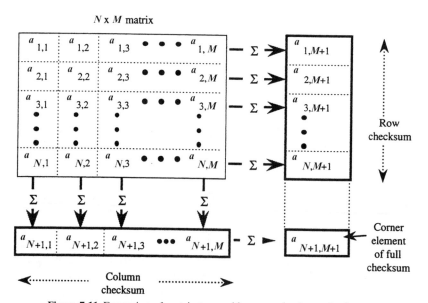

Figure 7-11 Expansion of matrix to provide row and column checksums.

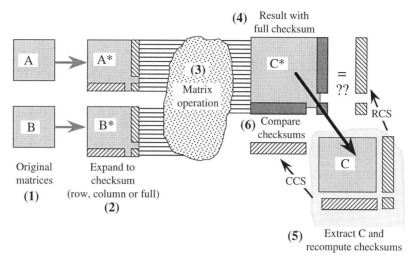

Figure 7-12 Matrix operation and steps in computation/error detection.

row and $\hat{c}_{j,M+1}$ in the extra column. The column checksums $c_{N+1,i}$ and the row checksums $c_{j,M+1}$ are recomputed from the embedded $N \times M$ matrix of the $(N + 1) \times (M + 1)$ full checksum output matrix C. An error is indicated when one or more of the recomputed checksum elements disagree with the corresponding received checksum elements. In addition, the error can be corrected since the computed difference provides the correction directly. This assumes that the result matrix has checksum elements following the matrix operation which agree (in the absence of faults) with computation of the checksum elements from that result matrix. Five major matrix operations satisfying this requirement are given in [36].

The checksum matrices provide a powerful fault tolerance capability since they provide detection, isolation, and correction of faulty data. This is illustrated in Figure 7-13. An internal error produces both a faulty row element and a faulty column element. The capabilities are generally as follows.

- An error results in a comparison difference for a row checksum and for a column checksum, as illustrated by the solid areas in Figure 7-13.
- The element in the resultant matrix which is faulty is given by the intersection of the row and column having erroneous checksums.
- Since the checksum errors are computable and the erroneous element is locatable, that element can be corrected by adding the difference in the checksum comparison.

The principal limitation associated with the checksum fault tolerance approach described above arises from deviations of computed numbers from the precisely correct values due to roundoff errors incurred during the computations. Such roundoff errors can cause errors in both the normal resultant matrix elements and the checksums of that resultant matrix. It is therefore necessary to define some

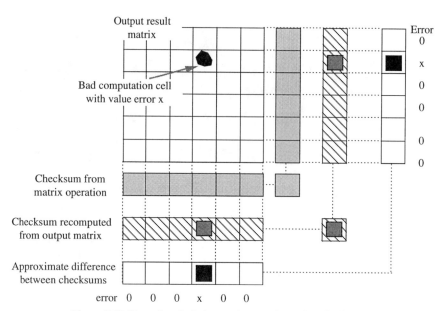

Figure 7-13 Detection, isolation, and correction using checksums.

threshold on checksum errors, below which the computation is assumed correct and above which the computation is assumed faulty [36].

Algorithm-based testing is reviewed in [37]. Algorithm-based fault tolerance for Fast Fourier Transform (FFT) computations, in this case encoding the data according to the properties of FFTs, is described in [38]. The general approach has obvious application to systolic array computations, which emphasize matrix and FFT-type computation algorithms, as discussed in [39].

7.3 GLOBAL RECONFIGURATION

A variety of formal results have been developed to clarify the complexity of reconfiguration of linear or two-dimensional arrays of N cells as N becomes large. The general approach assumes a regular array (linear or two-dimensional) as the target architecture. The physical architecture is similarly an array (generally two-dimensional) of physical cells, with the possibility that some of the physical cells are faulty. Capabilities are added such that the faulty cells can be bypassed by rerouting interconnections, creating from the physical set of functional cells the desired "logical" array. This section highlights such array reconfiguration for two cases: (1) a regular array of computational cells and (2) a memory array. Memory arrays routinely use reconfiguration to bypass faulty rows and/or columns, as discussed later. However, despite the considerable study of reconfigurable arrays of computational cells, it remains to be seen whether such approaches will see significant practical use.

Many of these techniques were originally developed from the perspective of research on wafer scale integration, where large arrays of computational cells could

7.3 ■ Global Reconfiguration

be readily achieved within the large area of an entire wafer. As VLSI technologies continue to advance, such arrays previously requiring larger silicon area will be realizable within the area of a single IC. If substantially larger arrays (requiring larger than normal IC chip areas) can be achieved at reasonable added cost, then fault tolerant arrays may play a substantial role. The general approaches reviewed in this chapter are representative of the techniques that have been explored (assuming a monolithic realization of the array). There are several practical "cost functions" that impact actual reconfiguration schemes. However, most of the theoretic results emphasize only simple models of performance such as interconnect length (impacting speed) and the number of interconnections (impacting area). In addition, somewhat stylized representations of the physical layout are used while developing the results.

7.3.1 Reconfigurable Arrays of Computation Cells

Figure 7-14 illustrates a representative model, used, for example, by Green and Gamal [43] for their early studies of reconfigurable arrays. All cells are identical (assumed throughout this section unless specified otherwise), allowing any cell to be substituted for a faulty cell. The cells (shaded regions) of a two-dimensional array are combined with interconnection structures, that is, *wiring channels* running between adjacent rows and between adjacent columns of the array and *switching sites* at the intersections of horizontal and vertical wiring channels. Each channel consists of a number of distinct interconnections (called *tracks*), allowing multiple paths (different source-destination connections) to flow between any pair of cells. The distance measure can be simplified by assuming that the width of a cell is equal to the width of a track, each with distance ℓ. The center-to-center distance between adjacent cells is then $(1 + k)\ell$, capturing the impact of both the cell and additional

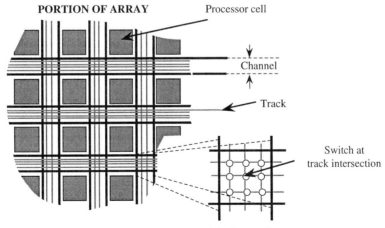

Figure 7-14 Simplified model of general array.

wiring between cells.[1] To construct a functional array of K cells, under conditions such that some cells of an array can be defective, requires a starting array of $N > K$ cells, with $N - K$ cells providing the redundancy needed to bypass faulty cells.

The Configurable Highly Parallel (CHiP) computer proposed by Snyder and his colleagues [44–49], has a physical architecture similar to the general architecture above. The CHiP computer was an efficient, circuit-switched parallel computer with static reconfiguration. The general architecture of the CHiP computer is illustrated in Figure 7-15 and consists of (1) a switch lattice providing a regular interconnection structure that can be customized using programmable switches, (2) a uniform array of identical processors whose input and output connections terminate on the switching nodes of the switch lattice and (3) external connections to the processor array provided at lattice switching sites along the lattice perimeter. The switch lattices can be adjusted for particular applications by adjusting the corridor width, the degree of the lattice, and the crossover capability. The *corridor width* is the number of switches separating two adjacent processors. The *degree* of the lattice is the number of data paths incident on a processor or switch, using the element (switch or processor) with the smaller number of incident data paths. The *crossover capability* is the number of distinct data paths that can be independently connected at a switch.

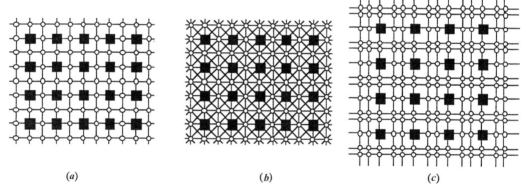

Figure 7-15 CHiP computer configurations specified by number of data paths per channel and the degree of the switching nodes on channel interconnections. (*a*) Degree 4 switches, 1 data path per channel. (*b*) Degree 8 switches, 1 data path per channel. (*c*) Degree 4 switches, 2 data paths per channel.

Figure 7-16 shows two typical examples of regular parallel processor networks (a mesh-connected array and a binary tree connected array of processors) implemented in the CHiP array. Several embeddings of standard networks within a mesh network have been demonstrated by Snyder and his colleagues, as well as by several others. The CHiP computer architecture provided a basis for extensive studies not only of the general structures of parallel computers but also of the software environments supporting such architectures [50]. For example, the reconfiguration

---

[1] It is conventional in descriptions of theoretical results to use a normalized distance, that is, to take $\ell \equiv 1$. With this normalization, the center-to-center distance between adjacent cells is $1 + k$. Such theoretical limits also generally assume that the fault probability p is $p = 1/2$.

7.3 ■ Global Reconfiguration

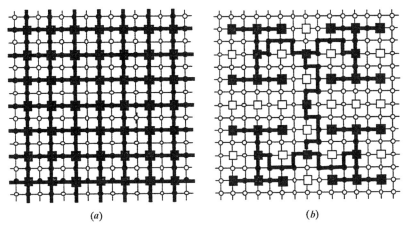

Figure 7-16 Lattice configuration examples. (*a*) Embedded mesh. (*b*) Embedded binary tree.

switches can be used to "program" a desired topology (tree, etc.) of cells, using the natural mesh organized array which would be seen in a VLSI layout. In this case (e.g., Figure 7-16), the reconfiguration is provided not for the purpose of fault tolerance but for the purpose of establishing a preferred set of interconnections (according to the preferred topology) among cells.

Next, we consider the two major classes of array, the linear array and the two-dimensional array. In the case of linear arrays, two alternative strategies are discussed, namely (1) the use of all functional cells when constructing the target array and (2) the use of most of the functional cells (discarding those functional cells whose use in the target array produces substantial degradations in performance). In the case of two-dimensional arrays, the size of the target two-dimensional array (assumed rectangular) will generally not exactly match the number of functional cells of a larger two-dimensional array with redundant cells. In this case, "most" of the functional cells of the two-dimensional array are used, creating the largest possible two-dimensional array topology. Another important conceptual approach discussed in this section is "self-reconfiguration" in which testing and reconfiguration are provided within the monolithic circuit.

7.3.1.1 Linear Arrays: All Functional Cells Used.

Green and Gamal [43] considered the linear array illustrated in Figure 7-17. The objective is to obtain a linear array of K functional cells, using an original array of $N > K$ cells. Let p be

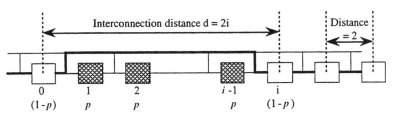

Figure 7-17 Linear array and reconfigured distance between connected cells.

the probability that a cell is defective. Then, $q = (1 - p)$ is the probability that the cell is functional. To obtain K functional cells, $N > K/(1 - p)$. A single *track* is sufficient in the linear array shown in Figure 7-17. The minimum distance of 2 corresponds to the assumption that the cell width and channel width are equal, with value unity, as discussed earlier.

The interconnection distance d in Figure 7-17 represents the additional interconnection length necessary to bypass faulty cells. For a given specific array and interconnection/driver technology, there is a maximum interconnection distance D for which the timing constraints for a given throughput rate can be satisfied. Given a maximum distance D, the maximum number of cells that can be bypassed is $N_{bypass} = (D/2) - 1$. The probability P_{reconf} of successful reconfiguration, subject to the maximum length constraint, is found to be [43]

$$P_{reconf} < N(1 - p)[1 - p^{D/2}]^{K-1} < \frac{K}{R} exp(-Kp^{D/2}) \qquad (7\text{-}1)$$

where $N \equiv K/R$ has been used. This result leads to significant conditions (see, for example, [43]) required to achieve a high probability of successful reconfiguration.

The reconfigurability limitations can be relaxed by providing a richer reconfiguration structure. Figure 7-18, where solid squares are defective cells and open squares are functional cells, illustrates the contrast between a serial array used for reconfiguration and a full two-dimensional array for embedding a serial array. In Figure 7-18a, the first row bypasses several defective cells and reconnects to the second row by extending the bypass interconnection to the rightmost edge, leading to a long line length, even though the next functional cell to be connected lies immediately below the first functional cell. Using a lattice of reconfiguration connections, such as shown in Figure 7-18b, we can reroute connections both horizontally and vertically, minimizing distances between adjacent functional cells of the final serial array.

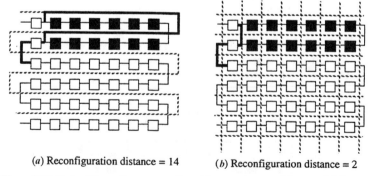

(a) Reconfiguration distance = 14   (b) Reconfiguration distance = 2

Figure 7-18 Embedding a linear, reconfigurable array in a mesh layout. (a) Reconfiguration distance = 14. (b) Reconfiguration distance = 2.

7.3 ■ Global Reconfiguration

The *patching method* [51, 52] shown in Figure 7-19 reconfigures a linear array, with a two-dimensional lattice of reconfiguration tracks, using all functional cells of the array and using, with probability $1 - \mathcal{O}(1/N)$, wires of maximum length $\mathcal{O}(\sqrt{\log N})$. The overall array of N cells in Figure 7-19 is first divided into square subarrays with $2 \log N$ cells and edge dimension $\sqrt{2 \log N}$ and with each subcell having a high probability of containing at least one functional cell. The first step is to connect (within each subcell) all the functional cells of that subarray. Next, connections between adjacent subarrays are made to complete configuration of the overall initial array (e.g., using successive left-to-right and right-to-left connections among subarrays, starting with the top row and proceeding to the bottom row of subarrays). Since the maximum wire length required to interconnect two successive subarrays is $3\sqrt{2 \log N}$, the maximum wire length in the completed linear array is $\mathcal{O}(\sqrt{\log N})$.

The increasing reconfiguration distance d with increasing N arises from local regions with a large number of faulty cells, combined with the constraint that all functional cells be used in the final linear array. Reconfiguration using most of the functional cells can be achieved with a constant distance d, independent of N, as considered in the next section.

7.3.1.2 Linear Arrays: Most Functional Cells Used.
The use of all functional cells may require some connections longer than desired. The approaches considered in this subsection identify and eliminate, from the set of cells used, those functional

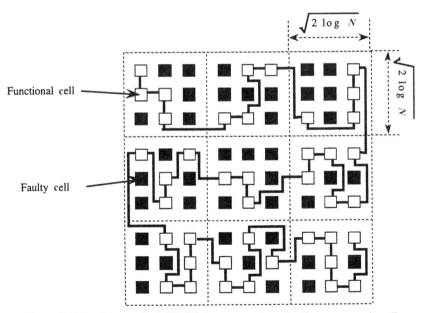

Figure 7-19 Patching method [51, 52] for construction of a linear array using all functional cells.

cells that would require excessively long interconnections. As a result, a physical array that cannot be reconfigured using all functional cells (due to excessive line length to some functional cells) can be successfully reconfigured. Two early results are summarized here, illustrating the general approach.

Leighton and Leiserson [51, 52] applied techniques used for the standard minimum weight spanning tree problem to identify and remove from the final array those functional cells that introduce longer than desired wiring paths. The overall procedure combines construction of a minimum weight spanning tree to the array of functional cells, eliminating problematic functional cells, followed by construction of a linear array from that tree. With probability $1 - \mathcal{O}(1/N)$, at least $1 - \varepsilon$ of the live cells on an N-cell wafer can be connected in a linear array using wires of length $\mathcal{O}(s\sqrt{\log_p \varepsilon})$ and channels of width 2. Here, p is the probability of a given cell being faulty, s is the edge length of each cell, and $1/N \leq \varepsilon \leq p < 1$.

Greene and Gamal [43] used a variant of the patching method, initially dividing the $\sqrt{N} \times \sqrt{N}$ physical array into N/b square arrays (blocks) as illustrated in Figure 7-20. Connections are then established to create a logic linear array (of $K = RN$, $R < 1$, cells) in this physical array such that the maximum wiring length is constant. The parameter b is chosen such that, with high probability, each block contains at least four functional cells. Blocks with fewer than four functional cells are treated as faulty. Reconfiguration proceeds first by connecting all the functional blocks (a percolation procedure that progressively forms connections from the current block to a neighboring block until a path is found which traverses all the functional blocks). After the input and output edges of each block in the step above are defined, the linear array is formed within each "functional" block by "snaking" the connection through all the functional cells of the block, as shown in Figure 7-20. Using this construction and for arbitrarily large N and any $R < (1 - p)$, we can connect a chain of length $K = RN$ from a $\sqrt{N} \times \sqrt{N}$ array with yield $1 - \mathcal{O}(1/N)$ and maximum wire length $d = [(9 \log([1 - p - R]/c)/\log(p))^{1/2}]$ for some

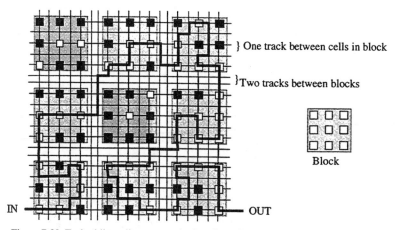

Figure 7-20 Embedding a linear array by forming subarrays and discarding unusable cells (either nonfunctional or inaccessible from neighboring cells) according to a percolation model.

constant $c > 1$. No more than two tracks are required in any channel between blocks, as shown in Figure 7-20.

These theoretical estimates not only provide useful general bounds on reconfigurability but also illustrate some of the powerful graph-oriented techniques that can form the basis of actual reconfiguration algorithms.

7.3.2 Two-Dimensional Arrays

Efficient embedding of a rectangular array of functional cells in a two-dimensional array with faulty cells is more complex than the linear array embedding discussed in the previous section. General evaluations based on particular reconfiguration approaches have been reported (e.g., [43, 51, 52]). Only two simple algorithms are summarized here.

A divide and conquer algorithm is described by [51, 52] using the functional cells of an $N \times N$ array to construct a functional $M \times M$ array ($M \leq N$). If the original array contains more than M^2 functional cells, then some of the functional cells are not used. The algorithm begins by recursively dividing the array into subarrays of size $\Theta(\log N)$.

Figure 7-21, which shows the physical 8×8 array as well as the 6×6 target array, illustrates the general approach [51, 52]. The technique recursively bisects the physical N-cell array, alternating between vertical and horizontal cuts. In Figure 7-21c, the vertical cut provides 19 functional cells on the left and 17 functional cells

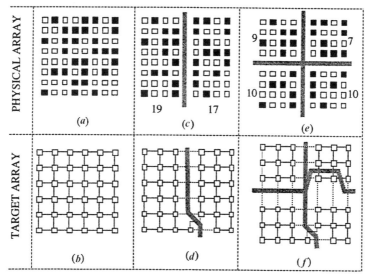

Figure 7-21 Divide-and-conquer algorithm example. (*a*) The original physical array with faulty cells (dark). (*b*) The target array constructed from the functional cells of the physical array. (*c*) Vertical bisection of the physical array. (*d*) Distorted vertical bisection of target array. (*e*) Horizontal bisection of physical array. (*f*) Corresponding horizontal bisection of target array.

on the right. A similar bisection of the target cell would require 18 cells on each side. Letting the cut of the physical array define which cells in the target array are to the left and right of the similar vertical cut of the target array, we see that the vertical cut of the physical array must be distorted to match the number of functional cells on each side of the cut of the target array. The bisections of the physical and target arrays continue until the target array has been divided into subarrays with $\Theta(\log N)$ cells. These distortions define which cells in the target array are "out-of-place" when mapped into the physical array. Next, the cells within the subarrays are connected to provide a mesh-connected, target subarray (a tree-of-meshes approach is described in [51, 52]). Note that not all functional cells within a physical subarray are necessarily used in the final target array. The cell interconnections achieved in this manner are shown as solid interconnection lines in the subarrays of the reconfigured array in Figure 7-22. Finally, connections between subarrays can be completed, as defined by the construction of the subarrays. Using this approach, with probability $1 - \mathcal{O}(1/N)$, we can construct a two-dimensional array from the functional cells of the $\sqrt{N} \times \sqrt{N}$ array using wires of length $\mathcal{O}(\log N \log \log N)$ and channels of width $\mathcal{O}(\log \log N)$.

Figure 7-22 Reconfigured array from example in Figure 7-21.

The wafer-scale integrated versions of the CHiP computer (i.e., the WASP computer [53]) shown in Figure 7-23 used a patching method. A two-level hierarchical reconfiguration strategy was used to organize the overall array as an array of modules, each module containing a set of neighboring processing elements (PEs) and associated switches. This allows local reconfiguration of individual modules to

7.3 ■ Global Reconfiguration

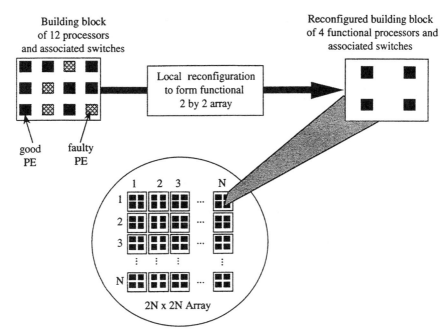

Figure 7-23 General organization of the WASP computer.

obtain functional modules that can then be interconnected without higher level, global reconfiguration. The building blocks consist of arrays of 12 PEs. The individual building blocks are reconfigured to obtain good 2×2 arrays. The overall wafer is then configured as a $K \times K$ array of such reconfigured building blocks, providing an overall $2K \times 2K$ array of processing elements.

These discussions have considered linear and nearest-neighbor connected two-dimensional arrays. Important cases of other arrays have also been studied. The Diogenes approach [54–58] is a general, graph-theoretic approach developed by Rosenberg and his colleagues for reconfiguration of general array topologies, and it continues to provide a foundation for more recent work [59].

7.3.3 Self-Reconfiguration Algorithms for Two-Dimensional Arrays

The above approaches have all assumed an external analysis of the faulty cells and execution of various algorithms to embed a functional array in the physical array, avoiding the faulty cells. In this section, we consider a very different approach, namely, an intelligent array of cells capable of reconfiguring itself. The early studies at the Politecnico di Milano on self-organizing (self-reconfiguring) two-dimensional arrays were particularly important in the development of reconfiguration approaches. Reviews of their approaches to self-reconfiguration, using both time- and space-redundancy, are provided in [60, 61]. A variety of reconfiguration switching/multiplexing elements have been considered and are not considered here. Details of various approaches can be found in [60–68].

In this section, the simplest of the spatial reconfiguration schemes is described.

The general approach is based on launching, from an edge of the array, a probe of the possibilities for alternative connections throughout the array combined with a means for a selection of a particular set of reconnections.

Figure 7-24 illustrates the adjacent cells, in a rectangular array with horizontal and vertical nearest neighbor connections, to which outputs of a functional cell can be connected. Unidirectional data flow (inputs from the left and the top and outputs to the right and to the bottom) is assumed for each cell, representing, for example, a wavefront or systolic processor structure. Cells to which the horizontal output from a cell can be connected comprise the horizontal *neighborhood cells* of the functional cell. In the examples considered here, the *horizontal neighborhood* is shown in Figure 7-24*a*, and includes three adjacent cells to the right. Various sets of nearby cells, consisting of three, four, or five cells, can be used to construct the *vertical neighborhood,* as illustrated in Figure 7-24*b*. The combination of the horizontal and vertical neighborhoods is labeled as $/H_n * V_n/$, giving the $/3*3/$, $/3*4/$ and $/3*5/$ structures corresponding to the combination of the horizontal neighborhood in Figure 7-24*a* with the 3, 4, or 5 cell vertical neighborhoods, respectively, in Figure 7-24*b*.

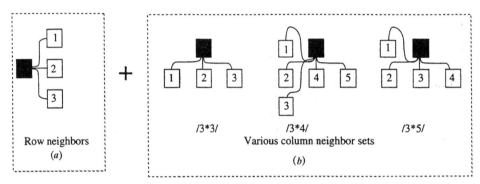

Figure 7-24 Cell neighborhoods.

Reconfiguration begins with a request from the input edges of the array (i.e., the top edge and left edge) to connect cells of those input edges to functional cells within the array.[2] It is assumed that all cells have been identified as functional or faulty and know their state.

Row extension is considered first. A functional cell (the *requesting cell*) requests, from each member of its horizontal neighborhood set, a response indicating whether the *contacted cell* can accept the requesting cell's horizontal connection. On receipt of requests (perhaps simultaneously from several cells), a *contacted cell* will be expected to acknowledge to the *requesting cells* whether or not it can accept a horizontal connection. The algorithm proceeds by requiring that the contacted cell not reply until it has determined whether it can extend its connection, causing requests from the left side to ripple through the array until reaching the right side,

---

[2] Rows are constructed from left to right, though a logical row may include cells from more than one physical row. Columns are constructed from top to bottom, with logical columns able to use cells from more than one physical column.

7.3 ■ Global Reconfiguration

at which point replies are propagated backwards toward the left edge. The specific actions in the example algorithm described here are as follows, for the horizontal connections.

1. A *contacted cell* responds to *requesting cells* that it can accept a horizontal connection *if that contacted cell can itself complete a horizontal connection to its horizontal neighborhood.* The first request to a *contacted cell* represents the propagation of the reconfiguration command launched at edges of the array to that *contacted cell.* The contacted cell, in turn, contacts cells in its horizontal neighborhood to extend its own connections. Not until it has received acceptance of its request can the *contacted cell* accept requests made to it. A highly dynamic sequence of events is therefore initiated by the edge-initiated prolong command, the reconfiguration process relaxing to a stable configuration as requests and responses to requests throughout the array evolve.
2. Each cell can receive requests to extend horizontal connections (i.e., extend rows) from several cells. It provides the same response (either "can accept a connection" or "cannot accept a connection") to all cells requesting connections. A contacted cell able to accept a horizontal connection is an *accepting cell.*
3. A *requesting cell* receives connection acknowledgments from all cells in its horizontal neighborhood, and more than one of those cells may be able to accept a horizontal connection. The *requesting cell* must therefore follow some set of rules on choosing which connection, among those allowed, to use. Each neighborhood cell (for both horizontal and vertical neighborhoods) in Figure 7-24 is shown with a priority number associated with the requesting cell.
   (a) The requesting cell seeking a horizontal connection selects the *accepting cell* with the lowest priority number.
   (b) The requesting cell then, having located a horizontal connection, cancels its request for connections to those horizontal neighborhood cells having a higher priority number than the selected cell.

Vertical connections (i.e., logical columns) are built similarly, with two differences relative to the steps above for horizontal connections.

- Rather than accepting the neighborhood accepting cell with the lowest priority number, the *requesting cell* selects the *accepting cell* in its vertical neighborhood with the highest priority number.
- Row and column interconnections originating from or terminating on different cells can cross only within a cell. (Note: *passing cells* receive row and column connections from a single cell and have output row and column terminating on a single cell, as illustrated below in the reconfiguration example).

Figure 7-25 illustrates an example of array reconfiguration, using the algorithm summarized above and the /3*4/ neighborhood structure. This figure shows the *logical* flow of horizontal (row) and vertical (column) connections among cells in the array. The actual routing of interconnection paths requires physical wire links and interconnection circuitry. Various physical interconnection and routing circuit examples are included in the reports by researchers at the Politecnico di Milano (e.g., [60]).

7.3.4 Replacement of Full Rows (Columns) by Spare Rows (Columns)

Here, we consider replacement of an entire row (column) containing a faulty cell. Such an approach has been used extensively in yield-enhanced, VLSI memory circuits (e.g. [69]) and other regular arrays (e.g. [70–74]). The approach is inherently inefficient in the sense that all good cells of a row (column) with one or more faulty cells are discarded. However, there are physically very straightforward techniques to implement the approach, providing a practical efficiency in several cases. The approach, for a large array with only a few spare rows and/or columns, can accommodate only a relatively small number of faulty cells. It is therefore important to

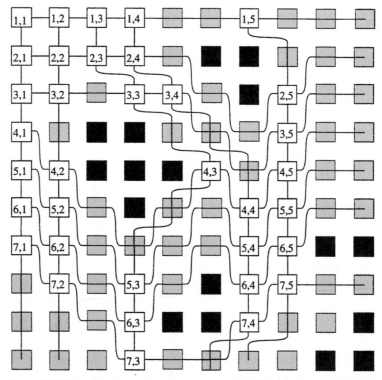

Figure 7-25 Example of reconfigured array using /3*3/ neighborhoods. Solid cells are faulty and unused. Shaded cells are functional but are not used in the reconfigured array. The output connections from cell (3,4) illustrate an example of the *passing cell*.

7.3 ■ Global Reconfiguration

develop algorithms that most efficiently allocate spare rows and/or columns to avoid (i.e., cover) as many faults as possible. Kuo and Fuchs [75] have reviewed two conventional reconfiguration algorithms (i.e., those of [76] and [77]) and suggested an improved reconfiguration algorithm based on the bipartite graph covering problem. Figure 7-26a shows the general structure of the array (with N rows and M columns, augmented by S_R spare rows and S_C spare columns). The example used in [75] to illustrate the reconfiguration approaches is shown in Figure 7-26b.

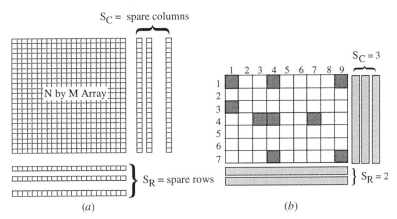

Figure 7-26 Spare row and column array structure and example used in text. (a) General array. (b) Example used in text.

A *bipartite graph* is a graph $\mathcal{G}(\mathcal{V}, E)$ whose vertex set $\mathcal{V}$ can be divided into two disjoint sets $\mathcal{A}$ and $\mathcal{B}$, with each edge in E having one end in $\mathcal{A}$ and the other end in $\mathcal{B}$. Let the bipartite graph be denoted as $(\mathcal{A}, \mathcal{B} : E)$. The collection of faulty cells in the rectangular array of Figure 7-26 can be labeled according to their indexed position, (x_i, y_i) for the ith fault, in the array. This maps directly into a bipartite graph structure by associating the set of faulty cell row indices $\mathcal{X} = (x_i)$ with the vertex set $\mathcal{A}$ and the set of faulty cell column indices $\mathcal{Y} = y_i$ with the vertex set $\mathcal{B}$ of the bipartite graph. Figure 7-27a illustrates this association. Each specific faulty cell is then defined by an edge of the bipartite graph, that is, representing the fault by an edge between the faulty cell's horizontal and vertical indices.

The problem of allocating spare rows and columns is related to the bipartite graph in an obvious way. Replacing a row having faulty cells with a spare row corresponds to removing that row's index (and the edges associated with that index) from the bipartite graph. When all the edges of the bipartite graph have been removed by using spare rows and/or columns, the faulty array has been repaired. The problem of allocating the minimum number (or lowest "cost" set) of spare rows and columns to bypass faulty cells is therefore the standard covering problem for the bipartite graph. More precisely, in a bipartite graph, a minimum vertex cover is exactly the minimum number of edges of the graph. The number of vertices in the minimum vertex cover is exactly the minimum number of spare rows and columns in the array for the system to be repairable [75].

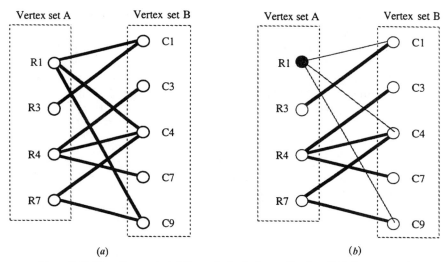

Figure 7-27 Bipartite graph. (*a*) Example for fault specification. (*b*) Edges covered by node R_1.

Other representative algorithms for allocating spare rows/columns include the following.

- The *greedy algorithm* [76] successively removes the remaining node with the highest degree (most edges). In special cases, this simple heuristic algorithm cannot reconfigure arrays that are, in fact, repairable by row/column substitution.
- An *exhaustive search* evaluates all possible row/column allocations, searching for the lowest cost solution [77]. While able to find minimum cost solutions, the exhaustive search is impractical unless the number of faulty cells is small.
- The *branch and bound* method [75] first determines whether sufficient rows and columns exist to allow repair of the array, using a polynomial time algorithm to solve the bipartite graph matching problem. Next, a branch and bound algorithm is performed to obtain the minimum cost allocation of rows and columns.
- A *heuristic method* was suggested by [75] to reduce the computation time needed to determine the row/column allocation. Reconfiguration of arrays up to 1024 × 1024, with $S_R = S_C = 20$ and 400 faulty cells using the branch and bound and the heuristic allocation algorithms are summarized in [75].

7.4 PHYSICAL SWITCH TECHNOLOGIES FOR RECONFIGURATION

This section reviews various physical technologies for implementing reconfiguration. A variety of schemes can be used, ranging from placement of electronic switches at reconfiguration points to physically altered interconnections. The following terminology [78] represents the variety of techniques that can be used.

7.4 ■ Physical Switch Technologies for Reconfiguration

- *Soft-configurable:* Alteration of interconnections using electronic switches that are programmed according to the distribution of faults in an array. The explicit switch states are generally stored in latches within the switch sites.
- *Vote-configurable:* Alteration of interconnections using (generally selector) switches according to local voting among redundantly operating sites, as in Triple Modular Redundancy (TMR) schemes. Here the switch states are locally defined by local fault-detection mechanisms.
- *Hard-configurable:* Physical alteration in interconnections by cutting links or physically adding links. The switch state is implicit in physical modifications to the interconnections and cannot be altered during service.
- *Firm-configurable:* Alteration of interconnections using structures such as floating gate switches with switch state implicit in the physical mechanism used to open or close switches. Programming is nonvolatile (e.g., e-beam programming of floating gate switch) but can be altered (e.g. by UV-erasure of the states and reprogramming). The settings are, however, "firm" in the sense that reprogramming requires application of specialized processes outside the range of conventional electronic in-situ programming.

The discussions below illustrate these general techniques.

7.4.1 Electronically Programmable Reconfiguration

Figure 7-28 illustrates a general, programmable electronic switch for reconfiguration of interconnections, illustrating some of the overhead functions associated with the switching function. In addition to the specific switching element, the state of the switch must be stored at the switch site (either in a nonvolatile memory

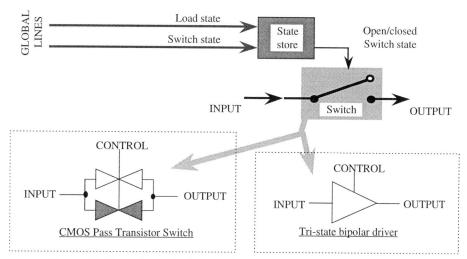

Figure 7-28 General model of electronic switch and two switch realizations (a CMOS pass transistor switch and a bipolar, tri-state driver).

element or, after loading, in a register associated with the switch). The programmed switch state information will typically have to be loaded through external interconnections. In Figure 7-28, this external loading is illustrated by two global lines, one providing the switch state and the other a control signal loading that state into the storage node.

The *overhead* associated with such electronic programmable switching of interconnections includes the *switch area overhead* of the switching circuit, the *control area overhead* of the state storage flip-flop/latch,[3] the *control interconnection area overhead* corresponding to distribution of the switch state to and loading that state into the storage node,[4] and the *signal delay overhead* imposed by the switch on the data signals being propagated down the reconfigured wire relative to a direct point-to-point interconnection between the reconfigured source/destination nodes. Physical restructuring, described later, avoids the control overhead of electronically programmable reconfiguration by performing a localized physical alteration of interconnections (opening and/or closing interconnection links).

The advantages of programmable electronic switching include the ability to correct in-service faults and full manufacturing compatibility with aggressive VLSI technologies (i.e., no additional fabrication steps such as laser restructuring sites). At the same time, there are defects that are difficult to cover using electronic reconfiguration. These include shorts in the power and ground distribution nets and defects in the clock distribution network. For these cases, physical restructuring may be preferred (or perhaps necessary).

Next, we summarize some representative switch designs that have been reported, illustrating the various techniques that can be used.

7.4.1.1 Independently Addressable Switch Sites.

Figure 7-29 shows the general structure of the array of control nodes associated with the reconfiguration switches of a reconfigurable mesh array of cells (using a CHiP-computer-styled layout) reported by Choi et al. [79]. Of particular interest here is the means by which fault tolerant access was provided to each switch site.

Programming the switches proceeds through the row and column "write-only" buses (i.e., buses that can be written from external sources but not from array cells, to avoid faulty cells contaminating the information on a control bus). Each switch has a unique address and circuitry to detect that address on configuration data passed along one of the two buses (i.e., either the column or row bus) intersecting that cell. This allows each switch to be programmed from either bus, providing tolerance of some faults. In particular, a switch is inaccessible only if both the row and column buses intersecting the cell are defective, as illustrated by the solid colored switch in Figure 7.29*b*.

---

[3] If, rather than a single wire, a parallel set of wires are simultaneously switched (as in a parallel data bus), then the control circuit area, which is independent of the number of wires simultaneously switched, becomes smaller relative to the switch area.

[4] The area associated with loading the switch state into all switching sites can be decreased by passing the switch state information along a pipelined serial data interconnection, simultaneously loading the state information along that chain into the switch sites in parallel (similar to the approach used in scan test designs).

7.4 ■ Physical Switch Technologies for Reconfiguration

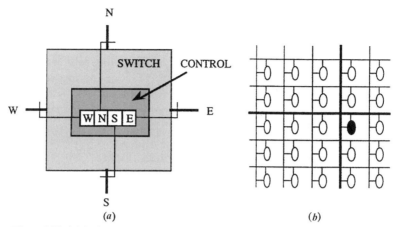

Figure 7-29 (*a*) Switch structure and (*b*) array programming organization in testable array suggested in [79].

7.4.1.2 Masking Faulty Cells. Figure 7-30*a* illustrates a general array structure described in [78]. Switches are represented by open circles, buses (allowing parallel-bit data paths) are represented by gray-colored lines, and "chips" are represented by solid squares. This structure is used to illustrate the "masking off" of faulty cells for applications that do not require a regular array of cells but that are implemented in an array form to achieve efficient layouts. Rather than using "reconfiguration" switches to retain a regular array of interconnected cells, switching cells instead "mask off" faulty cells or bus segments (i.e., simply delete them from the circuitry). This is illustrated in Figure 7-30*b* where a 5-input, 1-output multiplexer is used to implement the switch. Masking information (e.g., 1-bit per input bus) can be loaded into the switch to "turn-off" faulty elements. In addition to removing faulty elements, the switches can be used as part of a communications

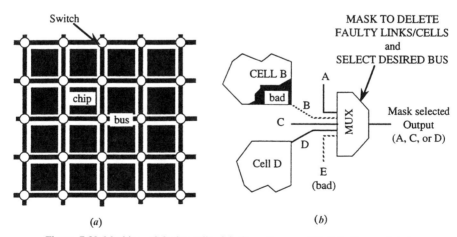

Figure 7-30 Masking of faulty cells. (*a*) Overall array. (*b*) Selection switch to deselect faulty cells.

network. Extensions discussed in [78] but not considered further here include general computer network communication functions such as routing with congestion and pipelining.

Such expansions of the function of the switching cells are costly for fine-grained array processing cells but become more practical as the granularity of the array processing cells increases. Indeed, the functionality of the switching function may have to expand as the array cell functionality expands to suit the requirements on the overall array architecture.

7.4.1.3 Four-Port Switch Examples. Figure 7-31 illustrates a typical construction of a multiport switch from simple pass transistor switching elements (open/closed switches) [64]. The pass transistor structure shown in Figure 7-31*b* illustrates the simple switch designs possible, in this case requiring only 1.5 pass transistors per port of the switch. Not shown in Figure 7-31 is the circuitry needed to hold the pass transistor states (3-bit register and decoding logic or 6-bit register). The individual pass transistors are set to either ON (closed) or OFF (open) states to create one of the five switch configurations shown in Figure 7-31*a*. In addition to the simplicity of such pass transistor structures, the switches are intrinsically bidirectional. These advantages may be offset, however, by the relatively large ON-state resistance of the pass transistors.

Chevalier and Saucier [80] describe a somewhat more complex design for a four-port switch, illustrated in Figure 7-32. The switch design here is designed to support not only reconfiguration of an array but also reliable propagation of test signals through the array. The switches and interconnection network must accordingly provide both routing for testing and routing for problem execution.

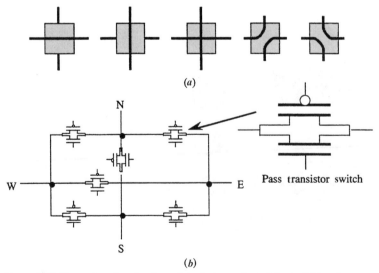

Figure 7-31 Example of a simple switch design, adapted from [64], using pass transistors. (*a*) Switch configurations. (*b*) Switch design.

7.4 ■ Physical Switch Technologies for Reconfiguration 253

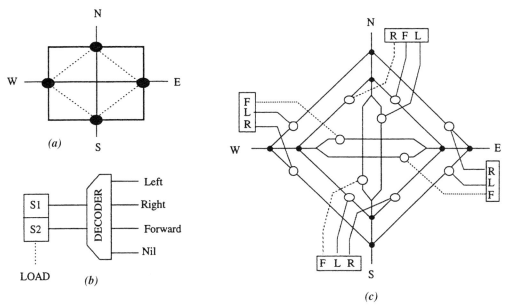

Figure 7-32 Flexible and testable switch design of Chevalier and Saucier [80]. (a) General switch model. (b) Control logic to store and control pass transistors associated with each port of the switch. (c) Organization of interconnections and pass transistors (open circles) of four-port switch, with port direction control blocks at each port.

The switch design in Figure 7-32c uses three pass transistors per port, in contrast to the design in Figure 7-31 which uses only 1.5 switches (pass transistors) per port. However, the increased number of switches per port facilitates the progressive probing of tests through the array. For example, entering from the "N" input, the interconnection can be passed directly to the "W," "S," or "E" port fully under control of the "N" port switches. This contrasts with the propagation of paths through the switching cell in Figure 7-31 where the global switch state must be set and where redundant tests of the switch are limited. The specific steps used to create paths through the switch lattice and the logical circuitry to program the switches are given in [80].

7.4.1.4 Crossbar Switch Example. Much of the preceding discussion has considered arrays of identical elements, with reconfigurations used to change the connections among those elements. An example of a fault tolerant array using a heterogeneous set of modules (with duplicates of each module to provide a working spare module as well as additional resources if multiple functional modules can be used) is described in [81]. The specific application combined a set of special-purpose, bit-serial signal processing functions (multipliers, adders, FIFOs, etc.) to provide a high-performance signal processor called *FIRST* (First Implementation of Real-time Signal Transforms). Connections among the various processing modules depend on (and are changed according to) the specific signal processing function implemented. Combining the programmable communications to "build" specific signal processing

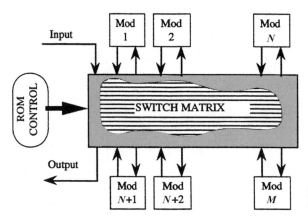

Figure 7-33 Crossbar switch interconnected collection of modules.

functions with the use of programmable communications for fault tolerance suggested a general switching fabric such as the crossbar switch such as shown in Figure 7-33. Tolerance of faulty modules is provided directly through the programmability of the crossbar switch and the duplicate modules, with the signal flow paths established only among the functional modules.

Figure 7-34 shows two crossbar switch organizations described in [81]. In Figure 7-34a, each output line from modules is connected to only a single horizontal line of the switch. This provides a significant reduction in the electronic circuitry of the switch (crosspoint switches and their local control circuitry). However, faults in the crossbar switch can make some connections among functional modules inoperable. The lack of redundancy within the switch, in this case, is clear since there is only one path between any given module's output and any target module input. An alternative design, requiring more switch and switch control circuitry than the nonredundant design is shown in Figure 7-34b. Here, not only the module input lines but also the module output lines are connectable through switches to each horizontal line. In the case of a local failure of a crossbar switch location or in a horizontal crossbar switch line, connections among modules can bypass the faulty portion of the switch. The actual circuit designs are presented in [81].

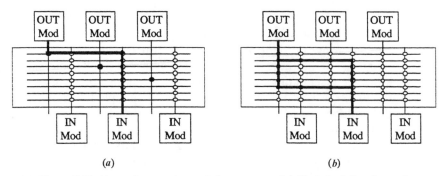

Figure 7-34 Alternative crossbar switch structures. (a) Dedicated line for each OUT module. (b) Programmable lines for OUT modules.

7.4.2 Physical Restructuring of Interconnections

Physical restructuring involves physical alteration of the circuit interconnection links through addition or deletion of connections between physical interconnections or through physical programming of switches controlling the interconnection paths. Physical restructuring provides two important advantages relative to use of electronic switches: namely, (1) additional on-chip area to "store" the state of the switch is not needed, and (2) smaller resistances are typically achieved at the switching site. Maintaining low interconnection resistance is important, for example, on clock distribution networks [82] and power distribution lines [82, 83].

7.4.2.1 Laser-Based Restructuring.
Cenker et al. [95] introduced repairable memory ICs to commercial practice, and fault tolerant designs have become standard practice in high-density silicon memory components [84–88]. Laser cutting of links (opening a link) is often used, providing a simple approach acting only on the surface layer of metallization.

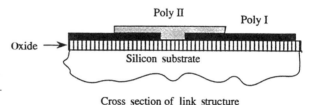

Figure 7-35 Planar polysilicon antifuse for laser closure.

Cross section of link structure

Laser closure of links is illustrated in Figure 7-35*b*. In this example, two closely spaced and heavily doped polysilicon (Poly1) lines can be joined by laser heating of the overlying, high-resistance polysilicon antifuse. The initial resistance of the open link can be in the range $10^8 \rightarrow 5 \times 10^9\ \Omega$. The laser-closed line has a moderately low resistance ($50 - 500\Omega$).

An example of general use of physical restructuring for VLSI/WSI (Wafer Scale Integration) is seen in the several studies of the RVLSI (Restructurable VLSI) program at MIT Lincoln Labs (e.g., [89–94]). Figure 7-36 shows a representative example of a reconfigurable cell, with connections made by closing connections between metal layers or broken by cutting metal on a layer.

Laser-programmed vertical links using amorphous silicon as the link insulator were initially used in the RVLSI effort [89–91]. The link structure is shown in Figure 7-37. The link is initially open but can be programmably closed using a low-power, CW argon laser, operated at between 1 and 2 watts. For use in standard CMOS VLSI technologies, the amorphous silicon layer introduces a nonstandard process and additional cost.

A laser-programmable diffused link structure, more compatible with CMOS technologies, was developed later within the RVLSI program [93, 94]. The diffused link structure is illustrated in Figure 7-38. The degenerately doped P-type implant regions are formed during source/drain doping of the N-channel MOS-FETs. The

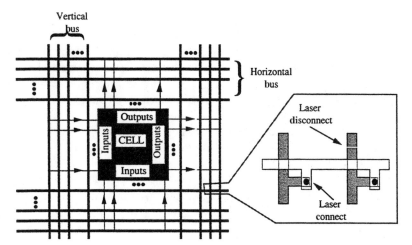

Figure 7-36 Planar polysilicon anti-fuse for laser closure.

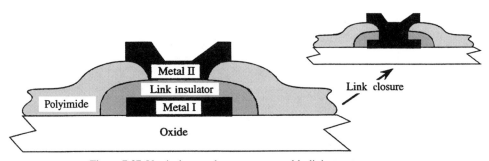

Figure 7-37 Vertical amorphous programmable link structure.

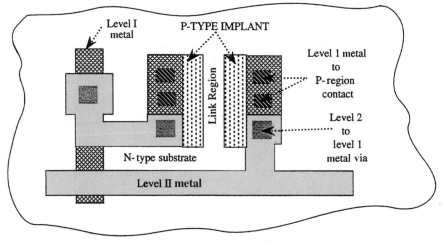

Figure 7-38 Laser diffused link structure.

two P-type implanted regions are separated by about 4 μm, with a width of about 15 μm. These regions, formed in the N-type substrate, correspond to series opposed pn diodes, with leakage currents of about 1 pA. The separation between the P-type implanted regions is closed by locally heating the substrate, causing local diffusion of the dopant which spreads across the gap. In [94], a 3.5 W, 1 msec argon laser pulse, focused to about 4 μm, is used. Three such laser pulses are required to form the 15 μm wide link shown in Figure 7-38. For a 1500 μm wide power link, 250 laser pulses are required to cover the gap. Link yields greater than 99.99% were obtained.

7.4.2.2 Electronically Field-Programmable Anti-Fuses. Field-programmable read-only memories (ROMs), written by applying programming voltages to "blow" normally open fuses (anti-fuses), are familiar electronic IC circuit functions. A similar approach can be used to physically and irreversibly add links between interconnection paths. Figure 7-39 illustrates the anti-fuse structure used to program interconnections on a passive silicon circuit board described in [96]. The two lower metal layers provide power distribution. The two upper metallization layers provide interconnection lines that can be linked by electrically switching the antifuse (the amorphous Si alloy sandwiched between the lines) from the as-fabricated OFF-state (with resistance greater than 200 MΩ) to a low resistance (<10Ω) ON-state. The switching is achieved by applying a voltage greater than the switching voltage to the anti-fuse. The voltage programming of the amorphous silicon "anti-fuse" illustrated in Figure 7-39 is about 20 volts.

7.4.2.3 Laser-Assisted Chemical Processing. The fuses and anti-fuse structures in the examples above are placed at points in the circuit where reconstruction of interconnections is to be performed. However, there may be interconnection lines that are difficult to convert to fault tolerant structures and that can exhibit

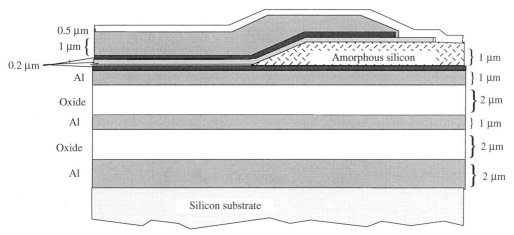

Figure 7-39 Electronically programmed amorphous silicon fuse.

defects at points anywhere along the interconnection. Examples include the power/ground distribution networks and the clock distribution network. In these cases, the interconnection area can be a significant portion of the circuit area. Faults in the power/ground and in the global clock distribution network would render the entire circuit faulty. In such cases, locally selective processing techniques provide the potential for repair.

Laser-assisted chemical processing uses the energy of light beams to enhance various chemical reactions, for example, for deposition or etching of materials. The laser beam is incident on a reaction chamber (e.g., [97]) containing the reactants (either liquid or gaseous) and accelerates or activates reactions at the points where the laser beam is incident. In this manner, the chemical reactions can be "written" directly on a local portion of the circuit surface. Chemical reactions can be laser-assisted either through generation of higher local temperatures as the substrate absorbs the radiation (*pyrolysis*) or through optical activation of reactions when reagents absorb the radiation (*photolysis*).

Several representative applications have been experimentally demonstrated, including the following.

> *Localized doping:* Localized doping of silicon, with linewidths as small as 1000Å by locally heating the substrate to high temperatures to rapidly diffuse impurities into the substrate.
>
> *Oxidation:* The oxidation rate can be thermally accelerated by heating the substrate in an oxygen ambient, allowing local growth of SiO_2 insulating layers.
>
> *Deposition:* Deposition rates can be accelerated either thermally or by optical activation of reactants. Deposition of interconnection lines to close links and etching of interconnect lines to open links have been demonstrated for possible use for repair of VLSI and WSI circuits.
>
> *"Maskless" lithography:* Local exposure of photoresist, photopolymerization of resist, or ablative photodecomposition removal of resist provides local lithographic patterning [84, 98]. Such "maskless" lithography is particularly useful for circuit customizing, avoiding the need for custom masks for each wafer [99–101].

7.4.2.4 Focused Ion Beams for Restructuring. Directing a beam of KeV energy ions onto a surface produces a number of effects that can be adapted to interconnection restructuring. Focused ion beam techniques and applications have been reviewed by Melngailis [102]. Focused ion beams have been used to cut interconnections [103–105] using a sputter etching mechanism. Secondary ions and electrons during sputter etching can be used to monitor the etching process. The use of ion milling to remove thin-film layers and then to deposit metal for vertical links has been described in [106].

7.4.2.5 Electron-Beam Programming of Floating Gates. Electron-beam (e-beam) programming of floating gate control logic provides a nonvolatile means of programming electronic switches for interconnection reconfiguration. The approach

7.4 ■ Physical Switch Technologies for Reconfiguration

described in [107] and illustrated in Figures 7-40 and 7-41a used a floating gate transistor. E-beam injection of charge onto the floating gate changes the transistor's threshold voltage, setting the transistor to either a low- or high-resistance state, depending on whether the transistor is N-type or P-type. The negative charge stored on the floating gate can be "erased" by exposure to UV radiation. Damage (and perhaps defect generation) to the substrate can degrade the programmability of the transistor. Conditions to avoid damage to the circuit by the e-beam exposure are discussed in [108].

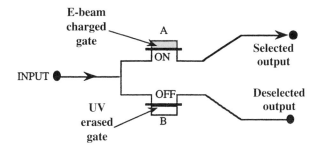

Figure 7-40 E-beam/UV programmed floating gate switch for reconfiguration.

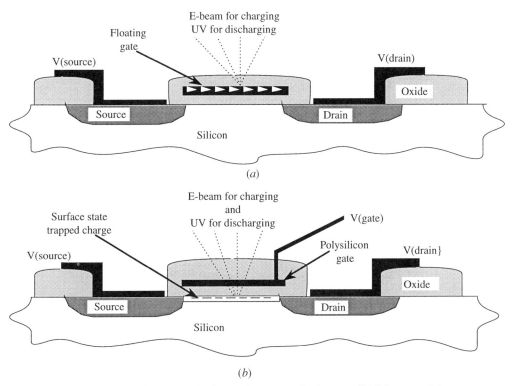

Figure 7-41 (a) E-beam injection of charge onto floating gate. (b) E-beam modulation of surface-state density.

A non-reprogrammable setting of a transistor by e-beam exposure can be obtained by modulating the surface-state density at the $SiO_2 - Si$ interface [109], as illustrated in Figure 7-41b, to change the device's threshold voltage.

7.5 SUMMARY

A rich set of approaches has been developed to provide fault tolerance of VLSI circuits and microelectronic systems. Most approaches are based on some form of redundancy (either added to the circuitry or intrinsically provided by the architecture). Architectural fault tolerance is clearly a very extensive topic, based on principles developed from applications ranging from large-scale reliable systems to large-area integrated circuits. The fault tolerant technologies (such as those used to repair faulty dynamic memory ICs) are used to repair components that emerge from manufacture as faulty. Several approaches to fault tolerance can also address the problem of failures in service (providing a "built-in-repair" capability). As advanced packaging techniques such as Multichip Modules (MCMs) and three-dimensional chip stacks make systems less repairable and as the VLSI components themselves become more complex and faster, fault tolerance can be expected to serve as an important adjunct to traditional techniques to minimize (or avoid altogether) faults and to provide in-service repair of systems. Just as built-in testing techniques have become increasingly necessary to provide practical components for insertion into microelectronic systems, so, too is built-in fault tolerance likely to become increasingly important.

REFERENCES

[1] P. K. Lala, *Fault-Tolerant and Fault Testable Hardware Design*, London: Prentice-Hall, 1985.

[2] B. W. Johnson, *Design and Analysis of Fault Tolerant Digital Systems*, Reading, MA: Addison-Wesley, 1989.

[3] D. P. Siewiorek and R. S. Swarz, *The Theory and Practice of Reliable System Design*, Bedford, MA: Digital Press, 1982.

[4] W. R. Moore, "A Review of Fault-Tolerant Techniques for the Enhancement of Integrated Circuit Yield," *Proceedings of the IEEE*, pp. 684–698, 1986.

[5] S. K. Tewksbury, *Wafer-Level System Integration*, New York: Kluwer Publishers, 1989.

[6] J. E. Brewer, L. G. Miller, I. H. Gilbert, J. F. Melina, and D. Garde, "A Single-Chip Digital Signal Processing Subsystem, *Proceedings of the 1994 IEEE International Conference WSI*, ed. by R. M. Lea and S. K. Tewksbury, Washington, DC: IEEE Computer Society Press, 1994.

[7] J. Carson, The Emergence of Stacked 3D Silicon and Its Impact on Microelectronic System Integration, *Proceedings of the 1996 International Conference on Innovative Systems in Silicon, IEEE*, pp. 1–8, Austin, TX, Oct. 1996.

[8] R. B. Boppana and S. Chalasani, "Fault-Tolerant Wormhole Routing Algorithms for Mesh Networks," *IEEE Transactions on Computers,* vol. 44, pp. 848–864, 1995.

[9] T. Lee and J. Hayes, "A Fault-Tolerant Communications Scheme for Hypercube Computers," *IEEE Transactions on Computers,* vol. 41, pp. 1242–1256, 1992.

[10] M. Slimane-kadi, A. Boubekaur, and G. Saucier, "Interconnection Networks with Fault-tolerance Properties," *Proceedings of the 1993 IEEE International Conference on Wafer Scale Integration,* pp. 213–222, San Francisco, Jan. 20–22, 1993.

[11] A. Varma and S. Chalasami, "Fault-Tolerance Analysis of One-Sided Crosspoint Switching Networks," *IEEE Transactions on Computers,* vol. 41, pp. 143–158, 1992.

[12] K. Roy and S. Nag, "On Routability of FPGAs under Faulty Condition," *IEEE Transactions on Computers,* vol. 44, pp. 1296–1305, 1995.

[13] N. J. Howard, A. M. Tyrell, and N. M. Allinson, "The Yield Enhancement of Field-Programmable Gate Arrays," *IEEE Transactions on VLSI Systems,* vol. 2, pp. 115–123, 1994.

[14] N. R. Saxena, C-W.D. Chang, K. Dawallu, J. Kohli, and P. Helland, "Fault-Tolerant Features in the HaL Memory Management Unit," *IEEE Transactions on Computers,* vol. 44, pp. 170–180, 1995.

[15] D. S. Phatak and I. Koren, "Complete and Partial Fault Tolerance of Feedforward Neural Nets," *IEEE Transactions on Neural Networks,* vol. 6, pp. 446–456, 1995.

[16] L. A. Belfore and B. W. Johnson, "Analysis of the Faulty Behavior of Synchronous Neural Networks," *IEEE Transactions on Computers,* vol. 40, pp. 1424–1428, 1991.

[17] N. Tsuda, "Rotary Spare Replacement Redundancy for Tree Architecture WSIs," in *Wafer-Scale Integration,* ed. by M. J. Little and V. K. Jain, pp. 83–89, Washington, DC: IEEE Computer Society Press, 1991.

[18] Y. Y. Chen and S. J. Upadhyaya, "An Analysis of a Reconfigurable Binary Tree Structure Based on Multilevel Redundancy," *Proceedings of the IEEE International Symposium on Fault Tolerant Computing,* pp. 192–199, 1990.

[19] Y. Y. Chen and S. J. Upadhyaya, "Reliability, Reconfiguration, and Spare Allocation Issues in Binary Tree Architectures Based on Multiple-level Redundancy," *IEEE Transactions on Computers,* vol. 42, pp. 713–723, 1993.

[20] A. D. Singh and H. Y. Youn, "A Modular Fault-Tolerant Binary Tree Architecture with Short Links," *IEEE Transactions on Computers,* vol. 40, pp. 882–890, 1991.

[21] A.S.M. Hassan and V. K. Agarwal, "A Fault Tolerant Modular Architecture for Binary Trees," *IEEE Transactions on Computers,* vol. 35, no. 4, pp. 356–361, 1986.

[22] M. B. Lowrie and W. K. Fuchs, "Reconfigurable Tree Architectures Using

Subtree Oriented Fault Tolerance," *IEEE Transactions on Computers,* vol. 36, no. 10, pp. 1172–1182, 1987.

[23] M. C. Howells, R. Aitken, and V. K. Agarwal, "Defect Tolerant Interconnects for VLSI," in *Defect and Fault Tolerance in VLSI Systems,* ed by I. Koren, New York: Plenum Press, 1989.

[24] T. Markas, D. M. Royals, and N. Kanopoulos, "Design and DCVS Implementation of a Self-checking Bus-monitor Unit for Highly Reliable Fault-Tolerant System Configurations," *IEEE Transactions on VLSI Systems,* vol. 2, no. 2, pp. 149–156, 1994.

[25] R. Leveugle, Z. Koren, I. Koren, G. Saucier, and N. Wehn, "The Hyeti Defect Tolerant Microprocessor: A Practical Experiment and Its Cost Effective Analysis," *IEEE Transactions on Computers,* vol. 43, pp. 1398–1406, 1994.

[26] S. Y. Kuo and W. K. Fuchs, "Fault Diagnosis and Spare Allocation for Yield Enhancement in Large Reconfigurable PLAs," *IEEE Transactions on Computers,* vol. 41, pp. 221–226, 1992.

[27] D. K. Pradhan and J. J. Stiffler, "Error Correcting Codes and Self-checking Circuits in Fault-Tolerant Computers, *IEEE Computer,* vol. 13, pp. 27–37, Mar. 1980.

[28] S. Lin and D. J. Costello, *Error Control Coding: Fundamentals and Applications,* Englewood Cliffs, NJ: Prentice-Hall, 1982.

[29] J. F. MacWilliams and N.J.A. Sloane, *The Theory of Error Correcting Codes,* Amsterdam: The Netherlands: North-Holland, 1977.

[30] C. L. Chen and M. Y. Hsiao, "Error Correcting Codes for Semiconductor Memory Applications: A State-of-the-Art-Review," *IBM Journal of Research and Development,* vol. 28, no. 2, pp. 124–134, 1984.

[31] J. Yamada, H. Kotani, J. Matsushima, and M. Inoue, "A 4-Mbit DRAM with 16-bit Concurrent ECC," *IEEE Journal of Solid-State Circuits,* vol. 23, pp. 20–26, 1988.

[32] C. H. Stapper and H-S. Lee, "Synergistic Fault-tolerance for Memory Chips," *IEEE Transctions on Computers,* vol. 41, pp. 1078–1087, 1992.

[33] P. Mazumder, "Design of a Fault-Tolerant Three-Dimensional Random-Access Memory with On-chip Error-Correcting Circuit," *IEEE Transactions on Computers,* vol. 42, no. 12, pp. 1453–1468, 1993.

[34] R. Blahut, *Theory and Practice of Error Control Codes,* Reading, MA: Addison-Wesley, 1983.

[35] E. R. Berlekamp, "The Technology of Error-Correcting Codes," *Proceedings of the IEEE,* vol. 68, pp. 564–593, 1983.

[36] K.-H. Huang and J. A. Abraham, "Algorithm-Based Fault Tolerance for Matrix Operations," *IEEE Transactions on Computers,* vol. C-33, pp. 518–528, 1984.

[37] J.-Y. Jou and J. A. Abraham, "Fault-Tolerant Matrix Arithmetic and Signal Processing on Highly Concurrent Computing Structures," *Proceedings of the IEEE,* pp. 732–741, 1986.

[38] J.-Y. Jou and J. A. Abraham, "Fault-Tolerant FFT Networks," *IEEE Transactions on Computers,* vol. C-35, pp. 548–561, 1988.

[39] J. A. Abraham, P. Banerjee, C.-Y. Chen, W. K. Fuchs, S.-Y. Kuo, and A.L.N. Reddy, "Fault-Tolerant Techniques for Systolic Arrays," *IEEE Computer,* pp. 65–75, July 1987.

[40] B. Vinnakota and N. K. Jha, "Diagnosability and Diagnosis of Algorithm-based Fault Tolerance," *IEEE Transactions on Computers,* vol. 42, no. 8, pp. 924–938, 1993.

[41] D. J. Rosenkratz and S. S. Ravi, "Improved Bounds for Algorithm-based Fault Tolerances" *IEEE Transactions on Computers,* vol. 42, no. 5, pp. 630–635, 1993.

[42] D. Gu, D. J. Rosendrantz, and S. S. Ravi, "Construction of Check Sets for Algorithm-based Fault Tolerance," *IEEE Transactions on Computers,* vol. 43, no. 6, pp. 641–650, 1994.

[43] J. W. Greene and A. El Gamal, "Configuration of VLSI Arrays in the Presence of Defects," *Journal of the ACM,* vol. 31, pp. 694–717, 1984.

[44] L. Snyder, "Overview of the CHiP Computer," in *VLSI 81,* ed. by John P. Gray, pp. 237–246, New York: Academic Press, 1981.

[45] L. Snyder, "Introduction to the Configurable Highly Parallel Computer," *IEEE Computer,* vol. 15, pp. 47–56, 1982.

[46] K. S. Hedlund, "Wafer Scale Integration of Parallel Processors," Ph.D. Thesis, Comp. Science Department, Purdue University, Dec. 1982.

[47] K. S. Hedlund and L. Snyder, "Wafer Scale Integration of Configurable Highly Parallel (CHiP) Processors," *Proceedings of IEEE International Conference on Parallel Proceedings,* pp. 262–264, 1982.

[48] K. S. Hedlund and L. Snyder, "Systolic Architecture—A Wafer Scale Approach," *Proceedings of the IEEE International Conference on Computer Design,* pp. 604–610, 1984.

[49] K. S. Hedlund, "WASP—a Wafer-Scale Systolic Processor," *Proceedings of the IEEE International Conference on Computer Design,* pp. 665–671, 1985.

[50] L. Snyder, "Parallel Programming and the Poker Programming Environment," *IEEE Computer,* pp. 27–36, 1984.

[51] T. Leighton and C. E. Leiserson, "Wafer-Scale Integration of Systolic Arrays," *IEEE Transactions on Computers,* vol. C-34, pp. 448–461, 1985.

[52] T. Leighton and C. E. Leiserson, "A Survey of Algorithms for Integrating Wafer-Scale Systolic Arrays," Technical Report MIT/LCS/TM-302, Laboratory for Computer Science, MIT, 1986.

[53] K. S. Hedlund, "The Design of a Prototype WASP Machine," in *Wafer Scale Integration,* ed. by G. Saucier and J. Trihle, Amsterdam, The Netherlands: Elsevier Science Publishers, pp. 89–97, 1986.

[54] A. L. Rosenberg, "The Diogenes Approach to Testable Fault-Tolerant Arrays of Processors," *IEEE Transactions on Computers,* vol. C-32, pp. 902–910, 1983.

[55] A. L. Rosenberg, "On Designing Fault-Tolerant VLSI Processor Arrays," in *Advances in Computing Research,* ed. by F. P. Preparata, vol. 2, pp. 181–204, 1984.

[56] A. L. Rosenberg, "A Hypergraph Model for Fault-Tolerant VLSI Processor Arrays," *IEEE Transactions on Computers,* vol. C-34, pp. 578–584, 1985.

[57] A. L. Rosenberg, "Graph-Theoretic Approaches to Fault-Tolerant WSI," in *Wafer Scale Integration,* ed. by C. Jesshope and W. Moore, pp. 10–23, Adam Hilger, 1986.

[58] A. L. Rosenberg, "Fault-Tolerant WSI Processor Arrays," Technical Report, Department of Computer Science, Duke University, Sept. 1987.

[59] K. P. Belkhale and P. Banerjee, "Reconfiguration Strategies for VLSI Processor Arrays and Trees Using a Modified Diogenes Approach," *IEEE Transactions on Computers,* vol. 41, pp. 83–96, 1992.

[60] R. Negrini and R. Stefanelli, "Comparative Evaluation of Space- and Time-Redundancy Approaches for WSI Processing Arrays," in *Wafer Scale Integration,* ed. by G. Saucier, and J. Trihle, Amsterdam: Elsevier Science Pubs., pp. 207–222, 1986.

[61] R. Negrini, M. G. Sami, and R. Stefanelli, "Fault-Tolerance Approaches for VLSI/WSI Arrays," *Proceedings of the IEEE International Phoenix Conference on Computers and Communications,* pp. 460–468, 1985.

[62] G. Gentile, M. G. Sami, and M. Terzoli, "Design of Switches for Self-Reconfiguring VLSI Array Structures," *Microprocessing and Microprogramming,* vol. 14, pp. 99–108, 1984.

[63] V. N. Doniants, S. Iori, M. Pellegrino, E. I. Pi'il, and R. Stefanelli, "Fault-Tolerant Reconfigurable Processing Arrays Using Bi-Directional Switches," *Microprocessing and Microprogramming,* vol. 14, pp. 109–115, 1984.

[64] V. N. Donaints, V. G. Lazarev, M. G. Sami, and R. Stefanelli, "Reconfiguration of VLSI Arrays: A Technique for Increased Flexibility and Reliability," *Microprocessing and Microprogramming,* vol. 16, pp. 101–106, 1985.

[65] A. Antola, R. Negrini, and N. Scarabottolo, "An Approach to Fault-Tolerance in Architectures for Discrete Fourier Transforms," *Microprocessing and Microprogramming,* vol. 18, pp. 275–288, 1986.

[66] R. Negrini, M. Sami, and R. Stefanelli, "Fault Tolerance Techniques for Array Structures Used in Supercomputing," *IEEE Computer,* pp. 78–87, Feb. 1986.

[67] F. Lombardi, M. G. Sami, and R. Stefanelli, "Reconfiguration of VLSI Arrays: An Index Mapping Approach," *Proceedings of the IEEE CompEuro,* pp. 60–65, 1987.

[68] F. Lombardi, D. Sciuto, and R. Stefanelli, "Functional Reconfiguration in Fixed-size VLSI Arrays," *Proceedings of the IEEE International Symposium on Circuits and Systems,* pp. 386–389, 1987.

[69] C.-L. Wey and F. Lombardi, "On the Repair of Redundant RAM's," *IEEE Transactions on Computer-Aided Design,* vol. CAD-6, pp. 222–231, 1987.

References

[70] F. Lombardi and D. Sciuto, "Algorithms for Delay-Bound Reconfiguration of Arrays," in *Wafer Scale Integration*, ed. by G. Saucier, and J. Trihle, pp. 197–206, 1986.

[71] J.A.B. Fortes and C. S. Raghavendra, "Gracefully Degradable Processor Arrays," *IEEE Transactions on Computers*, vol. C-34, pp. 1033–1044, 1985.

[72] M. Chean and J.A.B. Fortes, "The Full-Use-of-Suitable-Spares (Fuss) Approach to Hardware Reconfiguration for Fault-Tolerant Processor Arrays," *IEEE Transactions on Computers*, vol. 39, no. 4, pp. 564–571, 1990.

[73] F. Lombardi, M. G. Sami, and R. Stefanelli, "Reconfiguration of VLSI Arrays by Covering," *IEEE Transactions on CAD Integrated Circuits*, vol. 8, no. 9, pp. 952–965, 1989.

[74] T. A. Varvarigou, V. P. Roychowdhury, and T. Kailath, "Reconfiguring Processor Arrays Using Multiple-track Models: The 3-track-1-spare Approach," *IEEE Transactions on Computers*, vol. 42, no. 11, pp. 1281–1293, 1993.

[75] S-Y. Kuo and W. K. Fuchs, "Efficient Spare Allocation for Reconfigurable Arrays," *IEEE Design and Test*, pp. 24–31, Jan. 1987.

[76] M. Tarr, D. Boudreau, and R. Murphy, "Defect Analysis Speeds Test and Repair of Redundant Memories," *Electronics*, pp. 175–179, Jan. 12, 1984.

[77] J. D. Day, "A Fault-Driven Comprehensive Redundancy Algorithm," *IEEE Design and Test*, vol. 2, pp. 35–44, June 1985.

[78] M.G.H. Katevenis and M. G. Blatt, "Switch Design for Soft-Configurable WSI Systems," in *Wafer Scale Integration*, ed. by G. Saucier and J. Trihle, Amsterdam, The Netherlands: Elsevier Science Publishers, pp. 255–270, 1986.

[79] Y-H. Choi, D. S. Fussell and M. Malek, "Fault Diagnosis of Switches in Wafer-Scale Arrays," *Proceedings of the IEEE International Conference on Computer-Aided Design*, pp. 292–295, 1986.

[80] G. Chevalier and G. Saucier, "A Programmable Switch Matrix for the Wafer Scale Integration of a Processor Array," in *Wafer Scale Integration*, ed. by C. Jesshope and W. Moore, London: Adam Hilger, pp. 92–100, 1986.

[81] W. Chen, P. B. Denyer, J. Mavor and D. Renshaw, "Fault-Tolerant Wafer Scale Architectures Using Large Crossbar Switch Arrays," in *Wafer Scale Integration*, ed. by C. Jesshope and W. Moore, London: Adam Hilger, pp. 113–124, 1986.

[82] J. Fried, "An Analysis of Power and Clock Distribution for WSI Systems," in *Wafer Scale Integration*, ed. by G. Saucier and J. Trihle, pp. 127–137, 1986.

[83] R. M. Lea, "A WSI Image Processing Module," in *Wafer Scale Integration*, ed by G. Saucier and J. Trihle, pp. 43–58, 1986.

[84] R. T. Smith, J. D. Chlipala, J.F.M. Bindels, R. G. Nelson, F. H. Fischer, and T. F. Mantz, "Laser Programmable Redundancy and Yield Improvement in a 64K DRAM," *IEEE Journal of Solid-State Circuits*, vol. SC-16, pp. 506–514, 1981.

[85] B. F. Fitzgerald and E. P. Thoma, "Circuit Implementation of Fusible Redundant Addresses of RAMs for Productivity Enhancement," *IBM Journal of Research and Development*, vol. 24, pp. 291–298, 1980.

[86] C.-W. Chen, J.-P. Peng, M.Y.S. Shyu, M. Amundson, and J. C. Yu, "A Fast 32K × 8 CMOS Static RAM with Address Transition Detection," *IEEE Journal of Solid-State Circuits,* vol. SC-22, pp. 533–537, 1987.

[87] S. Asai, "Semiconductor Memory Trends," *Proceedings of the IEEE,* vol. 74, pp. 1623–1635, 1986.

[88] G. Nicholas, "Technical and Economical Aspect of Laser Repair of WSI Memory," in *Wafer Scale Integration,* ed. by G. Saucier and J. Trihle, Amsterdam: Elsevier Scientific Press, pp. 271–280, 1986.

[89] J. I. Raffel, A. H. Anderson, G. H. Chapman, K. H. Konkile, B. Mathur, A. M. Soares, and P. W. Wyatt, "A Wafer-Scale Integrator," *Proceedings of the IEEE International Conference on Computer Design,* pp. 121–126, Oct. 1984.

[90] J. I. Raffel, M. L. Naiman, R. L. Burke, G. H. Chapman, and P. G. Gottschalk, "Laser Programmed Vias for Restructurable VLSI," *Digest: IEEE International Device Meeting,* pp. 132–135, 1980.

[91] G. H. Chapman, J. I. Raffel, J. A. Yasaitis, and S. M. Cheston, "A Laser Linking Process for Restructurable VLSI," *Digest: OSA/IEEE Conference on Lasers and Electrooptics,* pp. 60, 62–63, 1982.

[92] G. H. Chapman, "Laser Linking Technology for RVLSI," in *Wafer-Scale Integration,* ed. by C. Jesshope and W. Moore, Adam Hilger Publishers, pp. 204–215, 1987.

[93] J. M. Canter, G. H. Chapman, B. Mathur, M. L. Naiman, and J. I. Raffel, "A Laser-Induced Ohmic Link for Wafer-Scale Integration," *IEEE Transactions on Electron Devices,* vol. ED-33, p. 1861, 1986.

[94] G. H. Chapman, J. I. Raffel, J. M. Canter and F. M. Rhodes, "Advances in Laser Link Technology for Wafer-Scale Circuits," *IFIP International Workshop on Wafer-Scale Integration,* Sept. 23–25, 1987, Brunel University.

[95] R. P. Cenker, D. O. Clemons, W. R. Huber, J. B. Petrizzi, F. J. Procyk, and G. M. Trout, "A Fault-Tolerant 64K Dynamic Random Access Memory," *IEEE Transactions on Electron Devices,* vol. ED-26, pp. 853–860, 1979.

[96] H. Stopper, "A Wafer with Electrically Programmable Interconnections," *Digest: IEEE International Solid-State Circuits Conference,* pp. 268–269, 1985.

[97] M. J. Mayo, "Photodeposition: Enhancement of Deposition Reactions by Heat and Light," *Solid State Technology,* pp. 141–144, Apr. 1986.

[98] D. C. Shaver, R. W. Mountain, and D. J. Silversmith, "Electron-beam Programmable 128-Kbit Wafer-Scale EPROM," *IEEE Electron Device Letters,* vol. EDL-4, pp. 153–155 (1983).

[99] P. W. Cook, S. E. Schuster, and R. J. von Gutfeld, *Applied Physics Letters,* vol. 26, p. 124, 1975.

[100] Y. C. Kiang, J. R. Moulic, W.-K. Chil, and A. C. Yen, "Modification of Semiconductor Device Characteristics by Lasers," *IBM Journal of Research and Development,* vol. 26, pp. 171–176, 1982.

[101] J. A. Yasaitis, G. H. Chapman, and J. I. Raffel, *IEEE Electron Device Letters,* vol. EDL-3, pp. 184–186, 1982.

[102] J. Melngailis, "Focussed Ion Beam Technology and Applications," *Journal of Vac. Science Technology B,* vol. 5, pp. 469–495, 1987.

[103] L. R. Harriott, A. Wagner, and F. Fritz, "Integrated Circuit Repair Using Focussed Ion Beam Milling," *Journal of Vac. Science Technology B,* vol. 4, p. 181, 1986.

[104] J. Melngailis, C. R. Musil, E. H. Stevens, M. Utlaut, E. M. Kellog, R. T. Post, M. W. Geis, and R. W. Mountain, "The Focussed Ion Beam as an Integrated Circuit Restructuring Tool," *Journal of Vac. Science Technology B,* vol. 4, p. 176, 1986.

[105] D. C. Shaver and B. W. Ward, "Integrated Circuit Diagnosis Using Focussed Ion Beams," *Journal of Vac. Science Technology B,* vol. 4, p. 185, 1986.

[106] C. R. Musil, J. L. Bartlet, and J. Melngailis, "Focussed Ion Beam Microsurgery for Electronics," *IEEE Electron Device Letters,* vol. EDL-7, pp. 285–287, 1986.

[107] P. Girard, B. Pistoulet, M. Valenza, and R. Lorival, "Electron Beam Switching of Floating Gate MOS Transistors," *IFIP International Workshop on Wafer Scale International,* Brunel University, Sept. 23–25, 1987.

[108] P. Girard, F. M. Roche, and B. Pistoulet, "Electron Beam Effects on VLSI MOS: Conditions for Testing and Reconfiguration," in *Wafer-Scale Integration,* ed. by G. Saucier and J. Trihle, Amsterdam, The Netherlands: North-Holland, pp. 301–310, 1986.

[109] D. C. Shaver, Proceedings of the Second CALTECH Conference on VLSI, p. 111, California Institute of Technology, Pasadena, CA, Jan. 1981.

8

Design for Test and Manufacturability

Adit Singh and D. K. Pradhan

8.1 INTRODUCTION

8.1.1 The Basic Problems of Testing

Testing remains a major challenge in the low-cost production of high-quality VLSI circuits. This chapter explores why test costs can at times add 50% or more to the production costs of complex IC packages and still fail to screen out all the defective parts. Although improved test generation techniques, along with design for testability, are commonly used approaches for improving test coverage as discussed in Chapter 3, here we consider the relationship between test coverage, yield, and test escape levels, and we also look at some new techniques for optimizing testing and screening high-quality circuits for special applications. Figure 8.1 depicts the typical testing scenario of an integrated circuit.

The testing of VLSI circuits is carried out in phases, beginning with dice on the wafers, followed in turn by the testing of packaged chips, boards, and subsystems. The success of any testing strategy depends on early recognition of defective parts; the longer the defect eludes detection, the more expensive the consequences. For example, a fault detected in the field is much more expensive to fix than if it were detected during product assembly. A key question is how best to test a circuit at each stage during manufacture.

It is important to recognize that no test can completely guarantee that a complex IC (or board or larger subsystem) is failure-free. Intuitively, one simple way to test a digital circuit is to apply all possible binary input combinations and verify that all outputs are correct.

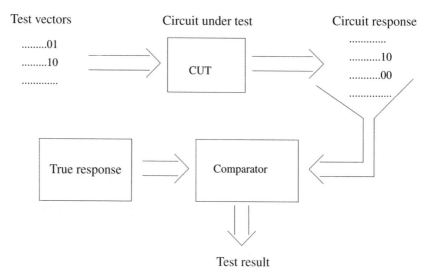

Figure 8-1 Testing a circuit.

But this is impractical for real ICs. For a circuit, for instance, with three binary inputs and thus eight possible input combinations, such exhaustive testing is trivial. However, even modest VLSI circuits now have several dozen inputs. For a circuit with, say 64 inputs, there are approximately 10^{19} different input combinations. At a test application rate of 1 MHz (10^6 combinations per second), applying all combinations will take 10^{13} seconds or 10^7 years! In reality, a complete test would take even longer, since most circuits contain internal memory elements with many different internal states. An exhaustive test would require all input combinations to be applied for each internal state. Clearly, testing a circuit exhaustively is impractical.

Thus, in reality it is practical to apply only a very small fraction of all possible inputs to test a circuit. Obviously, such a test can never be complete and will always fail to detect some types of failures. The best that can be done is to develop tests that go after the most likely failures, those caused by the most commonly occurring defect mechanisms. Targeting the most common failures screens out the largest number of defective parts for a given test cost. [8, 9, 10].

8.2 TESTING FOR STUCK-AT FAULTS

Unfortunately, defect statistics are a function of many inputs such as circuit details, fabrication process, and packaging technology. These statistics are not only difficult to compile accurately but often also vary greatly from run to run, and even lot-to-lot. For this reason, test engineers have traditionally simplified the problem by targeting a logical abstraction, the set of all "stuck-at" faults, for test generation. The stuck-at fault model assumes that in a digital circuit, real physical defects manifest themselves at the logic level by forcing some signal line in the circuit to be permanently stuck at the high-level logic value (stuck-at-1) or low-logic value

8.2 ■ Testing for Stuck-at Faults

(stuck-at-0). Now if all real stuck-at faults would screen out all defective circuits, such a test could be much shorter than the exhaustive test.

Consider, as a simple example, the four-input NAND gate in Figure 8.2.

There are 10 possible stuck-at faults in the circuit, corresponding to each of the five signal lines stuck high or stuck low. A $\langle 1111 \rangle$ input applied to the gate will detect faulty behavior due to any input line stuck-at-0 by giving an incorrect output (1) in the faulty circuit when compared to the expected output (0) for the good circuit. (Note that our goal here is only to detect a faulty circuit, and not to identify which specific stuck-at fault has occurred.) This input combination will also detect a stuck-at-1 fault on the output line. Input $\langle 0111 \rangle$ will detect input A stuck-at-1 as well as the output stuck-at-0. To detect the three remaining stuck-at faults, we need $\langle 1011 \rangle$ to detect B stuck-at-1, $\langle 1101 \rangle$ to detect C stuck-at-1, and $\langle 1110 \rangle$ to detect D stuck-at-1. Thus, applying the five input combinations $\langle 1111, 0111, 1011, 1101, 1110 \rangle$ and comparing the observed output with the expected value will always detect a faulty circuit with a stuck-at type failure. If all defects manifest themselves as stuck-at faults, we therefore need to apply only 5 input combinations rather than 16 in the exhaustive test to completely test the circuit. More importantly, observe from the above discussion that an N-input NAND will need only $N + 1$ input combinations for the complete stuck-at test instead of the 2^N for the exhaustive test. For large N, this is a dramatic reduction in test set size. Although it is generally not always possible to reduce the exponential test complexity to linear for more complex circuits, testing only for stuck-at faults can greatly reduce the size of the test set when compared to exhaustive testing.

Figure 8-2 Four-input NAND gate.

For these reasons, the stuck-at fault model is by far the most common fault model employed for test generation in practice. The goal usually is to obtain a test set with as close to 100% coverage (detection) of all single stuck-at faults as possible. Using Automated Test Pattern Generation (ATPG) techniques [4], we can generally obtain test sets with single stuck-at fault coverage of 95% or better for most combinational circuits. However, generating the test sets can be computationally intense for some large circuits. Sequential circuits have the added problem of setting and observing the internal flip-flop states while exercising the circuits; such memory elements are usually not directly accessible from the IC pins. Generating tests with high single stuck-at fault coverage for sequential circuits often requires some testability support built-in at design time. Common design for testability approaches includes test point insertion and scan techniques. The latter approach essentially transforms the sequential circuit to a combinational one for testing by making the flip-flops externally controllable and observable. Most complex designs today employ at least some design for testability to allow high-coverage testing for single stuck-at faults. Typically, some "functional test vectors" are also added to the ATPG-generated test patterns to exercise the intended functionality of the design.

8.2.1 Limitations of the Stuck-at Fault Model

For the past three decades, the stuck-at fault model has proven to be an adequate model for targeting test generation. This is because in bipolar and older N-type Metal Oxide Semiconductor (NMOS) technologies, many realistic physical defects and failures, such as opens and shorts, did in fact result in stuck-at values on signal lines. Furthermore, for most realistic circuits, complete testing for stuck-at faults requires thousands of test vectors. Exercising the circuit with this large test set also provides a good chance to detect other types of faults that may not specifically result in stuck type behavior.

With the increasing complexity of VLSI circuits, however, and the widespread move to CMOS technology over the past decade, there is concern that this classical fault model may no longer be adequate. In other words, tests generated to detect all stuck-at faults may fail to detect many actual faults in real circuits. This concern stems from the recognition that realistic CMOS defects often do not result in stuck-at behavior.

Figure 8.3 shows an open fault (line break) in a CMOS NOR gate. Assume that as a result of this open, the nMOS transistor at the end of the line can no longer be turned on and is therefore permanently off. Normally, for any input combination, the output of a static CMOS gate is either connected through conducting nMOS transistors to ground (if the output is low), or through pMOS transistors to VDD if the output is high. For a $\langle 10 \rangle$ input, however, because the open gate transistor will not turn on, the output of this faulty gate is electrically isolated (the load at the output of CMOS circuits is purely capacitive) and will "float." As a result, the output node will retain the charge and voltage from the previous input

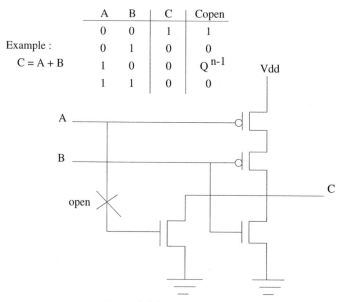

Figure 8-3 Stuck open fault.

condition. Thus, this open fault results in the combinational circuit displaying sequential behavior. Whether or not a stuck-at test detects the fault can depend on the order of test application. If the gate output is charged to 1 by the vector before the test input is applied, then an incorrect output may be observed. However, from the NOR truth table it is obvious that a 0 initial value is significantly more likely for a random ordering of vectors. Moreover, even if a two-vector test is deliberately applied, a glitch may still occur during the transition between the two vectors that can discharge the output and invalidate the test. Experience with CMOS indicates considerable glitching activity in circuits. Obtaining hazard-free two-vector tests for such open faults, so-called robust tests, is often impossible.

We have thus seen that a stuck-at test may fail to detect a common class of CMOS failures such as opens. A similar problem arises with shorts or bridging faults.

A bridging fault is caused by a short between two leads. The effect of such a fault depends on the technology and on how the lines are being driven. In general, if the high voltage is dominant, then the bridging fault will result in an OR function on the two signals; if low voltage is dominant, an AND function results. Thus, in TTL, where the low-level voltage is dominant, the two bridged lines will perform an AND function; both bridged lines take a logic low unless both are high. In CMOS, depending on the strength and combination of the driving transistors (which depends on the input), either line may dominate, resulting in complex functional behavior that can defeat a stuck-at test.

Finally, common CMOS physical failures such as opens or gate-oxide shorts can also result in delay faults such that the circuit outputs are slow to switch. With the increasing emphasis on VLSI performance, such faults must now also be tested for. The delay fault coverage of typical stuck-at fault test sets is generally quite poor. Unfortunately, because of charge-sharing effects within CMOS circuit nodes which can significantly affect switching delays, reliable worst-case delay fault testing of CMOS remains an open problem.

Despite this limitation, industry practice remains centered on testing to high stuck-at fault coverage. Such test sets are often enhanced with functional or random vectors and sometimes run at operational speeds to provide some delay testing. Critical paths may also be tested with specially tailored delay tests. Although in theory many realistic failures can escape such tests, the large number of test inputs applied often results in the random detection of many nontargeted faults. This was well illustrated in a recent experiment [7] which found that a compact ATPG-generated test set with 90% single stuck-at fault coverage was less effective in screening out defective circuits than a larger functional test that covered only 80% of the single stuck-at faults. Thus, stuck-at fault coverage is not always a good measure in evaluating the quality of a test. Even a test with 100% stuck-at fault coverage cannot ensure that significant numbers of defective parts do not pass the test and are shipped out to customers. Unfortunately, no better measure of test effectiveness has as yet been developed. Moreover, given the complexity of the CMOS testing problem, a single test quality measure that is both practical and comprehensive with respect to realistic CMOS faults appears unlikely. This is an important factor contributing to the continuing popularity of the stuck-at model.

One should keep in mind that high test coverage is usually not a goal in itself.

The ultimate objective of production testing is to minimize the number of defective parts that pass testing and result in field failures. In the next section we explore the relationship between test coverage and the defect level in the parts declared good by a test.

8.3 TEST COVERAGE AND DEFECT LEVELS

The defect level [14], also called the field reject ratio [1, 11], can be expressed as

$$DL = \frac{\text{Number of bad chips passed by the test}}{\text{Total number of chips passed by the test}}$$

Observe from this expression that the defect level will depend on both the test coverage and the yield of the incoming parts. If most incoming parts are good, that is, high incoming yield, then even a relatively poor test that screens out only some of the bad parts can still give low defect levels. On the other hand, if most incoming parts are bad, then the test must have high coverage to ensure that the number of bad parts passed are not a significant fraction in comparison to the few good parts.

Williams and Brown [14] first studied the relationship between defect levels, yield, and fault coverage. Their work assumes that stuck-at faults on signal lines are the only type of faults that can exist in the circuit. However, a circuit may have more than one stuck-at fault since individual faults are all independent and all have an equal probability of occurrence P. Let c be defined as the single stuck-at fault coverage—that is, the fraction of all single stuck-at faults covered by a test set. Observe that, even if only stuck-at faults are considered, this is, in general, different from the probability that a test detects a faulty chip, unless the likelihood of multiple faults on a chip is very low.

Let n be all possible stuck-at faults in the chip or circuit, of which m are detected by the test. Then $c = \frac{m}{n}$.

The yield Y is the probability of no faults in the chip.

$$Y = (1 - P)^n$$

Now the probability of having exactly k faults in the chip

$$= \binom{n}{k}(1-p)^{n-k}p^k = \frac{n!}{k!(n-k)!}(1-p)^{n-k}p^k$$

If m out of the n faults are detected by the test, the probability of not detecting at least one fault when k faults are present [7]

$$= \begin{cases} \dfrac{\binom{n-k}{m}}{\binom{n}{m}} & \text{for } n - k - m \geq 0 \\ 0 & \text{for } n - k - m < 0 \end{cases}$$

8.3 ■ Test Coverage and Defect Levels

The probability of failing to detect any of the faults and thus accepting a chip containing k faults, when the test detects m out of the n possible faults =

$$= \begin{cases} \binom{n}{k}(1-p)^{n-k}p^k \dfrac{\binom{n-k}{m}}{\binom{n}{m}} & \text{for } n-k-m \geq 0 \\ 0 & \text{for } n-k-m < 0 \end{cases}$$

Thus, the probability of accepting a chip with one or more defects

$$p_a = \sum_{k=1}^{n-m} \binom{n}{k}(1-p)^{n-k}p^k \dfrac{\binom{n-k}{m}}{\binom{n}{m}}$$

In [14] this is shown to reduce to

$$p_a = (1-p)^m - (1-p)^n$$

Since the probability that a chip is good is $(1-p)^n$, the defect level is given by

$$DL = \frac{p_a}{p_a + (1-p)^n} = 1 - y^{1-m/n}$$

Recall that coverage $c = \frac{m}{n}$

Thus

$$DL = 1 - y^{1-c}$$

Figure 8.4 from [14] shows a plot of defect level versus fault coverage.

The figure shows that even if stuck-at faults were the only faults that existed in circuits, achieving defect levels of a few tens (or even hundreds) of parts per million, as commonly called for in the semiconductor industry, requires near complete fault coverage for large circuits. How such low defect levels can be achieved in the presence of the many other failure types that can be realistically expected in CMOS technology remains a difficult challenge and is the focus of much ongoing research.

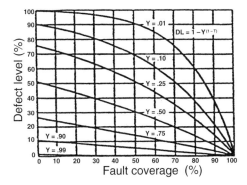

Figure 8-4 Defect level as a function of yield and fault coverage.

8.4 QUALITY SCREENING BASED ON DEFECT CLUSTERING

Recent work has recognized a major problem with attempting to achieve very low defect levels by improving the stuck-at fault coverage alone: the reduced defect detection efficiency of the test set as the stuck-at coverage approaches 100%. This is because for high stuck-at fault coverage values, most defects that exhibit stuck-at behavior are already detected; the test escapes are mostly defects that display other failure modes. Thus, improving stuck-at fault coverage in such a situation does not yield a corresponding reduction in test escapes.

This problem of eliminating uncommon defect types not specifically targeted by the test set is an important one in applications, such as flight control systems, where very low defect levels are required. Extensive testing with random inputs is often attempted, but unless the circuits are small, a very small fraction of the input space can be exercised in this way. Any improvement in defect levels is minimal.

Recently, a new statistical approach, based on the clustering of defects on semiconductor wafers, has been developed to screen out circuits with possible undetected failures. This yields significantly improved defect levels in the screened circuits, even for fault classes that are not explicitly targeted by the test set applied. This approach can also be used to screen out many potential "infant mortality" failures, eliminating the need for the burn-in of bare die in Multichip Module (MCM) applications.

Defect clustering information was first exploited in test optimization [12]. If c_{true} is the true defect coverage of the test, that is, that fraction of all faulty chips actually detected by the test, and Y the yield of the manufacturing line, then the fraction of chips passed by the test is $Y + (1 - c_{\text{true}})(1 - Y)$, and the defect level

$$DL = \frac{(1 - c_{\text{true}})(1 - Y)}{Y + (1 - c_{\text{true}})(1 - Y)}$$

Note from this expression that for the same test coverage, the defect level can vary greatly depending on the yield. For example, for a true defect coverage of 99%, that is, test transparency $TT = 0.01 (1\%)$, a 90% yield gives a defect level of 0.0011, whereas a 10% yield gives a defect level of 0.083. Thus, chip lots with low yield must be much more thoroughly tested to achieve the same defect levels. This observation, along with the fact that defects on a wafer are not uniformly distributed but are known to exhibit clustering, has been used to optimize wafer-probe testing. The test optimization technique takes advantage of the fact that most of the good dies on a wafer are found adjacent to other good dies, while defective dies are similarly clustered. Thus, if the state of some or all of the neighbors of a given die are known, we can obtain a better estimate of its expected yield than just the average yield observed for all the dies on the wafer. Now if dies on the wafer are tested in order by rows or columns, for any die under test, test results for four of its eight immediate neighbors are already available. These are used by the test algorithm to obtain an improved yield estimate for the die, which is in turn used to optimize the test applied to the die. Given an average test cost per die, to minimize the overall defect levels, dies with mostly bad neighbors (therefore low expected yields) can be more extensively tested than those with mostly good neighbors. This wafer-probe test optimization scheme has shown a potential for improving defect levels by a factor of 2 to 3 for the same average test cost.

8.4.1 Binning for Low Defect Levels

Defect clustering information can also be used to screen for quality levels while testing packaged chips. Normally, all spatial defect distribution information from the wafer is lost once the dies are diced and packaged. However, this information can be captured by assigning each of the good dies to one of up to nine different bins depending on how many of its eight neighbors were found faulty during wafer probe testing. (Dies failing the wafer probe test are, of course, discarded.) The distinction between these different bins is then retained through the chip packaging process and is used in the subsequent phase of packaged chip testing.

The following simple example illustrates some of the ways in which this defect clustering information captured by binning the chips can be used to obtain improved defect levels. While the defect statistics used in the example have been selected primarily to illustrate our main idea, the numbers used are quite typical of real manufacturing processes and give some indication of the magnitude of improvement that can be expected. This example motivates the more formal analysis presented in the following sections.

Consider a fabrication run with an overall die yield of 50%. Suppose we assign the dies on the wafers into just two classes: those with three or fewer faulty neighbors, class C_1, and those with four or more faulty neighbors, C_0. Assume that the die yields for the two classes are $Y_1 = 80\%$ and $Y_0 = 20\%$, and that the number of dies in the two classes are equal. Also, the test transparency for the wafer-probe test is $TT_w = 0.01$. The dies that test good from each of the two classes are binned in B_1 and B_0, respectively. Note that because the number of faulty dies that test good is small, B_1 will contain approximately 80% of the chips that test good, and B_0 the remaining 20%.

The defect levels for the two bins can be computed as follows:

$$Y_0 = 20\%: DL_0 = \frac{0.8 * 0.01}{0.2 + (0.8 * 0.01)} = 0.0385$$

$$Y_1 = 80\%: DL_1 = \frac{0.2 * 0.01}{0.8 + (0.2 * 0.01)} = 0.0025$$

Again, because defect levels are small, the overall defect level for the two bins combined can be approximated by

$$Y_0 DL_0 + Y_1 DL_1 = 0.0077 + 0.002 = 0.0097$$

Observe that the poor defect level for the bin B_0 (obtained from the low-yield class of dies) is by far the major contributor to the overall defect level. Clearly, the next phase of testing must be biased toward testing the chips in bin B_0 if the overall defect level is to be efficiently minimized. Note again that the number of chips in B_0 is only about 20% of the total. This makes it much easier to test this bin more extensively. For example, even if we double the test effort expended on the chips in B_0, as compared to those in B_1, the overall test cost will only increase by about 20%. This fact—that because of defect clustering relatively few chips with many bad neighbors will test good at wafer probe testing—ensures that bins with high defect levels will generally be lightly populated. This is an important factor contributing to the effectiveness of test optimization based on defect clustering information.

Another important way to improve defect levels based on this clustering information is to discard altogether the chips in the high-defect-level bins. This would immediately reduce the defect level in our example by almost a factor of 4, from 0.0097 to 0.0025. The cost of this is the loss of 20% of the chips, which may be acceptable in some applications. (The actual cost may be lower, since dies to be discarded will not need to be packaged. Chips in the high-defect-level bins can also be sold separately at lower cost.)

This hypothetical example has shown the potential for obtaining reduced defect levels by using the defect clustering information captured by binning the chips. Although we have only considered two bins in the above example for simplicity, it is obvious that more bins (up to the maximum of nine possible) will yield even better results. In the next section, we present an analytical model to study the effectiveness of such an approach. The model is based on the widely accepted negative binomial model for defect distribution on the wafer and follows the theoretical development of Koren and Pradhan [5].

8.4.2 Analysis

The negative binomial distribution has two parameters: d, an average defect density and α, a clustering parameter [13]. The smaller the value of α, the greater the clustering of defects. As $\alpha \to \infty$, the clustering reduces to randomness; that is, there is no clustering. These parameters can be estimated by observing many wafers and using the usual statistical estimation techniques. Typically, α is found to range from 0.2 to 2.0. Choose the clustering and density parameters so that $\Pi(x)$ is the probability of having x defects in a random nine-die window on the wafer, as shown in Figure 8.5. As in [5], we assume that these defects are uniformly distributed among the nine dies.

Figure 8-5 The nine-die square.

Under the negative binomial distribution, the probability of having x defects on a prespecified area (in our case, the nine-die square) of the wafer is given by

$$\Pi(x) = \frac{\Gamma(x + \alpha)}{x!\,\Gamma(x)} \frac{(d/\alpha)^x}{(1 + (d/\alpha))^{\alpha+x}}$$

8.4 ■ Quality Screening Based on Defect Clustering

where Γ is most conveniently defined by the recursion

$$\Gamma(x) = \begin{cases} (x-1)\Gamma(x-1) & \text{if } x > 1 \\ \int_{t=0}^{\infty} e^{-t} t^{x-1} dt & \text{if } 1 \leq x \leq 0 \end{cases}$$

After some minor algebraic manipulations, we reduce the expression for $\Pi(x)$ to an easily computable form:

$$\Pi(x+1) = \frac{x+\alpha}{x+1} \frac{d/\alpha}{1+d/\alpha} \Pi(x)$$

$$\Pi(0) = (1 + d/\alpha) - \alpha$$

We seek to determine, from the analytical model, (a) the defect levels of dies n of whose neighbors have been found during wafer-probe testing to be faulty, and (b) the improvements in defect level possible by further testing.

We will make extensive use of Bayes's Law of conditional probability, that is,

$$P(A|B) = \frac{P(A \cap B)}{P(B)}$$

Everything that follows in this section refers to a nine-die square in Figure 8-5 mentioned earlier. We shall develop statistics for the central die in the nine-die square to obtain results for a random die on the wafer.

We will begin by defining some events.

$A(\ell)$: The central die passes the wafer test; ℓ of its neighbors are found to be faulty by the wafer test.
C: The central die is faulty.
$D(i)$: Exactly i of the nine dies in the square are faulty.
$E(j)$: The nine-die square has exactly j defects in it.
$G(i)$: Exactly i of the eight outer dies in the square are faulty.

We seek $P[C|A(\ell)]$, which is the probability that the central die is faulty, given that it and $8 - \ell$ of its eight neighbors have passed the wafer test, and the ℓ remaining neighbors have failed the wafer test. We have, using Bayes's Law:

$$P[C|A(\ell)] = \frac{P[A(\ell)|C] P[C]}{P[A(\ell)]}$$

$$= \frac{\sum_{i=0}^{8} P[A(\ell)|C \cap G(i)] P[C \cap G(i)]}{P[A(\ell)]}$$

(8-1)

To evaluate the right-hand side of the above equation, we write Equations (8-2)–(8-11). These are generally applications of Bayes's Law and need no explanation. The following diagram illustrates the use of these equations to obtain Equation (8-1).

$$1 \Leftarrow \begin{cases} \begin{matrix} 3 \\ 5 \end{matrix} \Leftarrow \begin{cases} 2 \Leftarrow 6 \Leftarrow 7 \\ 9 \Leftarrow 6 \Leftarrow 7 \end{cases} \\ 8 \Leftarrow \begin{cases} 3 \\ 4 \\ 10 \Leftarrow 3, 9 \\ 11 \Leftarrow 9, 10 \end{cases} \end{cases}$$

In the following, recall that TT_w is the test transparency of the wafer test.

$$P[C|G(i)] = \frac{C \cap G(i)}{P[G(i)]}$$

$$= \frac{P[C \cap G(i)|D(i+1)]P[D(i+1)]}{P[G(i)|D(i)]P[(D(i)] + P[G(i)|D(i+1)]p[D(i+1)]} \quad (8\text{-}2)$$

$$= \frac{(i+1)P[D(i+1)]}{(9-i)P[D(i)] + (i+1)P[D(i+1)]}$$

$$P[A(\ell)|C \cap G(i)] = \binom{i}{\ell}(TT_w)^{i-\ell+1}(1-TT_w)^\ell \quad (8\text{-}3)$$

$$P[A(\ell)|\overline{C} \cap G(i)] = \binom{i}{\ell}(TT_w)^{i-\ell}(1-TT_w)^\ell \quad (8\text{-}4)$$

$$P[C] = \sum_{i=0}^{8} P[C|G(i)]P[G(i)] \quad (8\text{-}5)$$

$$P[D(i)] = \sum_{j=1}^{\infty} P[D(i)|E(j)]P[E(j)] \quad (8\text{-}6)$$

$$P[D(i)|E(j)] = \begin{cases} 0 & \text{if } i > j \\ 0 & \text{if } 0 = i < j \\ 1 & \text{if } i = j = 0 \\ 1 & \text{if } i = j = 1 \\ P[D(i)|E(j-1)](i/9) + \\ P[D(i-1)|E(j-1)](10-i)/9 & \text{otherwise} \end{cases} \quad (8\text{-}7)$$

$$P[A(\ell)] = \sum_{i=0}^{8} P[A(\ell)|C \cap G(i)] \, P[C \cap G(i)] \quad (8\text{-}8)$$

$$P[G(j)] = \left(1 - \frac{j}{9}\right) P[D(j)] + \frac{j+1}{9} P[D(j+1)] \quad (8\text{-}9)$$

$$P[C \cap G(i)] = P[C|G(i)] \, P[G(i)] \quad (8\text{-}10)$$

$$P[\overline{C} \cap G(i)] = P[G(i)] - P[C \cap G(i)] \quad (8\text{-}11)$$

We now compute the defect levels that result from running a test of average cost per chip T on the chips that have passed the wafer test and have been diced. As mentioned earlier, we sort them according to how many of their eight neighbors were detected by the wafer test to be faulty, so that bin i contains all chips i of whose neighbors were faulty.

What is the fraction of the chips, f_i, that pass the wafer test that are in bin i? This can be written down by inspection from the definition of $P[A(\ell)]$: recall that it is the probability that the central die has passed the wafer test, but that exactly $8-\ell$ of its neighbors failed and were discarded.

$$f_i = \frac{P[A(i)]}{\sum_{i=0}^{8} P[A(i)]}$$

The ratio

$$g_i = \frac{\text{Number of good dies in bin } i}{\text{Number of dies in bin } i} = 1 - P[C|A(i)]$$

Also, we have:

$$\frac{\text{Number of good dies in bin } i}{\text{Number of dies in bin } i} = \frac{P[\overline{C} \cap A(i)]}{P[\overline{C}]} = \frac{\{1 - P[C|A(i)]\} P[A(i)]}{P[\overline{C}]}$$

Let t_i denote the length of the test applied to a chip in bin i, $i = 0, \ldots, 8$. Let $TT_a(T)$ denote the test-transparency of a test of length t. We can then write the overall defect level after test to be:

$$DL(t_0, \cdots, t_8) = \frac{\sum_{i=0}^{8} f_i (1 - g_i) \, TT_a(t_i)}{\sum_{i=0}^{8} f_i g_i + \sum_{i=0}^{8} f_i (1 - g_i) \, TT(t_i)}$$

Given an average test time per chip, T, how should it be allocated to the various bins? To minimize the defect level, we have to minimize the number of bad chips that escape the test; that is, we must minimize

$$\Pi(t_0, \cdots, t_8) = \sum_{i=0}^{8} f_i (1 - g_i) \, TT_a(t_i)$$

subject to the constraint that

$$\sum_{i=0}^{8} f_i t_i = T$$

This is a standard optimization problem and can be solved, for example, by the method of steepest descent [6].

8.4.3 Test Transparency Function

The preceding analysis assumes a function $TT_a(t)$ that relates true test transparency (with respect to all defect types) to test cost t. To obtain a better understanding of the test transparency function, we need to consider the testing process in more detail.

Since it is practically impossible to exhaustively test most VLSI circuits, there is virtually no limit to the amount of testing that can be performed on a chip. A decision is always made as to when the chip has been tested "enough." This is typically based on a cost/benefit analysis (often informal because of incomplete information) of additional testing. Of course, for testing to be cost effective, any sequence of tests should be so ordered that for any test cost the test transparency is the minimum achievable.

Let us assume that wafer-probe testing and packaged chip testing are two phases of the same testing process, characterized by the same test transparency function. Thus, repeating the wafer-probe test at chip test time will not uncover additional failures. This implicitly assumes that the packaging process does not introduce additional faults in the circuits. From a practical viewpoint, this does not introduce any significant errors in the analysis since packaging yields are relatively high, and the kinds of failures introduced in packaging are readily caught in subsequent testing.

A simplistic measure of test cost is the size of the test set applied to the chip. While this approach does provide some measure of the tester time required, it ignores the set-up time associated with presenting the chip to the tester probes, which can often be comparable to the test execution time. Moreover, to achieve low defect levels, complex VLSI circuits may need to employ a number of different types of tests in sequence, each providing an incremental improvement in test transparency. Thus, for example, classical stuck-at fault tests may be followed by test sets for stuck-open faults, delay faults, or IDDQ testing for bridging faults, stuck-on transistors, and oxide shorts. In general, each of these different test types can require its own specialized test equipment and have its own set-up times and execution rates, resulting in differing per vector test costs. Clearly, then, the number of vectors in the test set, by itself, is not a good measure of test costs, and other, more general, test cost measures should be considered.

Now consider the test transparency function. Some recent studies have shown that for low values of test transparency, $TT_a(t)$ is nearly a linear function of test cost [3]. This model was used for die test optimization in [12]. However, in general, the linear model is too optimistic in projecting improvements in test transparency with increased test costs, and the exponential function better captures the general trend of the test transparency/test cost relationship. To numerically study some of the optimization trade-offs analyzed in the previous section, we therefore model test transparency as an exponential function of test cost. Thus,

$$TT(t) = e^{-t/10}$$

This relationship implies that the cost of reducing test transparency by a constant factor, say a half, at any point in the testing process is the same. Thus, it costs the same to reduce test transparency from 1% to 0.5% as it does to reduce it from 5% to 2.5%. *Note that we employ this exponential model only to obtain some numerical estimates of potential improvements in defect levels.* The analysis in the previous section is completely general, however, and can be used in conjunction with any other test transparency function that may better characterize a given circuit. Furthermore, the defect-level improvements projected by this cost model are pessimistic when compared with those for a linear test transparency function.

8.4 ■ Quality Screening Based on Defect Clustering

8.4.4 Numerical Results

We now present some numerical results that show the range of improvements in defect levels that are possible if defect clustering information is used to optimize testing. Unless otherwise mentioned, we use $TT_a(t) = e^{-t}$ and $TT_w = 0.05$.

Figure 8-6 plots the ratio of the defect levels achieved for a given test cost for equal testing of all chips and optimum testing based on the binning information. The method of steepest descent was used. The plots are for yields in the range of 20% to 60% and α between 0.2 (highly clustered defects) and 1.0 (not very clustered). As expected, the results indicate that the improvement in defect level is greater for greater clustering. The improvement also increases with decreasing yields: this is because for high yields, the poor-quality bins are very lightly populated and have a smaller influence on defect levels. The figures suggest that for relatively large VLSI chips having low to moderate yields, a three- to fivefold improvement in defect levels can be reasonably expected. Note that this improvement does not strongly depend on the actual defect level achieved.

The effect of wafer-probe test transparency on this defect level is shown in Figure 8-7. With α equal to 0.4, it costs about 50% more to achieve a defect level of 0.001 with equal testing when compared with optimal testing. Again, this difference increases with greater clustering (smaller α) and lower yields.

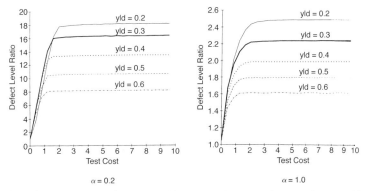

Figure 8-6 Defect level (equal-length) versus defect level (proposed approach).

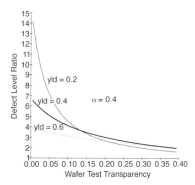

Figure 8-7 Cost under equal testing versus cost under optimal testing.

8.5 APPLICATIONS

Perhaps the most powerful use of the defect clustering information in obtaining very low defect levels is by considering only the best bins. Figure 8-8 plots defect levels that are obtained from the remaining bins when some of the bins are dropped from consideration. To show the cost of doing this, the plot displays the percentage of good chips contained in the bins that are discarded. If all but the test bin is discarded, the defect level improvement can easily be well over a factor of 10. Such low defect levels may never be attainable without binning in many cases, because of the difficulty of obtaining tests with sufficiently low test transparency. Furthermore, binning can screen out unusual defect-based failure modes that are not targeted by the test set. Clearly, this is very significant for critical applications requiring high-quality components.

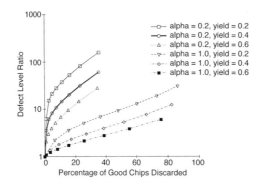

Figure 8-8 Effects of discarding poorer bins.

With the maturing of MCM and other advanced packaging techniques, bare die sales are becoming a significant growth market for semiconductor manufacturers. In many MCM packages, detecting and replacing failed dice is prohibitively expensive. Bare die customers therefore demand the same high quality available through extensive test and burn-in in packaged parts, but they are generally unwilling to pay any extra costs. Unfortunately, burning-in bare die is expensive, requiring placement on and removal from temporary die carriers. Die manufacturers are therefore looking for ways to screen out potential burn-in failures without the need for an actual burn-in step. IDDQ testing is one proposed solution; elevated quiescent current in CMOS appears to be a predictor for burn-in failures. The binning approach provides a more general screen for eliminating a majority of the defective dice that escape wafer-probe testing. Again, it can potentially screen out not only likely burn-in failures, but also other defects whose faulty behavior is not targeted by the fault models employed by the test generation programs. A combination of IDDQ testing and binning, in addition to traditional wafer-probe testing, may be able to deliver high-quality "known good die" without the need for burn-in. Dice from the poorer bins can be packaged, tested (with burn-in if required), and sold in the traditional way.

REFERENCES

[1] D. V. Das, S. C. Seth, P. T. Wagner, J. C. Anderson, and V. D. Agrawal, "An Experimental Study on Reject Ratio Prediction for VLSI Circuits: Kokomo Revisited," International Test Conference, 1990.

[2] M. H. DeGroot, *Optimal Statistical Decisions,* New York: McGraw-Hill, 1970.

[3] R. B. Elo, "An Empirical Relationship Between Test Transparency and Fault Coverage," International Test Conference, 1990.

[4] P. Goel, "An Implicit Enumeration Algorithm to Generate Tests for Combinational Logic Circuits," *IEEE Transactions on Computers C-30,* vol. 3, Mar. 1981.

[5] I. Koren and D. K. Pradhan, "Introducing Redundancy into VLSI Designs for Yield and Performance Enhancement," *Proc. FTCS-15,* 1985.

[6] D. L. Luenberger, *Introduction to Linear and Nonlinear Programming,* Reading, MA: Addison-Wesley, 1973.

[7] P. C. Maxwell and R. C. Aitken, "The Effect of Different Test Sets on Quality Level Prediction."

[8] E. J. McCluskey, "IC Quality and Test Transparency," International Test Conference, 1988.

[9] Phil Nigh, Private Communication.

[10] S. C. Seth and V. D. Agrawal, "On the Probability of Fault Occurrence," in *Defect and Fault-Tolerance in VLSI Systems,* New York: Plenum, 1989.

[11] S. C. Seth and V. D. Agrawal, "Characterizing the LSI Yield Equation from Wafer Test Data," IEEE Transactions on Computer-Aided Design, vol. CAD-3, 1984.

[12] A. D. Singh and C. M. Krishna, "On Optimizing Wafer-Probe Testing for Product Quality Using Die-Yield Prediction," International Test Conference, 1991.

[13] C. H. Stapper, "Correlation Analysis of Particle Clusters on Integrated Circuit Wafers," *IBM Journal of Research and Development,* vol. 31, no. 6, 1987.

[14] T. W. Williams and N. C. Brown, "Defect Level as a Function of Fault Coverage," *IEEE Transactions on Computers,* vol. C-30, 1981.

9

Testing Solutions for MCM Manufacturing

Yervant Zorian

9.1 INTRODUCTION

Today's need for denser packaging technologies is dictated mainly by systems requiring smaller physical sizes or higher system performances. Multichip Module (MCM) technology is a key technology that meets the current miniaturization and performance demands. Advanced consumer products, such as personal communicators and military equipment, need MCMs primarily to achieve smaller size systems, whereas telecommunication equipment and high-speed computers are considered performance-driven applications. They require MCMs to achieve operations with very high speeds.

In addition to the size and performance improvements, MCMs offer other advantages when compared to single-die devices performing the same function. Examples of such advantages are the possibilities of using multivendor components, mixing different process technologies (bipolar and CMOS, digital and analog, etc.), and reducing cost and development time.

For the most part, today's MCM technology has not yet settled into a set of standard materials and techniques the way printed circuit board and surface-mount technologies have. MCMs today consist of single packages containing multiple bare dies or discrete components built with different configurations and sizes connected to substrates using different types of chip-to-substrate attachment techniques, such as wire bond, TAB, or flip-chip [5, 19, 34]. Multiple-substrate technologies are currently being used to provide high-density interconnect circuits for signals, power, and ground [5, 19]. Examples of such substrate technologies are co-fired ceramic (MCM-C), which is a hybrid circuit technology with thick-film screen printing; laminates (MCM-L), which is an advanced form of printed circuit board technology and is used for midrange performance and low-cost needs; and deposited thin-film over silicon or metal base (MCM-D), similar to an integrated circuit manufacturing process. These technologies are used for high-performance and density requirements, but they also have an associated high cost [5]. As for substrate-to-board attachment, peripheral input/output, such as Grid Arrays, and surface input/output, such as Land Grid Array technologies, are in use [5]. Figure 9-1 shows an MCM consisting of four bare dies attached via wire bonding to an MCM-C substrate.

Even though there is hardly any standardization in the MCM technology, and the MCM manufacturers often use proprietary techniques, the current problems associated with MCM technology are known to most existing manufacturers. Some of the most important of these problems are:

The unsatisfiable quality of unpackaged chips, that is, bare dies

The unavailability of their simulation models

The low yield of MCM assembly

The high cost of MCM manufacturing

The complexity of the rework process, by which defective components can be diagnosed and replaced

Adequate solutions for testability, repairability, and electrical modeling are crucial to the advancement of MCM technology.

Figure 9-1 MCM with four bare dies attached via wire bonding to an MCM-C substrate.

The problems of bare die quality, MCM assembly yield, and defective component identification are, in fact, test and diagnosis problems. Specifically, they are due to limitations in conventional testing techniques. For instance, the conventional test at the wafer level, as performed by most IC suppliers, is limited to a simple parametric test and a low-speed functional test to verify if a chip is alive. Therefore, it does not often meet the quality levels required by MCM manufacturers. In general, if a bare die is found to be defective after assembly onto the MCM substrate, either the substrate is scrapped with the rest of the chips or the MCM is repaired by removing the bad chips and replacing them with presumably good ones. Both alternatives are expensive and undesirable. Hence, test strategies that result in providing bare dies with high quality before mounting them onto the MCM substrates have to be adopted. An approach based on structured testability, as presented in this chapter, can be very effective in helping to solve this problem.

Another such problem is testing MCMs after assembly. In this case, the conventional test approach, that is, using in-circuit testing (i.e., bed-of-nails), faces two major obstacles. One is the difficulty of accessing internal nodes in a module, due to high chip density and small interconnection line dimensions [5, 19]. The other is due to the speed limitations of automatic test equipment, hence, the difficulty in having the performance test at the MCM level. The structured testability solution discussed in this chapter helps overcome these problems.

Yet another issue is the ability to diagnose failed dies or substrates at the module level. The MCM manufacturing process requires isolating the defective components. Hence, the module level test approach needs to provide diagnostic capabilities during the MCM repair process.

This chapter discusses the current practices in MCM testing and emphasizes a test approach based on structured testability that can help resolve the limitations of conventional testing. Since this is a testability approach, first it has to be incorporated in the design and then utilized for manufacturing tests at different assembly levels throughout the MCM fabrication process.

The chapter is organized as follows: Section 9.2 discusses the MCM testing problem. Section 9.3 analyzes the structured testability approach and the associated test and diagnostic procedures. Section 9.4 describes the BIST and boundary-scan incorporation in the chip design and their use during wafer testing. Section 9.5 addresses the testability needs for MCMs, and the design considerations to obtain a self-testing MCM. Finally, Section 9.6 summarizes the chapter and provides some concluding remarks.

9.2 MCM TESTING PROBLEM

The MCM production flow can be divided into four major nonoverlapping processes, as shown in Figure 9.2: the process of fabricating the wafer; the production of individual bare dies; the fabrication of substrates; and the assembly of bare dies and substrates to compose MCMs. MCM testing takes place during each of these four processes. The test-related activities are represented by shaded boxes in Figure 9.2. They can be divided into four sets of activities that correspond to the MCM

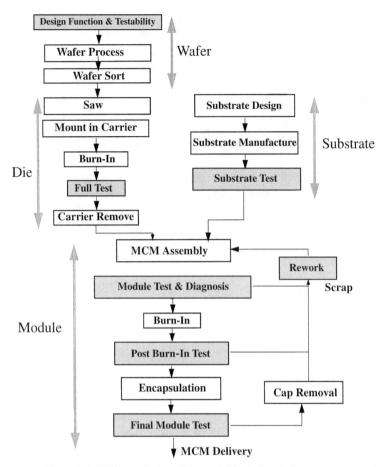

Figure 9-2 MCM production flow model including test processes.

production processes. They are the wafer test, bare die test, substrate test, and assembled module test and rework. Each of these test processes faces certain challenges and requires adoption of specific strategies to meet the MCM quality requirements and improve its yield.

9.2.1 Wafer and Bare Die Test

Ideally, chip-level testing should be done by chip suppliers. The chip supplier traditionally performs a simple wafer sort, which consists of running a structural integrity test, called a functional test, at low speeds and an input/output parametric test. Both tests are performed at the wafer level. The full test, which consists of comprehensive performance and reliability components, is done after the chip is packaged [2]. However, since MCM technology requires bare dies to be attached directly to the substrate, MCM manufacturers ask the chip suppliers to provide "known good dies" [6, 7]. A known good die is defined as a confidence level that ensures a die to be fully functional over specifications and a temperature range. This means that, in order to provide higher quality and higher reliability, the test performed on bare dies has to include the performance and reliability components, which are normally done at the packaged chip level [2].

The low-speed functional and input/output parametric tests for bare dies will continue to be performed at the wafer level. The input/output parametric test verifies that finished dies meet input/output voltage and leakage specifications. Using regular wafer probes for this test has been a standard procedure [3].

The functional test for packaged chips has a stuck-at fault coverage usually in the range of 90%. Even though this coverage should be raised for packaged chips in order to improve the yield of IC production, improving the fault coverage is far more important for bare dies. This is because the MCM yield is a composite function of the individual yields of the bare dies it contains. Hence, in order to meet the MCM yield requirements, a known-good-die fault coverage will probably be above 99+% [5, 19]. This can hardly be obtained for today's complex chips if structured testability techniques such as BIST are not used during the design cycle of a chip [1]. This is a major change over conventional chip design.

The performance test is needed to detect delay-type faults that are not manifested during the conventional low-speed wafer-level tests. Bare dies often need to be tested for such faults prior to MCM assembly. If an MCM application is density-driven rather than performance-driven, performance testing of dies may not be as crucial. But for performance-driven applications, at-speed testing of dies is necessary. The existing wafer probes have some speed limitations. Even though technically it is possible to improve the probes for full high-speed capabilities, this could be an expensive task [15]. A number of tester-based techniques that provide performance testing at wafer level are in development [2, 5, 19, 26]. A different approach to this problem is packaging samples from each wafer lot and testing the samples for performance, based on the fact that an individual wafer has a maximum spread of only a few percent [6]. However, this approach does not have good precision and cannot be adopted for critical MCM applications in terms of performance. Another solution would be to mount a bare die in a (sacrificial or permanent) carrier and apply the performance test to it. A number of carrier-based solutions have been proposed [2, 8, 26, 32]. A preferable solution would be to invoke a built-in at-speed test in the bare die [39]. This requires minimal improvement in the high-speed capabilities of existing tester probes. Such a built-in at-speed testing solution is discussed in Section 9.3.

The complexity of bare die testing differs from one method of attachment to the other. The flip-chip attach, where the die bond pads are distributed all over the active area of the die, and the wire-bonded dies cannot be effectively tested at-speed by conventional test techniques until the assembly [37]. On the other hand, the dies designed for tape automated binding (TAB) attachment can be tested for performance, by conventional test techniques before entering the assembly process after the inner lead bonding. Hence, the defective TAB dies can be removed or repaired prior to assembly. This maximizes MCM test yields. However, TAB devices have poor repairability and are very expensive; they may only be justified in volume operations. Devices that are designed for TAB and wire-bond attachment are generally more difficult to replace than flip-chip devices. Because of their maximum interconnect density, minimum requirements for substrate real-estate, maximum performance, heat dissipation capability, and repairability, flip-chip would seem to be the ideal attachment choice. The major drawback of flip-chip is the difficulty of applying performance testing.

Another yield-related issue is the chip-level reliability test, that is, burn-in. Burn-in is required for chips to identify if they have latent failures. Burn-in is usually done on packaged dies and can also be performed on TAB dies. For dies designed with other types of attachments, in order to guarantee bare die reliability, chip suppliers need further development in burn-in techniques [32]. Several experimental approaches based on temporary die packaging have been reported, for example, [2, 26, 27]. Most of these approaches are still on the expensive side.

In summary, the wafer testing for MCM bare dies faces three major difficulties: the first is the expected known good die quality which necessitates test vector sets with very high fault coverages; the second is the speed limitations of existing wafer probes, hence the problem with performance testing; and the third is the problem of burn-in associated with the required reliability test.

9.2.2 Substrate Testing

It is important to test the substrates for electrical integrity prior to attaching the dies. A substrate defect hurts MCM yield as much as, or more than, poor yielding die. Substrate repair is often not possible, and components are sometimes damaged during the removal process. Hence, failure to detect a substrate defect can be extremely expensive.

The usual substrate testing failure types include opens and shorts. Substrate testing is done either by flying probes, such as two moving probes using resistance as a base to detect faults in a net, or a single moving probe measuring the capacitance of each node [5, 19]. Mechanical probes are slow for substrates with a large number of nodes. They may damage the substrate and have serious problems with dense package geometries. Contactless probing, such as electron-beam probing, is much faster than mechanical probing [4]. The Voltage Contrast Electron-Beam test is an attractive technique for high-throughput contactless testing of unpopulated packaging substrates [5, 19].

Results have shown that it is possible to effectively test today's MCM substrates with the existing mechanical probing or contactless probing techniques [4].

9.2.3 MCM Assembly Testing and Repair

The testing of MCMs is a difficult task, owing to both the complexity of MCMs and the fact that no established infrastructure on MCM testing exists [18]. MCM testing combines the complexities of chip and printed circuit board testing [5, 7, 20, 37]. An MCM test, which meets the quality requirements, has to ensure that all dies are properly connected; that they are still functionally correct; and that the MCM as a device meets its performance specifications. This requires an interconnect test, a full functional test, and a performance test.

Because of limited accessibility to MCM circuitries (accessibility other than MCM primary input/output pins may be nonexistent), it is impossible to test the interconnects and the chips with conventional test techniques. Therefore, it is essential to incorporate testability features during the functional design of the MCM chips.

The interconnect test detects mechanical assembly defects and does not identify functional failures. On the other hand, the full functional at-speed test of the dies reveals defective dies. The at-speed test is usually less important with the printed circuit board testing, because the packaged chips in this case are fully tested. As a result of MCM's bare die handling and bonding processes, there is a chance of damaging the dies. Hence, we need to run a thorough test again after mounting the dies on the substrate [28, 39]. In addition, thermal stresses are present while the module is powered, which can also result in damage to the individual dies. These steps imply that the MCM receives the same testing done to a single-chip IC.

If one or more tests detect failures, an MCM may enter a repair cycle. As shown in Figure 9.2, the repair cycle sometimes includes cap removal, if the repair is done after final test and rework. Understandably, the cost effectiveness of the repair depends on the MCM technology adopted. For instance, a failed MCM using laminate-type substrates may be scrapped because of their relatively low cost. But in general, the test procedure must include, or be followed by, a diagnostic procedure in order to identify the defective element in the MCM. This problem is addressed in the following section.

9.3 A STRUCTURED TESTABILITY APPROACH

Based on the above analysis of the MCM testing problems, this section presents a structured testability approach and discusses the associated test procedures for bare die testing and MCM assembly test and diagnosis.

The testability approach in based on implementing the Built-in-Self-Test (BIST) and module level Boundary-Scan in the MCM design. Both BIST and Boundary-Scan are testability techniques; that is, they are meant to improve the controllability and observability of a circuit under design to make it easily testable [39]. The use of both techniques requires early planning, since they impact the chip design process [40].

The BIST capability allows a circuit to test itself. The circuit could be a chip, an MCM, a board, or a system. Various schemes of chip-level BIST [1, 14, 24, 30] and module-level BIST [40] have been used in the last decade. The BIST schemes needed for the MCM technology are those that provide very high fault coverage [39]; examples are presented in Section 9.4. Most of these schemes run at nominal speed and typically achieve the required fault coverages.

Boundary-Scan is a general testability strategy. The IEEE/ANSI standard 1149.1-1990 [12] is a set of guidelines accepted industrywide to implement Boundary-Scan. Hence, for the rest of the chapter the term *Boundary-Scan* is used as a reference to the above standard. The architecture of Boundary-Scan allows associating memory cells with each input and output of every die so that known signals can be sent across interconnections and captured for observation [11, 18]. The

Boundary-Scan architecture provides a single Test Access Port (TAP) to each chip, through which all the test-related instructions and data can be transferred. Thus, we recommend executing BIST through the Boundary-Scan TAP.

The incorporation of BIST and Boundary-Scan in a chip and techniques to obtain very high fault coverages at the chip and the module-level test are covered in Sections 9.4 and 9.5. For the rest of this section, we will analyze the benefits of having BIST and Boundary-Scan in bare dies, and utilize them in bare die and MCM assembly testing.

9.3.1 The Bare Dies Test Procedure

The test applied on bare dies prior to the module assembly process consists of the following three operations: the parametric test; the functional at-speed test; and the reliability test. The parametric test operation, as discussed earlier, already exists in the conventional wafer test; thus, it does not need to be addressed here.

The second operation, that is, the functional at-speed test, becomes possible owing to the incorporation of BIST in the design of the chip under test. In fact, the two main problems of wafer testing, namely, obtaining a functional test with very high fault coverage and running it at system speed for the performance test, are greatly simplified through use of BIST. The execution of this operation, that is, running BIST, is autonomous. It only needs the applications of Boundary-Scan standard's RUNBIST instruction [12, 18, 28, 39] through the Boundary-Scan TAP to start the BIST execution, and upon completion, the BIST response needs to be scanned out for evaluation. This is also done through the Boundary-Scan TAP [13].

The use of BIST and Boundary-Scan impacts the reliability test operation. Here, during burn-in, BIST can be kept active on the die under test. This needs only the Boundary-Scan TAP interface. It allows continuous monitoring of the response status for additional in-process information.

9.3.2 The Assembled MCM Test and Diagnosis Procedure

To perform a typical test for MCMs with BIST and Boundary-Scan, the following five-step procedure needs to be followed: integrity and identity check; interconnect test; functional at-speed chip test; MCM performance test; and MCM parametric test.

9.3.2.1 Integrity and Identity Check. This step verifies the integrity of Boundary-Scan circuitry prior to its use in subsequent tests. The Boundary-Scan standard provides a certain method to perform this test, which yields relevant diagnostic information regarding the location of a failure, if any [12, 28]. Figure 9-3 shows an MCM design composed of four ICs. Each IC contains its corresponding Boundary-Scan facilities. As an example, the BSR (Boundary-Scan Register) of IC 1 is illustrated in Figure 9-3. The scan chain that connects all four ICs is represented by a thick line starting from TDI (Test Data Input) and ending with TDO (Test Data Output) [12]. The remaining Boundary-Scan signals (TMS, TCK, BCE) are distributed to all ICs in parallel [13].

9.3 ■ A Structured Testability Approach

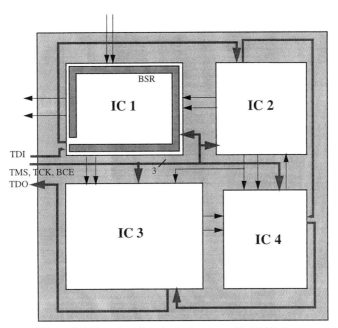

Figure 9-3 Boundary-Scan including in an MCM Design.

Following the integrity test, a test to check the identity of each die will be performed. The Boundary-Scan standard allows permanent storage of chip-level ID codes [12]. Each MCM die will have its ID code, which will be read through the Boundary-Scan TAP and be compared with a reference value. This test detects if a wrong die is mounted on the substrate or a die is not oriented properly. This is especially important with MCMs since several of the chips on an MCM may have the same die footprint [23]. Hence, it is almost impossible to rely on visual inspection. The output of this test carries diagnostic information, which identifies the wrong or incorrectly oriented die(s).

9.3.2.2 Interconnect Test. This test checks proper interconnection of dies with the substrate. The existence of the Boundary-Scan chain in the module, consisting of the Boundary-Scan registers of each die and the connections between these registers through the substrate, creates a virtual electronic bed-of-nails [11, 18] built into the module. This bed-of-nails is independent of the module's density. Boundary-Scan latches provide the stimulus and can sense across all interconnects. This permits testing the opens or shorts between inputs/outputs of two dies and between inputs/outputs of a die and the inputs/outputs of MCMs [28]. The interconnect test needs to use detection algorithms with high fault coverages. It is usually possible to automate the generation of test vectors based on such algorithms and hence greatly reduce the effort needed to develop interconnect test vectors.

Interconnect tests based on the Boundary-Scan can even provide diagnostic information identifying the faulty net and the type of fault. This information is very useful for the repair process. Even though a substrate is usually unrepairable, the die to substrate attachments can be corrected by replacing the die. Section 9.5 addresses the issue of interconnect testing for dies without Boundary-Scan.

9.3.2.3 Functional At-Speed Chip Test. This operation can be realized by the chip-level BIST run. This is similar to the one performed during the bare die test. This test should be repeated in order to detect the failures that may have occurred during the handling and mounting processes of bare dies. The test vectors in this operation are generated in the chips autonomously, due to BIST. Hence, there is no need to access each pin in order to run the functional test. The access needed is to the power and ground lines and TAP of each die, and that is possible through the Boundary-Scan chain of the MCM [39]. The autonomous generation of test vectors has an additional benefit for MCM manufacturers. Since nearly all MCM users must rely on outside suppliers for dies, substrates, and interconnection-related components, and since information may need to pass through several companies before reaching the final MCM assembler, BIST is considered a very advantageous test approach. It is a built-in capability and hence transparent to such transfers. Figure 9-4 shows an MCM design with BIST in its individual dies. As an example, the Test Pattern Generator (TPG) and Output Data Evaluator (ODE) of IC1 are illustrated.

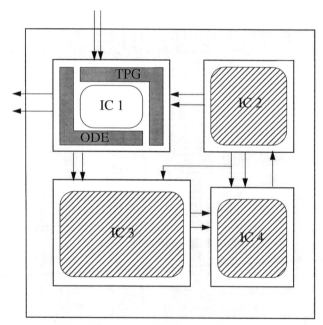

Figure 9-4 Bare Die BIST in an MCM Design.

This operation also provides diagnostic information. The analysis or automatic comparison of the BIST responses can identify which chip has failed. Similar to any additional feature, BIST causes additional on-chip real estate. However, the benefits of using BIST seem to dominate. Today, various chip suppliers are using BIST in their manufacturing tests [1, 14, 30]. Those suppliers need to provide information to MCM manufacturers who are using their chips regarding the BIST execution and its fault coverage. This will allow the usage of existing BIST capabilities during MCM assembly test. If some of the chips in an MCM are not designed with BIST capabilities, such as RAM chips, other measures can be taken into account in order to test them during the module test. Possible measures are discussed in Section 9.5.

9.3.2.4 MCM Performance Test. The MCM performance test is often an at-speed test on the assembled MCM level. It verifies whether the finished module meets performance requirements, and it tests for propagation delay times (path delays), including chip delays and substrate routing delays. This test is crucial, even though the performance test is applied at the chip and substrate levels. Some studies argue that, if certain design guidelines are followed, the module-level performance test can be avoided [23]. But in this case, the test requires automatic test equipment with high pin count and high-speed capabilities, which are in general unavailable or expensive.

The test vectors required for performance testing are developed by the MCM designer. Similar to developing them for ICs or boards, the MCM designer has to write the test vectors or generate them using module-level simulation, if the behavioral models of the chips are available.

If any performance-related problems are detected, it is possible to obtain diagnostic information to locate the problem by capturing the state of the Boundary-Scan cells (or snapshots). The availability of Boundary-Scan architecture in the chips is extremely useful in this case since the Boundary-Scan sample mode [12] can be used to take snapshots [23]. Such snapshots are taken at timed events and scanned out through the Boundary-Scan chain for analysis of module propagation data between chips. If the MCM is meant for performance-driven applications, then a multichip BIST scheme [40] provides an attractive solution for the module-level performance test problem.

9.3.2.5 MCM Parametric Test. The parametric test verifies that the MCM meet its parametric specifications. The Boundary-Scan chain can also be useful in the case of the input/output parametric test, by allowing proper input/output initialization to predefined values [23].

These five test operations can be applied during the MCM assembly process or after the assembly completion. In-process testing is preferable in order to identify the faulty element(s), and perform repair before the final MCM lid is put on.

As stated, the repair cycle includes faulty element isolation and rework. The diagnostic information, obtained during each one of the five testing operations, is utilized to isolate the faulty element. For instance, the BIST signature identifies the faulty chip, and the interconnect test based on the Boundary-Scan identifies the failed net, and so on. The rework involves replacing the faulty element, and a retest involves the rerun of all five operations.

9.4 CHIP-LEVEL STRUCTURED TESTABILITY INCLUSION

The structured testability approach presented in Section 9.3 emphasizes BIST and Boundary-Scan in the MCM chips. This section discusses a process of incorporating BIST and Boundary-Scan features in an MCM chip design.

The creation of chips with BIST and Boundary-Scan requires three major stages. The first stage is the planning stage. It consists of partitioning the chip based on a divide and conquer approach to structural blocks [39], as in Figure 9-5. Here the example chip is composed of several RL (random logic) blocks, RAMs, ROMs, Multipliers, and RFs (register files). Following the partitioning, the schedule for BIST execution is planned, according to the specifications of the partitioned blocks, such as the number of identical blocks, their clock domain distributions, the floor-planning, and the BIST power dissipation of the partitioned blocks. The BIST planning is completed by identifying the BIST hardware sharing and determining the BIST scheduling profile [38].

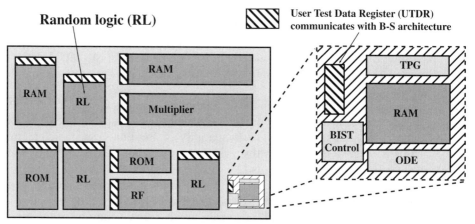

Figure 9-5 IC-Level BIST Incorporation: Divide and Conquer.

The second stage in the process is the BIST incorporation, which is basically the BIST hardware generation state. This starts with adopting an appropriate BIST scheme for each block type. Typically, a chip consists of random logic blocks and regular structure blocks, as in Figure 9-5. Depending on the performance considerations, area overhead limitations, and fault coverage requirements, a BIST scheme is selected for each block. The fault coverage is the dominant factor here. As stated in the last two sections, a very high fault coverage is needed if a chip is to be built into an MCM.

The last stage of the process includes recombining all blocks with BIST, adding the BIST control network, and inserting Boundary-Scan facilities [1].

BIST schemes, in general, need hardware facilities for Test Pattern Generation (TPG), as well as hardware facilities for Output Data Evaluation (ODE), as shown in Figure 9-5. Typically, BIST schemes detect single stuck-at faults in the functional circuitry. Most schemes can detect some other faults as well. The ideal BIST scheme would be a generic one that would be applicable to any block in the chip. However, such a scheme may not be realizable, insofar as digital chips today consist of blocks of different types of structures, device densities, and fault models associated with each. For example, embedded memory blocks like RAMs or ROMs consist of structures, which are much denser than random logic blocks, and have distinct sets of faults, in addition to single stuck-at faults. Hence, specific BIST schemes are needed to test each type of block, in order to achieve very high fault coverages.

The existing BIST schemes used for random logic blocks, in general, can be divided into two categories: pseudorandom-based schemes and exhaustive test-based schemes [1, 40]. For instance, Circular BIST [24] and Partial Scan BIST [16] belong to the first category. The fault coverages obtained depend on the BIST scheme and the circuit under test. Typically, random logic BIST schemes can achieve >98% single stuck-at fault coverages by using some enhancement techniques [16, 24].

On the other hand, the fault coverages of regular structures depend on the fault models and test vector sets used. In general, regular structures are grouped as read/write type memory blocks, such as RAMs, RFs, or FIFOs (first in, first out); read-only type memory blocks, such as ROMs or PLAs; and combinational blocks, such as Multipliers or ALUs. For each type of structure, a specific BIST scheme is adopted. Neither the pseudorandom pattern test nor exhaustive pattern test schemes are appropriate to test regular structure blocks, for two reasons. First, both schemes are based on the stuck-at fault model; hence, they do not take into consideration all other types of faults like transition, coupling, retention, stuck-open, and crosspoint, which are manifested in memory blocks. Second, these approaches are structure-independent. They cannot take advantage of the regularity of such structures to provide a set of test vectors [35].

An optimal BIST scheme for a regular structure might be based on the deterministic test approach. The set of test vectors under such an approach will depend on the given structure and its appropriate fault model. The deterministic test algorithm in such a scheme covers faults in all other surrounding blocks, in addition to the faults in the regular cells of the core. For instance, in a Register File, the BIST algorithm must detect all the faults in the memory cell matrix, the stuck-at faults in the decoders, the input/output registers, and address register [35]. Examples of such regular structure BIST schemes, with 99+% fault coverages, can be found in [21] for RAMs, [36] for ROMs, [35] for Register Files, [9] for Multipliers, [10] for Data Paths, and [31] for FIFOs.

The execution of BIST in each block is performed autonomously, except in the cases where BIST facilities have been shared between blocks. This means that a block with BIST contains, in addition to the facilities needed to generate test vectors and evaluate output data, a controller to execute its BIST session, BRC (BIST Resource Controller). This controller will have the role of interfacing with the Boundary-Scan [29, 38]. In fact, the BIST controllers for most blocks can be interconnected to execute a predefined BIST schedule. An example of such a BIST control scheme can be found in [38].

One major advantage of BIST, in addition to obtaining very high fault coverages, is its ability to run at system speed. This is ideal for performance tests, as discussed earlier.

9.5 MODULE-LEVEL TESTABILITY NEEDS

With BIST and Boundary-Scan incorporated into the chips, most of the testability facilities are ready for the module-level tests. In fact, all the BIST and Boundary-Scan capabilities will be directly accessible from the MCM boundaries, after connecting the four Boundary-Scan lines (TDI, TDO, TMS, TCK) [12] to the dedicating primary MCM input/output pins.

This section discusses some additional testability needs that a module may have. Three such testability additions are described: an addition of BIST capability to test some chips without BIST facilities in their designs; an addition of Boundary-Scan capability around chips without Boundary-Scan; and creating MCM-level self-testing capability.

Examples of chips, which are used extensively in MCMs and have no BIST capabilities, are off-chip memories. In an MCM, where all the chips have BIST, it is desirable to provide self-testing of the memory arrays, too. This can be obtained either by adding a dedicated chip with BIST and other memory-oriented capabilities, such as the MCERT chip [25], or by specifically designing BIST facilities in the logic of a neighboring chip, such as the ASIC(s) controlling the memory array [1]. The BIST facility design can be based on the regular structure BIST schemes described in Section 9.4. With both cases, BIST can be executed during the functional test operation, and the exact faulty memory can be identified and replaced. The same concept can also be used with other types of off-the-shelf chips.

Even though numerous commercial chips offer Boundary-Scan in their products today, at least for some time, the problem of having off-the-shelf chips without Boundary-Scan will be faced. In the case of memory chips, which are unlikely to have Boundary-Scan capabilities, the issue is to provide interconnect testing in order to check the interconnects between the memory chips and the controlling ASIC(s). In this case, there is no need to run the memory BIST algorithms with high complexities. A less powerful algorithm with memory pin fault isolation capability, typically with lower complexity, can be applied using the Boundary-Scan chain of the controlling ASIC(s) [23]. Another approach for chips that do not have Boundary-Scan is to enhance the controllability/observability by adding to the MCM certain dedicated circuits for this purpose, such as the Boundary-Scan strip [33]. This is to directly control and observe the non-Boundary-Scan portions of the logic by a virtual Boundary-Scan chain placed in a neighboring chip. Another practical approach is documented in [17]. This approach allows simultaneous testing of non-Boundary-Scan logic clusters and Boundary-Scan interconnects. It uses the same test sequences for fault detection and fault diagnosis processes.

The last testability addition to be mentioned in this section is for the purpose of creating MCM level self-testing. With the BIST logic incorporated into the chips and their primary inputs/output, an MCM can have module level BIST. An adequate scheme, which puts the entire module in a self-test mode, is presented in [40]. This scheme augments the conventional single-chip BIST approach and provides detection of static and dynamic faults. It also identifies failed elements, that is, bad dies or substrate. This multichip self-test scheme is based on a pseudorandom test approach and uses a multisignature evaluation technique. The hardware designs of multichip and single-chip self-test blocks are combined under a common architecture. The scheme proposes a set of design configurations to create this architecture. This BIST solution [40] provides a reliable static and dynamic test at the module level as well as the bare die level. The multichip BIST can be reused at all levels of test from module testing to board, system, and field test.

With or without using a Boundary-Scan controller, the module testing effort is extremely simplified. The first three module testing operations will not require sophisticated test equipment, if the proposed testability strategy is used. Low-cost Boundary-Scan based testers will be sufficient to perform these detection and diagnosis operations. Moreover, use of BIST and Boundary-Scan avoids ad hoc efforts and promotes standardization, which is essential for a smooth process and speedy time to market.

9.6 CONCLUSION

An MCM testability strategy can be considered universal since it is independent of MCM's functionality, adopted technology, or complexity. This strategy needs to be considered during the early product design stages and helps meet the MCM quality requirements. It impacts the test process at the wafer level, module level, and beyond. Among the resulting benefits of this strategy are achieving very high fault coverages on the chip level due to the usage of effective BIST schemes; obtaining the performance test at the wafer level by running BIST at-speed; performing the interconnect test without pin access problems by using Boundary-Scan; allowing in-process self-testing and concurrent monitoring for its results during the burn-in test; and finally, obtaining a reduced MCM repair interval by using the diagnostic information provided at each stage to isolate the faulty module elements.

REFERENCES

[1] V. D. Agarwal, C. J. Lin, P. W. Rutkowski, S. Wu, and Y. Zorian, "Built-In Self-Test for Digital Integrated Circuits," *AT&T Technical Journal,* pp. 30–39, Mar./Apr. 1994.

[2] M. Andrews et al. "MCC Phase I KGD Report: Consortia for Known Good Die (KGD)," *MCC,* Feb. 1994.

[3] J. Bond, "Test Dominates MCM Assembly," *Test & Measurement World,* pp. 59–64, Mar. 1992.

[4] R. Schmid, R. Schmitt, M. Brunner, O. Gessner, and M. Sturm, "Electron Beam Probing—A Solution for MCM Test and Failure Analysis," in *Multi-Chip Module Test Strategies,* Boston: Kluwer Academic Publishers, p. 55, 1997.

[5] D. A. Doane, and P. D. Franzon, *MultiChip Module Technologies and Alternative—The Basics,* New York: van Nostrand Reinhold, ch. 13, pp. 615–658, 1993.

[6] R. A. Fillion, R. J. Wojnarowski, and W. Daum, "Bare Chip Test Techniques for Multichip Modules," *Proceedings of the 40th EIA/IEEE Electronic Component Technology Conference,* p. 554, May 1990.

[7] A. Flint, "MCM Test Strategy Synthesis Forms Chip Test and Board Test Approaches," in *Multi-Chip Module Test Strategies,* Boston: Kluwer Academic Publishers, p. 65, 1997.

[8] A. E. Gattiker, W. Maly, and M. E. Thomas, "Are There Any Alternatives to Known Good Die?," *Proceedings of MCM Conference,* pp. 102–107, Mar. 1994.

[9] D. Gizopoulos, A. Paschalis, and Y. Zorian, "An Effective BIST Scheme for Booth Multipliers," *Proceedings of the International Test Conference,* pp. 824–833, Oct. 1995.

[10] D. Gizopoulos, A. Paschalis, and Y. Zorian, "An Effective BIST Scheme for Data Paths," *Proceedings of the International Test Conference,* Oct. 1996.

[11] S. C. Hilla, "Boundary-Scan Testing for MultiChip Modules," *Proceedings of the IEEE International Test Conference,* pp. 224–231, 1992.

[12] IEEE Standard Test Access Port and Boundary-Scan Architecture, IEEE Std. 1149.1-1990, IEEE Standards Office, NJ, May 1990.

[13] N. T. Jarwala, "Designing 'Dual Personality' IEEE 1149.1 Compliant Multi-Chip Modules," in *Multi-Chip Module Test Strategies,* Boston: Kluwer Academic Publishers, p. 77, 1997.

[14] *Proceedings of the IEEE International Test Conference,* pp. 445–446, Oct. 1994.

[15] J. A. Jorgenson and R. J. Wagner, "Analyzing the Design-for-Test Techniques in a Multiple Substrate MCM," *Proceedings of 12th IEEE VLSI Test Symposium,* pp. 360–365, Apr. 1994.

[16] D. C. Keezer, "Bare Die Testing and MCM Probing Techniques," *Proceedings of the IEEE Multi-Chip Module Conference (MCMC),* pp. 20–23, Mar. 1992.

[17] C. J. Lin, Y. Zorian, and S. Bhawmik, "PSBIST: A Partial-Scan based Built-In Self-Test Scheme," *Proceedings of the IEEE International Test Conference,* pp. 507–516, Oct. 1993.

[18] M. Lubaszewski, M. Marzouki, and M. H. Touati, "A Pragmatic Test and Diagnosis Methodology for Partially Testable MCMs," *Proceedings of IEEE MCM Conference,* pp. 108–113, Mar. 1994.

[19] C. M. Maunder and R. E. Tulloss, *The Test Access Port and Boundary-Scan Architecture,* Los Alamitos: IEEE Computer Society Press, 1990.

[20] G. Messner, I. Turlik, J. W. Balde, and P. E. Garrou, *Thin Film Multichip Modules, A Technical Monograph of the International Society for Hybrid Microelectronics,* ch. 13, pp. 487–592, 1992, Virginia.

[21] P. R. Mukund and J. F. McDonald, "MCM: The High-Performance Electronic Packaging Technology," *IEEE Computer Magazine,* Special Issue, Apr. 1993.

[22] M. Nicolaidis, "An Efficient Built-in Self-Test Scheme for Functional Test of Embedded RAMS," *Proceedings of the 15th International Symposium on Fault-Tolerant Computing,* pp. 118–123, June 1985.

[23] R. H. Parker, "Bare Die Test," *Proceedings of the IEEE Multi-Chip Module Conference (MCMC),* pp. 24–27, Mar. 1992.

[24] K. E. Posse, "A Design-for-Testability Architecture for Multichip Modules," *IEEE International Test Conference,* pp. 113–121, Oct. 1991.

[25] M. M. Pradhan, E. O'Brien, S. L. Lam, and J. Beausang, "Circular BIST with Partial Scan," *Proceedings of the IEEE International Test Conference,* pp. 719–729, Washington, DC, Sept. 1988.

[26] P. Raghvachari, "Circuit Pack BIST from System to Factory—The MCERT Chip," *Proceedings of the International Test Conference,* p. 641, Oct. 1991.

[27] P. Roebuck et al., "Known Good Die: A Practical Solution," *ICEMM Proceedings,* pp. 177–182, 1993.

[28] L. Roszel, "MCM Foundry Test Methodology and Implementation," *Proceedings of the IEEE International Test Conference,* pp. 369–372, Oct. 1993.

[29] K. Sasidhar, A. Chatterjee, and Y. Zorian, "Optimal Multiple Chain Relay Testing Scheme for MCMs on Large Area Substrates," *Proceedings of the IEEE International Test Conference,* Oct. 1996.

[30] H. N. Scholz, R. E. Tulloss, C. W. Yau, and W. Wach, "ASIC Implementations of Boundary-Scan and BIST," *8th International Custom Microelectronics Conference,* pp. 43.0–43.9, London, U.K., Nov. 1988.

[31] T. Storey, "A Test Methodology for VLSI Chips on Silicon," *Proceedings of the International Test Conference,* pp. 359–368, Oct. 1993.

[32] A. J. van de Goor and Y. Zorian, "An Effective BIST Scheme for Ring-Address FIFOs," *Proceedings of the International Test Conference,* Oct. 1994.

[33] B. Vasquez, D. VanOverloop, and S. Lindsey, "Known-Good-Die Technologies on The Horizon," *Proceedings of the IEEE VLSI Test Symposium,* pp. 356–359, Apr. 1994.

[34] H. Whittemore, "nCHIP Module Level Test—An Experience Report," *Proceedings of the MCM Test Advanced Technology Workshop,* Sept. 1994.

[35] M. A. Zimmerman, "The Technology of Molded Multichip Modules," *AT&T Technical Journal,* pp. 73–83, Sept./Oct. 1993.

[36] Y. Zorian, "A Structured Approach to Macrocell Testing Using Built-In Self-Test," *Proceedings of the IEEE Custom Integrated Circuits Conference,* pp. 28.3.1–28.3.4, Boston, 1990.

[37] Y. Zorian and A. Ivanov, "An Effective BIST Scheme for ROMs," *IEEE Transactions on Computers,* vol. 41, no. 5, pp. 646–653, May 1992.

[38] Y. Zorian, "A Universal Testability Strategy for Multi-Chip Modules Based on BIST and Boundary Scan," *Proceedings of IEEE International Conference on Computer Design,* Cambridge, pp. 59–66, 1992.

[39] Y. Zorian, "A Distributed BIST Control Scheme for Complex VLSI Devices," *Proceedings of the 11th IEEE VLSI Test Symposium,* pp. 4–9, Atlantic City, 1993.

[40] Y. Zorian, "A Structured Testability Approach for Multi-Chip Modules Based on BIST and Boundary-Scan," *IEEE Transactions on Components, Packaging and Manufacturing Technology, Part B: Advanced Packaging,* vol. 17, no. 3, pp. 283–290, Aug. 1994.

[41] Y. Zorian and H. Bederr, "An Effective Multi-Chip BIST Scheme," in *Multi-Chip Module Test Strategies,* Boston: Kluwer Academic Publishers, p. 87, 1997.

Index

A

Acceptability regions, 166–171
   methods of approximation of, 168–171
Accepting cells, 245
Algorithm-based fault tolerance, 232–234
Application mode test, 15
Architectural fault tolerance, 217–260
ATE (Automatic Test Equipment), 10, 44
Automated Test Pattern Generation (ATPG)
      programs, 57, 271
Automatic Test Equipment (ATE), 10, 44
Averaging procedure, 182

B

Bare dies test procedure, 294
Base material, 17
"Bathtub"-like failure rate, 10–11
Bayes's Law of conditional probability, 279
BCH (Bose-Chadhuri-Hocquenghem) code, 232
BER (bit error rates), 221
Bernoulli trials, 133, 174
Bias term, 202
Binning for low defect levels, 277–278
Bipartite graph, 247
Bipolar Junction Transistor (BJT)
   empirical statistical model, 163
BIST (Built-in-Self-Test), 293–301
Bit error rates (BER), 221
BJT (Bipolar Junction Transistor)
   empirical statistical model, 163

Block error correcting codes, 229–232
Block parity error correcting codes, 228–229
Boolean polygons, 127
Bose-Chadhuri-Hocquenghem (BCH) code, 232
Bose-Einstein statistics, 89
Boundary-scan, 293–294
Bounding hyper-planes, 177
Branch and bound method, 248
Breaks, 122
Bridging faults, 52, 122, 124
Built-in-self-test (BIST), 293–301
Burn-in, 292

C

Cadence *Edge* system, 131
CAD (Computer-Aided Design) tools, 46
Capability indices, 207–208
Capacitors, deep-trench, 64
Catastrophic fault model, 73–75
Catastrophic fault probabilities, 5
Catastrophic faults, 122
Catastrophic fault simulation, 122
Centers of Gravity method, 183–184
Characterization text, 14
Check bits, 229
Chemical Vapor Deposition (CVD) technique, 16
Chip area, 85–86, 103–104, 114
CHiP (Configurable Highly Parallel) computer, 236

Chip costing, low-yield cutoff and, 109–110
Chip-level structured testability inclusion, 298–300
Chips, spatial clustering within, 137
Cholesky decomposition-based model, 173–174
Circuit abstraction, 3
Circuit-level fault simulation, 146
Circuit model issues, 137
Circuit performance, 160
Circuit performance faults, 122
Circuit Under Test (CUT), 45
Circuit variables, 158, 159–160
Cluster behavior, 35–37
Clustered-random defect distributions, 103
Cluster factor, 98
Cluster parameter, 93
CMOS circuits, 51
CMOS defects, mapping of, at stuck-on faults, 50–51
CMOS logic gate, 53
CMOS physical failures, 273
Comb-meander-comb structure, 27, 28
Common factors, 164, 205
Complementary indicator function, 168
Computation cells, reconfigurable arrays of, 235–241
Computer-Aided Design (CAD) tools, 46
Conducting material defects, 32
Confidence interval, 133
Configurable Highly Parallel (CHiP) computer, 236
Contacted cells, 244–245
Contamination, 100–101, 121
   mapping to defects, 125–126
Convolution function smoothing, 196–197
Correlated defects, 86
CP factor, 30, 31
CPK factor, 30, 31
Critical area, 68–69, 96–97, 103–104, 115
   computation of, 106–107
   defined, 106, 124–125
   effective, 125
   fault probability prediction and, 121–146
   in layout, 2
   memory redundancy and, 112–113
   in product design, 111–113
   subcircuit metal layer indicating, 105
   in yield calculation, 113
Critical area calculation
   geometrical methods for, 126–134
   Monte Carlo methods for, 132–135
Critical volumes, 136
Crossbar switch example, 253–254
Current, quiescent, 63
CUT (Circuit Under Test), 45
Cutting-plane approximation, 170, 171
CVD (Chemical Vapor Deposition) technique, 16

D

DDD (Defect Density Distribution), 70
Deep-trench capacitors, 64
DEFAM system, 135–136
Defect center, 124
Defect clustering, quality screening based on, 276–283
Defect clustering factor, 98
Defect density, 32–33, 100
   acceptable, 109
   term, 102–103
   for yield models, 100
Defect density derivation, 102–103
Defect Density Distribution (DDD), 70
Defect-fault relationship, 66–68
Defect levels, 274–275
   fault coverage versus, 275
   low, binning for, 277–278
Defect Limited Yield (DLY), 110
Defect mechanisms, 19–21
Defect modeling, 28–39
   of global defects, 29–31
   of local defects, 31–39
Defect monitoring, 21–28
Defect monitors, 4
Defect-Oriented Testing, 65
Defect radius, 69
Defects, 121
   classified, 2
   contamination mapping to, 125–126
   defined, 101
   faults versus, 46
   footprint, 136, 138
   killer, 87
   mapping to faults, 126–137
      hierarchical, 138–142
   random, 86–94
   spatial distribution of, 103
   three-dimensional, 136–137
Defect sensitive design, 68
Defect sensitive layout, 68–69

Defect sensitivity, 2
  layout-related, 68–69
Defect signatures, 145
Defect simulation, 122
Defect size distribution, 33–35, 101–102
Deformations, 121
Delay fault models, 61–63
Delay faults, 61
Delay testing, 61
Deposition, 16–17
Designable parameters, 159
Design Centering, 158, 176
  performance space-oriented, 178–180
  simplicial approximation-based, 176–177
  statistical, 180
  and Tolerance Assignment, 158
  worst-case distance-driven, 177–178
Designer's specifications, 167
Design failures, 11–12
Design for manufacturability, 145–146
Design for Quality (DFQ), 5, 158, 201–209
Design for Testability (DFT), 71, 145–146
Design for Test (DFT) methods, 45–46
Deterministic noise, 197
Deterministic optimization algorithms, 182–183
Development flow of integrated circuits, 12–15
Device integration, 44
Device mismatches, 163
DFQ (Design for Quality), 5, 158, 201–209
DFT (Design for Testability), 71, 145–146
DFT (Design for Test) methods, 45–46
Diffusion technique, 16
Digital Signal Processing (DSP), 220
Direct yield enhancement methods, 158
Discrete active RLC circuits, 161
Disturbances, 121
Divide and conquer algorithm, 241
DLY (Defect Limited Yield), 110
Dopant material, 16
Doped areas, manufacturing, 15–16
Dot throwing, 122
Double comb structure, 27, 28
Double-sided gradient estimator, 199
"Drive-in" step, 16
DSP (Digital Signal Processing), 220
Dummy optimization variables, 159

E

Electrical deformations, 67
Electrical test, 145
Electron-beam programming of floating gates, 258–260
Electronically field-programmable antifuses, 257
Electronically programmable reconfiguration, 249–254
Electronic systems, known-faulty components in, 218–219
Engineering faults, 2–3
Enhancement transistors, 54
Environmental impact, 20
Equipment failure, 19
Error correcting codes, 227–232
Exhaustive search, 248
Expected quality loss function, 202
External bridges, 52

F

FA (Factor Analysis), 163
FABRICS simulator, 199–201
Factor Analysis (FA), 163
"Fail" points, 183, 186
Failure analysis, 144–145
Failures
  faults mapping to, 142
  in-service, 220–221
  in Time Standardized (FITS), 11
Fault classes, 48–49
Fault collapsing, 49–50
Fault coverage, defect level versus, 275
Fault-defect relationship, 66–68
Fault density, 86
Fault detectability, 146
Fault Detection Size (FDS), 62
Fault diagnosis, 22
Fault dominance, 48–49
Fault equivalence, 48
Fault excitation test vector, 54
Fault-hidden cells, 222
Fault hierarchy, 2–3
Fault-known cells, 222
Fault modeling
  functional-level, 60–61
  layout-level, 60
  levels of, 46–65
  logic-level, 47–52
  objectives of, 45–46
  transistor-level, 53–60
Fault-modeling complexity, 2
Fault models, 45, 46

Fault probability prediction, critical area and, 121–146
Fault resolution, 48
Faults, 122
   defects mapping to, 126–137
      hierarchical, 138–142
   defects versus, 46
   defined, 101
   intermittent, 65
   mapping to failure, 142
   temporary, 64–65
   transient, 64–65
Fault tolerance, 5, 6, 217–218
   algorithm-based, 232–234
   local, 223–234
   selective, 221–223
Fault tolerant busses, 227
Faulty cells, masking, 251–252
FDS (Fault Detection Size), 62
Field Programmable Gate Arrays (FPGAs), 223
Field reject ratio, 274
Firm-configurable techniques, 249
FITS (Failures in Time Standardized), 11
Focused ion beams for restructuring, 258
Footprint defects, 136, 138
Four-port switch examples, 252–253
FPGAs (Field Programmable Gate Arrays), 223
Functional circuit performance, 157
Functional description, 13
Functional faults, 21, 122
Functional-level fault modeling, 60–61
Functional tests, 43
Functional yield, defined, 85
Functional yield modeling, 85–115
Functional yield models, 4–5, *see also* Yield models
Fuzzy event, 203

G

Gamma model, 92–94, 103
Gate delay fault model, 62
Gate-oxide monitors, 25–26
Generalized Loss Function (GLF), 202
Generalized Measure, 203
Generalized negative binomial distribution, 93
Geometrical deformations, 67
Geometrical design centering, 176
Geometrical methods for critical area calculation, 126–134
GLF (Generalized Loss Function), 202

Global defects, 2, 20, 67, 86
   defect modeling of, 29–31
   monitoring, 22–24
Global disturbances, 121
Global fault reporting, 141–142
Global mismatch models for integrated circuits, 164
Global net names, 141
Global reconfiguration, 234–248
Global yield loss, 99
Gradient averaging, 193
Gradient estimators, 199
Greedy algorithm, 248

H

Halo, 140
Hamming codes, 230
Hamming distance, 229–230
Hard-configurable techniques, 249
Hard defects, soft defects versus, 37–39
Hard performance faults, 122
Hard yield, 85
HCDB (hierarchical chip database), 139–140
Hessian matrix, 187
Heuristic method, 248
Hierarchical chip database (HCDB), 139–140
Hierarchical circuit extraction of nonoverlapping layout areas, 140–141
Hierarchical defect to fault mapping, 138–142
   of nonoverlapping layout areas, 141
Horizontal neighborhood, 244
Huber function, 180
Human errors, 19
Hyper-planes, bounding, 177

I

ICs, *see* Integrated circuits
IDDQ testing, 284
IFA, *see* Inductive Fault Analysis
Image evaluation, 25
Implantation technique, 16
Implementation flow of integrated circuits, 13
Importance sampling yield estimation, 185
Income Index Maximization, 203–204
Independently addressable switch sites, 250–251
Indicator function, 168
Indirect yield enhancement methods, 158, 176
Inductive Fault Analysis (IFA), 52, 65–77, 144
   basic concepts of, 69–71
   experiment on analog circuits, 75–76

experiment on embedded DRAM, 73–75
experiment on SRAM, 71–72
experiment on standard cells, 72–73
graphical representation of, 70
practical experiences with, 71–76
strengths and weaknesses of, 77
Initializing test vector, 54
In-line monitoring, 24–25
In-Line Quality Control, 201
In-service failures, 220–221
Integrated circuits (ICs)
complete development flow of, 3
complexity of, 1–2
development flow of, 12–15
implementation flow of, 13
large-area, 219–220
manufacturing defects in, 11–21
manufacturing process for, 15–19
market developments for, 9
price of, 9–10
quality and reliability of, 10–11
testing, 43
Interconnect delay, 62
Interconnection distance, 238
Interconnections, physical restructuring of, 255–260
Interconnect monitors, 26–28
Intermittent faults, 65
Internal bridges, 52

J

Jacobian matrix, 206, 209
Joint Test Association Group (JTAG), 6

K

Kerf, 104
Killer defect density, 96
Killer defects, 87
KLA image inspection system, 25
Known-faulty components in electronic systems, 218–219

L

Large-area integrated circuits, 219–220
Large-sample derivative-based methods
for discrete circuits, 184–188
for integrated circuits, 188–191
Large-sample heuristic methods for discrete circuits, 183–184
Laser-assisted chemical processing, 257–258
Laser-based restructuring, 255–257
Layout
critical areas in, *see* Critical area
nonoverlapping areas in, *see* Nonoverlapping layout areas
Layout-level fault modeling, 60
Layout-related defect sensitivity, 68–69
Layout rules for SOP fault detection, 58–59
Leakage fault models, 63–64
Level-by-level defect density, 96
Level set, 161
Linear arrays, 237–241
Linear shrink, 114
Local, term, 67
Local defects, 2, 20, 86
defect modeling of, 31–39
monitoring, 24–28
Local disturbances, 121
Local fault tolerance, 223–234
Local mismatch models for integrated circuits, 164
Local mismatch variables, 164, 165
Local yields, 189
Logic-level fault modeling, 47–52
Long feedback loop, 22
Low-yield cutoff, chip costing and, 109–110

M

Manufacturability, design for, 145–146
Manufacturing, 13–14
Manufacturing defects, 12
in integrated circuits, 11–21
Manufacturing process, for integrated circuits, 15–19
Manufacturing process disturbances, 66–67
Manufacturing yield, 157
Manufacturing yield enhancement, process optimization for, 199–201
Market developments for integrated circuits, 9
Mask artwork, rectilinear, 127
Masking faulty cells, 251–252
Masks, 15
Matrix operations, 232–234
Maxwell-Boltzmann statistics, 87–88
MC, *see* Monte Carlo *entries*
MCMs, *see* Multichip Modules
Mean-based capability indices, 208
Meander structure, 27, 28
Measured yield, 108
Memory redundancy, critical area and, 112–113

Merge tiles, 140
Metal Oxide Semiconductor (MOS) processing steps, 3
Metal patterns, 17
Mismatch model, 166
Modular redundancy, 223–227
Module-level testability needs, 300–301
Monitoring
  global defects, 22–24
  in-line, 24–25
  local defects, 24–28
  parameter, 23–24
Monte Carlo (MC) algorithm, 172–173
Monte Carlo methods for Critical area calculation, 132–135
MOS (Metal Oxide Semiconductor) processing steps, 3
Multichip Modules (MCMs), 2, 7, 221, 276, 284, 287
  assembled, test and diagnosis procedure for, 294–297
  assembly testing and repair of, 292–293
  chip-level structured testability inclusion for, 298–300
  functional at-speed chip test for, 296–297
  integrity and identity check for, 294–295
  interconnect test for, 295–296
  manufacturing of, testing solutions for, 287–301
  module-level testability needs for, 300–301
  parametric test for, 297
  performance test for, 297
  structured testability approach for, 293–297
  substrate testing of, 292
  testing problem with, 289–293
Multi-objective optimization problem, 202
Murphy models, 91–92

N

NAND gate open defects, 51–52
Negative binomial distribution, 278
Negative Binomial Model, 36–37, 103
Negative photoresists, 17
Neighborhood cells, 244
NFET (N-type Field Effect Transistor), 56
Nominal point minimax centering, 178
Noncatastrophic fault model, 73–75
Noncorrelated defects, 86
Nonoverlapping layout areas
  hierarchical circuit extraction of, 140–141
  hierarchical defect to fault mapping of, 141
  identification of, 139–140
Nonrobust tests, 63
Norm body, 161
N-type Field Effect Transistor (NFET), 56

O

ODOS (One-Dimensional Orthogonal Search) technique, 170, 171, 184
Off-Line Quality Control, 201
One-Dimensional Orthogonal Search (ODOS) technique, 170, 171, 184
Open-circuit faults, 122
Open defects, 51–52
Opens, 122
Oxide areas, manufacturing, 16–17

P

Package costs, 10
Packaging limitation, 43–44
Parameter monitoring, 23–24
Parameter value, effective, 29–31
Parametric circuit performance, 157
Parametric faults, 20–21, 122
Parametric Sampling technique, 187
Parametric yield, 157, 171–175
Parametric yield optimization, 157–210
Parts per million (PPM), 10
Passing cells, 245–246
Passive discrete RLC elements, 161–162
Passive integrated RLC elements, 162
"Pass" points, 183, 185
Patching method, 239
Path delay fault model, 62–63
Pattern recognition algorithm, 108
PCA (Principal Component Analysis), 163
PCB (printed circuit board), 218–219
PCMs (Process Control Monitors), 3, 22–23
Performance capability index, 208
Performance space, 176
Performance space-oriented design centering, 178–180
PFET (P-type Field Effect Transistor), 56
Photolithography, 15
Photoresists, 17
Physical mismatch modeling, 166
Physical restructuring of interconnections, 255–260
Physical switch technologies for reconfiguration, 248–260

Physical Vapor Deposition (PVD) technique, 17
POF (Probability of Failure), 104, 106, 125
Poisson model, 90, 103
Poisson probability distribution function, 97
Poisson statistics, 88
Poisson yield model, 35
Polysilicon patterns, 17
Positive photoresists, 17
Potential capability index, 207
POV (Propagation of Variance-covariance) formula, 206–207
PPM (parts per million), 10
Price of integrated circuits, 9–10
Price yield model, 92
Primitive tiles, 139
Principal Component Analysis (PCA), 163
Principal random variables, 159, 163
Printed circuit board (PCB), 218–219
Probability of Failure (POF), 104, 106, 125
Probing dimensions, 104
Process Control Monitors (PCMs), 3, 22–23
Process diagnosis and monitoring, 144–145
Process instabilities, 20
Process-level disturbances, 159
Process optimization for manufacturing yield enhancement, 199–201
Product design, critical area in, 111–113
Production line testing, 145
Production test, 14
Production yield, 10, 13
Propagation of Variance-covariance (POV) formula, 206–207
Propagation of variance method, 205–209
P-type Field Effect Transistor (PFET), 56
PVD (Physical Vapor Deposition) technique, 17

Q

Quality, term, 10
Quality classes, 203–204
Quality demands, 44
Quality improvement, role of testing in, 44–45
Quality screening based on defect clustering, 276–283
Quiescent current, 63

R

Radial Exploration of Space, 170, 171
Random Access Memory (RAM), testing, 61
Random defects, 86–94
Randomness testing, 108
Random number generators, 172
Random perturbations, 196
Random variables, 159
Realistic, term, 65
Reconfigurable arrays of computation cells, 235–241
Reconfiguration
  electronically programmable, 249–254
  global, 234–248
  physical switch technologies for, 248–260
Rectilinear mask artwork, 127
Redundancy, 97
Redundancy analysis, 143–144
Reed-Solomon code, 232
Regression function, 181
Reliability demands, 44
Repair, difficulty of, 221
Requesting cells, 244–245
Research directions, 146
Reticle map, 107
Ring-oscillator, 23
Robust tests, 63
Robust test sequence, 55

S

SAF, *see* Stuck-at Fault model
Scaling applications, yield models for, 113–115
Scan technique, 14
Scribe, 104
Seeds model, 92
Seeds yield model, 89
Selective fault tolerance, 221–223
Self-reconfiguration algorithms for two-dimensional arrays, 243–246
Semiconductor yield, 2
Sensitivity, 128
  relative, 207
Sensitivity function, 38
Sequential Quadratic Programming technique, 191
Short circuit, 2
Short-circuit faults, 122
Short defects, 52
Short-loop defect monitors, 4
Short-loop monitor defect density, 98
Short-loop monitors, 96
Shrink, linear, 114
Significant, term, 31

Silicon area, 10
Simplicial approximation, 170, 171
Simplicial approximation-based design centering, 176–177
Simulator time, 160
Simulator variables, 159–160
Single active device modeling for discrete and integrated circuits, 162–164
Single-pattern SOP fault detection procedure, 57
Single-sided gradient estimator, 199
Single-Stuck-Line (SSL) faults, 48
Singular Value Decomposition (SVD), 209
Size distribution, defect, 33–35, 101–102
SL (Stuck-Line) fault model, 47
Small-sample stochastic approximation-based methods
   for discrete circuits, 191–196
   for integrated circuits, 196–199
Smoothing parameter, 203
Soft-configurable techniques, 249
Soft defects, hard defects versus, 37–39
Soft errors, 64
Soft faults, 64, 122
Soft performance faults, 122
Soft yield, 85
SON, *see* Stuck-on faults
SOP, *see* Stuck-open faults
Spacing dimension, 111
Spatial clustering within chips, 137
Spatial distribution of defects, 103
Spatial yield distributions, 110
SPC (Statistic Process Control) programs, 30
Spin-on technique, 17
Spot defects, 2, 67, 121, 122
Sputtering, 17
SSL (Single-Stuck-Line) faults, 48
Stack map, 107
Stapper model, 93
Statistical analysis, 158
Statistical circuit design, problems and methodologies of, 158
Statistical design centering, 180
Statistical modeling, 161
Statistical models, 159
Statistical yield optimization methods, 180–201
   case study, 199–201
   large-sample derivative-based methods
     for discrete circuits, 184–188
     for integrated circuits, 188–191

large-sample heuristic methods for discrete circuits, 183–184
large-sample versus small-sample methods, 181–182
problem classification, 180–181
small-sample stochastic approximation-based methods
   for discrete circuits, 191–196
   for integrated circuits, 196–199
using standard deterministic optimization algorithms, 182–183
Statistic Process Control (SPC) programs, 30
Stochastic Approximation procedure, 192
Structural testing, 45
Stuck-at Fault (SAF) model, 14, 47–48
   shortcomings of, 51
Stuck-at Faults, testing for, 270–274
Stuck-Line (SL) fault model, 47
Stuck-on (SON) faults, 50
   mapping of CMOS defects at, 50–51
Stuck-open (SOP) faults, 50
   layout rules for detection of, 58–59
Subcircuit metal layer indicating critical area, 105
Surface-integral yield, 188–189
Surfscan technique, 25
Susceptible sites, 127
SVD (Singular Value Decomposition), 209
Systematic defects, 86
Systematic error, 182
System-level faults, 3

T

Taguchi approach, 5
Taguchi measure, 202, 206
Taguchi "on-target" design, 202
Technology parameter, 101
Temporary faults, 64–65
Testability, design for (DFT), 71, 145–146
Test costs, 10
Test error, Type I and Type II, 30
Test generation, 144
Testing, 2, 14–15, 45, 269
   basic problems of, 269–270
   effectiveness of, 7
   role of, in quality improvement, 44–45
   solutions for MCM manufacturing, 287–301
   for stuck-at faults, 270–274
Test sequence, robust, 55
Test transparency function, 281–282

Test vectors, 54
Thermal oxidation, 16
Three-dimensional defects, 136–137
Three-pattern test, 56
Timing, 61
Timing problems, 12
TMR (Triple Modular Redundancy), 223–227
Tolerance body, 161
Tolerance region, 161
Total Quality Management (TQM), 45
Transient errors, 6
Transient faults, 64–65
Transistor-level fault modeling, 53–60
Transistors, enhancement, 54
Transistor stuck-on fault model, 59–61
Transistor stuck-open fault model, 53–59
Triple Modular Redundancy (TMR), 223–227
Two-dimensional arrays, 241–243
    self-reconfiguration algorithms for, 243–246
Types of test errors, 30

U

Union method, 104

V

Validation test, 14
Variability minimization, 201
Variability term, 202
Varimax factor rotation, 163
Vectors, 159
Vertical neighborhood, 244
Very High Description Language (VHDL), 13
Very Large-Scale Integration (VLSI) systems, *see* VLSI *entries*
VHDL (Very High Description Language), 13
Virtual artwork method, 104
Virtual artworks, 129–130
Virtual fabrication line, 122
Virtual factory, 121
VLSI Layout Simulation for Integrated Circuits (VLASIC), 71
VLSI (Very Large-Scale Integration) systems, 1
Volume-integral yield, 188–189
Vote-configurable techniques, 249
Voter circuit, 224

W

Wafer and bare die test, 290–292
Wafer defects, 19

Wafer map, 107
Wafer Scale Integration (WSI), 6
Wafer yield distribution, 109
Windowing technique, 107–108
Worst-case distance-driven design centering, 177–178
WSI (Wafer Scale Integration), 6

X

XLASER system, 128–133

Y

Yield
  local, 189
  manufacturing, 157
  measured, 108
  parametric, 157, 171–175
  surface-integral, 188–189
  volume-integral, 188–189
Yield calculation, critical area in, 113
Yield distributions, 110–111
  spatial, 110
Yield-driven routers, 145–146
Yield enhancement
  indirect methods of, 175–180
  manufacturing, process optimization for, 199–201
Yield gradient, 185–186
Yield management, 1
Yield model derivations, 89–94
Yield models
  analytical, 124
  applications of, 108–115
  area term in, 103–104
  Class I, 94–95
  Class II, 95–96
  Class III, 96–99
  Class IV, 99
  classes of, 94–99
  components of, 99–108
  defect density for, 100
  defect density term in, 100–103
  definition of commonly used terms in, 100–101
  for scaling applications, 113–115
  yield loss in, 107–108
Yield optimization
  objective of, 175–176
  statistical methods of, *see* Statistical yield optimization methods

Yield prediction, 142–143
Yield probability function, 180
Yield simulation, 122
Yield statistics, basic, 86–94

Yield stimulation, 5

Z

Zadeh fuzzy set theory, 203

About the Editors

José Pineda de Gyvez received the degree in electronic systems engineering from the Technological Institute of Monterrey, Mexico; the M.Sc. degree from the National Institute of Astrophysics, Optics, and Electronics, Mexico; and the Ph.D. degree from the Eindhoven University of Technology, The Netherlands, in 1982, 1984, and 1991, respectively. From August 1986 through February 1991, he was a junior scientist with the Foundation for Fundamental Research on Matter, The Netherlands, working on deterministic methods to compute the critical areas of integrated circuit layouts.

In 1995 Dr. Pineda was a guest editor of a special issue on advanced yield modeling in the *IEEE Transactions on Semiconductor Manufacturing* and associate editor for technology from 1994 to 1996. From 1995 to 1997, he served as associate editor for cellular neural networks in the *IEEE Transactions on Circuits and Systems: Part I*. Dr. Pineda is currently an associate professor in the Department of Electrical Engineering and holds a joint faculty appointment with the Department of Computer Science. His accomplishments include the Eta Kappa Nu Outstanding Professor Award in 1993 and the Ernest A. Baetz '47 College of Engineering Faculty Fellow Award in 1997. He was also listed in *Who's Who Among America's Teachers* in 1996.

Dhiraj Pradhan is currently a visiting professor with the Department of Electrical Engineering at Stanford University, Stanford, California. He is on leave from Texas A&M University, College Station, Texas, where he is the COE Endowed Chair Professor in Computer Science.

Prior to joining Texas A&M, he served as professor and coordinator of Computer Engineering at the University of Massachusetts, Amherst. He also held positions at Oakland University, Michigan; University of Regina in Canada; and Stanford University, California.

Dr. Pradhan has made contributions to the fields of VLSI CAD and test, fault-tolerant computing, computer architecture, and parallel processing research with major publications in journals and conferences over the last 25 years. He has served as guest editor of special issues in prestigious journals, such as *IEEE Transactions.* Currently he is an editor for several journals, including *IEEE Transactions and JETTA.*

Dr. Pradhan has served as General Chair and Program Chair for various conferences and has received several Best Paper Awards, including the 1996 IEEE Transactions on CAD. A fellow of the IEEE, Dr. Pradhan was a recipient of Germany's Humboldt Distinguished Senior Scientist Award. He was also the recipient of the 1997–98 Fulbright–Flad Chair in Computer Science. Dr. Pradhan is co-author and editor of several books, including *Fault-Tolerant Computing: Theory and Techniques, Vols. I and II* (Prentice Hall, 1996), and *IC Manufacturability: The Art of Process and Design Integration* (IEEE Press, 1996).